Technological Innovation for a Dynamic Economy

(Pergamon Policy Studies—50)

Pergamon Policy Studies on Science and Technology

Starr/Ritterbush *SCIENCE, TECHNOLOGY AND THE HUMAN PROSPECT*

Related Titles

Encel/Ronayne *SCIENCE, TECHNOLOGY AND PUBLIC POLICY*

Joyner *A GUIDE TO OFFICIAL WASHINGTON FOR OVERSEAS BUSINESS*

Liebling *U.S. CORPORATE PROFITABILITY AND CAPITAL FORMATION*

Tasca *U.S.-JAPANESE ECONOMIC RELATIONS*

Ways *THE FUTURE OF BUSINESS*

ON SCIENCE AND TECHNOLOGY

Technological Innovation for a Dynamic Economy

Edited by
Christopher T. Hill
James M. Utterback

Published in cooperation with the
Center for Policy Alternatives, M.I.T.

Pergamon Press
NEW YORK • OXFORD • TORONTO • SYDNEY • FRANKFURT • PARIS

Pergamon Press Offices:

U.S.A.	Pergamon Press Inc., Maxwell House, Fairview Park, Elmsford, New York 10523, U.S.A.
U.K.	Pergamon Press Ltd., Headington Hill Hall, Oxford OX3 0BW, England
CANADA	Pergamon of Canada, Ltd., 150 Consumers Road, Willowdale, Ontario M2J, 1P9, Canada
AUSTRALIA	Pergamon Press (Aust) Pty. Ltd., P O Box 544, Potts Point, NSW 2011, Australia
FRANCE	Pergamon Press SARL, 24 rue des Ecoles, 75240 Paris, Cedex 05, France
FEDERAL REPUBLIC OF GERMANY	Pergamon Press GmbH, 6242 Kronberg/Taunus, Pferdstrasse 1, Federal Republic of Germany

Library of Congress Cataloging in Publication Data

Massachusetts Institute of Technology. Center for
Policy Alternatives.
Technological innovation for a dynamic economy.

(Pergamon policy studies)
Bibliography: p.
Includes index.
1. Technology and state—United States. 2. Technological innovations—United States. I. Hill, Christopher T. II. Utterback, James M., 1941-
III. Title.
T21.M37 1979 338.4'7'60973 79-18357
ISBN 0-08-025104-8
ISBN 0-08-025103-X pbk.

ON SCIENCE AND TECHNOLOGY

Technological Innovation for a Dynamic Economy

Edited by
Christopher T. Hill
James M. Utterback

Published in cooperation with the
Center for Policy Alternatives, M.I.T.

Pergamon Press
NEW YORK • OXFORD • TORONTO • SYDNEY • FRANKFURT • PARIS

Pergamon Press Offices:

U.S.A.	Pergamon Press Inc., Maxwell House, Fairview Park, Elmsford, New York 10523, U.S.A.
U.K.	Pergamon Press Ltd., Headington Hill Hall, Oxford OX3 0BW, England
CANADA	Pergamon of Canada, Ltd., 150 Consumers Road, Willowdale, Ontario M2J, 1P9, Canada
AUSTRALIA	Pergamon Press (Aust) Pty. Ltd., P O Box 544, Potts Point, NSW 2011, Australia
FRANCE	Pergamon Press SARL, 24 rue des Ecoles, 75240 Paris, Cedex 05, France
FEDERAL REPUBLIC OF GERMANY	Pergamon Press GmbH, '6242 Kronberg/Taunus, Pferdstrasse 1, Federal Republic of Germany

Library of Congress Cataloging in Publication Data

Massachusetts Institute of Technology. Center for
Policy Alternatives.
Technological innovation for a dynamic economy.

(Pergamon policy studies)
Bibliography: p.
Includes index.
1. Technology and state—United States. 2. Technological innovations—United States. I. Hill, Christopher T. II. Utterback, James M., 1941-
III. Title.
T21.M37 1979 338.4'7'60973 79-18357
ISBN 0-08-025104-8
ISBN 0-08-025103-X pbk.

This volume is based on research supported by and published with the permission of the United States Department of Commerce.

Contents

FOREWORD ix

ACKNOWLEDGEMENTS xiii

CHAPTERS

1 TECHNOLOGICAL INNOVATION: AGENT OF GROWTH AND CHANGE, Christopher T. Hill 1

Introduction 1
Technological Innovation and the Economy 3
Technological Innovation and the Quality of Life 28
The Conditions for Technological Innovation 32
Policy Implications 34
Notes 35

2 THE DYNAMICS OF PRODUCT AND PROCESS INNOVATION IN INDUSTRY, James M. Utterback 40

Introduction 40
A Dynamic Model of Product and Process Change 42
Major Product Innovations 46
Major Process Innovations 50
Incremental Improvements in Products and Processes 53
Innovative Challenges and Defensive Responses 55
Summary of Policy Options Arising From Findings on Industrial Innovation 57
Notes 59

3 THE SLOWDOWN IN PRODUCTIVITY ADVANCES: A DYNAMIC EXPLANATION, Burton H. Klein....66

Introduction 66
How Major Productivity Advances Occur 72
Understanding the Declining Rate of Productivity Advance 100
Restoring a Dynamic Economy: Policy Implications 107
Notes 115

4 TECHNOLOGICAL INNOVATION AND THE DYNAMICS OF THE U.S. COMPARATIVE ADVANTAGE IN INTERNATIONAL TRADE, Edward M. Graham....118

Introduction 118
International Trade, Comparative Advantage, and Technology-Intensive Products 118
Explanations for U.S. Comparative Advantage in World Trade of New Technology-Intensive Products 127
Changes in Global Patterns of Technological Innovation and in the U.S. Position in the International Economy 134
Policy Options for the United States Government 145
Notes 152

5 ENVIRONMENTAL, HEALTH, AND SAFETY REGULATION AND TECHNOLOGICAL INNOVATION, Nicholas A. Ashford, George R. Heaton, Jr., and W. Curtiss Priest....161

A Framework for Viewing the Effect of Regulation on Technological Change 161
The Effects of Regulation on Technological Change 167
Designing Regulatory Policy to Encourage Innovation 190
Summary and Policy Implications for Government and Special Interest Groups 204
Notes 210

6 LABOR, PRODUCTIVITY, AND TECHNOLOGICAL INNOVATION: FROM AUTOMATION SCARE TO PRODUCTIVITY DECLINE, Clinton C. Bourdon....222

Introduction 222
Trends in Productivity and Technology 225
Microanalysis of Productivity Change 243
Public Policy and Adjustment Assistance 250
Notes 252

7 DIRECT GOVERNMENT FUNDING OF RESEARCH AND DEVELOPMENT: INTENDED AND UNINTENDED EFFECTS ON INDUSTRIAL INNOVATION, Paul Horwitz................255

Introduction 255
R&D Funding Policy in the United States 257
R&D Funding Policy in Foreign Countries 261
Rationale for Government Support of R&D 268
Effects of R&D Funding on Technological Innovation 272
Policy Implications 285
Notes 288

8 POLICIES AND PROGRAMS OF GOVERNMENTS DIRECTED TOWARD INDUSTRIAL INNOVATION, J. Herbert Hollomon............292

Introduction 292
Actions of Governments that Affect Innovation Generally 294
Direct Intervention in the Innovation Process 303
Ameliorating the Adverse Effects of Innovation 309
Socioeconomic Climate and Innovation 313
Conclusions and Policy Implications 313
Notes 316

9 SUMMARY AND POLICY IMPLICATIONS, Christopher T. Hill and James M. Utterback................................318

The Central Issue: A Dynamic Economy 318
The Nature of the Innovation Process 319
Encouraging Technological Innovation 321
Policy Considerations 322
In Conclusion 328

THE AUTHORS 331

ABSTRACTS 333

INDEX 339

Foreword

The introduction of new and improved products, processes, and services to meet the changing needs of society provides an important basis for increasing the productivity and wealth of nations and for enhancing the health and welfare of their citizens. The new displaces the old -- coal substitutes for wood; oil for coal. The modern automobile is the horseless carriage. Old industies may disappear; regions of countries may become depressed; companies may fail; and workers may lose their jobs. At times, the widespread production and use of new goods have consequences that adversely affect society, while other innovations may alleviate these same adverse effects. Yet technological innovation is fundamental to the growth of economies. All nations -- rich and poor -- are concerned about conditions that foster it, and, at the same time, are anxious to assuage the possible harmful consequences of widespread use of new technology. Thus, most countries are now encouraging innovation and are developing the technology which undergirds the process.

Except for a few scholars and concerned individuals, few people in the United States paid attention to these matters until recently. Several factors explain the lack of attention. There was the spectacular success of new weapons development during World War II, when scientists were mobilized and liberally supported without concern for costs. After World War II, the United States was the strongest nation on earth -- politically, militarily, and economically. There was a pent-up demand for consumer goods which had been unavailable during the war. Europe and Japan had to be rebuilt. Most observers believed that the vigorous R&D effort in the United States would automatically produce new goods and services needed by the American people for rapid improvement of their health, welfare, and quality of life. The technical progress in the United States and the astonishing advances in nuclear energy, aircraft, and computers (spin-offs directly or indirectly from the space and defense programs) caused other nations to become apprehensive about the "technological gap" between their countries and the United States, and inspired them to intensify support for research and to create special "science" ministries. Although it usually provided unreliable predictions and ignored the details by which industrial goods and commercial services are conceived, produced, and distributed, macroeconomics dictated the country's policies. There was little concern with the details of industrial practices or with the system of providing goods themselves -- it was enough to assure demand.

In the early 1960s, a number of people, mostly with industrial experience, joined by a few scholars learned in industrial practices, began to question the wisdom of the prevailing attitudes. As an Assistant Secretary of Commerce, I encouraged the analysis of the significance of the innovation process -- the well-known Charpie Report was a result. Little action resulted in the United States in response to the recommendations of that report. However, other nations began to understand and to appreciate the importance of innovation. They began to perceive that R&D was only one part of a complex process required to bring new technology to practical and widespread use.

By the 1970s, it became clear that something was amiss. Improvements in productivity in the United States were, and had been for twenty years, less rapid than those of every major Western country, except the United Kingdom. The imbalance of trade of the United States with the rest of the world had become huge. (Even though Japan and West Germany import relatively more oil than the United States, they maintain a favorable balance of trade.) The people of four industrial nations (West Germany, Sweden, Switzerland, and Norway) have become more affluent than the people of the United States. Americans have become concerned with environmental deterioration and threats to human health and safety. They have noted the declining quality of U.S. products -- dramatized by the recall of millions of faulty automobiles, and, more recently, by the failure of a nuclear reactor. The enormous R&D effort of the United States, still nearly one-half of the world's total, has not produced the results its proponents had earlier proclaimed.

In 1978, President Carter called for a major Administration study and for policy recommendations relative to industrial innovation. The direction of the study was assigned to Secretary of Commerce Juanita Kreps, and Jordan Baruch, Assistant Secretary of Commerce, was delegated to manage the project. Public and government task forces were formed and special studies were solicited. The Center for Policy Alternatives at the Massachusetts Institute of Technology was asked to prepare eight reports on various aspects of industrial innovation by the Commerce Department. These policy studies, produced to meet the short-term schedule of the Administration, were not to be in-depth, scholarly treatises. Rather, they were designed to identify and delineate the principal characteristics of the innovation process, its consequences, and the ways governments deal with it. The revised and edited reports are the chapters of this book.

In many ways, the authors attack conventional thinking. They represent and elaborate viewpoints and ideas often opposite to those generally accepted. The authors believe that rivalry and competition, venture, risk, entrepreneurship, and creative engineering are crucial to the innovation process. They further believe that challenge and response are essential for a dynamic society, that there must be rewards for the risk taker, and that protection of the status quo will only stifle innovation. Of course, some will be hurt by change, and any resulting adversity must be shared by the society. It is important to help people adjust to new opportunities and not be bound by the old. Regulations intended to restore the environment and to protect human health and safety, discriminatingly applied, can be a spur to innovation.

The authors believe that unless the old perspectives and approaches are changed, the U.S. competitiveness will be further weakened. It is incumbent on our country to reverse the direction of past practices that have failed. I hope that this book will serve to stimulate political discussion of these important issues.

J. Herbert Hollomon

Acknowledgments

The first eight chapters of this book were originally prepared as a series of papers under contract to the U.S. Department of Commerce. We wish to thank Jordan J. Baruch, Assistant Secretary of Commerce for Science and Technology, and his staff, Theodore Schell and Theodore Schlie, for their support in carrying out this work. The views expressed herein are those of the authors and not of the U.S. Government, the Department of Commerce, the Department of the Treasury, or any of their officers or employees.

A number of people reviewed the chapters in various stages, and we are grateful for their many constructive comments and suggestions. In addition to several anonymous reviewers, we wish to thank Zeev Bonen, Al Brown, Robert Gilpin, Edward Greenberg, Sidney Hess, Franklin P. Huddle, Robert P. Morgan, Merton J. Peck, J. David Roessner, Nathan Rosenberg and George White for providing reviews of the early drafts, often under very short notice.

Our editor, Rebecca Packard, provided immeasurable help in converting very rough manuscripts to the final chapters. Inaccuracies and obfuscations remain despite her best efforts, rather than because of them. We wish to thank the staff of the Center for Policy Alternatives, and especially our secretary, Betsi Irving, for taking care of the myriad details necessary to prepare camera-ready copy. We also thank David Kagan and Amy Tighe, our research assistants, for their help in reviewing the papers and preparing the index, and Jeff Heehs and Robert Kaplan for production assistance.

Christopher T. Hill
James M. Utterback
Cambridge, Massachusetts
July 1979

1 Technological Innovation: Agent of Growth and Change

Christopher T. Hill

INTRODUCTION

Technological Innovation for a Dynamic Society

Technological innovation is a major agent of growth and change in society. Development and adoption of new technology are the primary means by which scientific advances are adapted to meet human needs and to expand human opportunities. The use of new technology is one of the main engines of economic growth and can help to control inflation. It enables major improvements to be made in the productivity of human effort and in the efficiency of the use of the Earth's resources. Technological innovation changes the character and nature of work. It provides an important basis for the competition in trade, in ideas, and in military strength among nations. Technological innovation is a significant force for social change because it changes human values and alters the character of human relationships. Use and abuse of new technology are at the root of many of the acute and chronic problems of our society such as the growth of the population, the exhaustion of resources, the mismatch of job opportunities and worker skills, the release of toxic materials, the growing influence of terrorism, and the proliferation of nuclear weapons. Paradoxically, new technology can also contribute to solving these problems.

The people and the government of the United States should be vitally concerned about the condition of technological innovation in their country and around the world. Better understanding is needed of the relationship between technological innovation and the factors mentioned above to ensure that technologies are developed at a rate and in a direction that best support the goals of our nation. Government policies and programs can have profound effects on technological innovation. They need to be designed, adopted, and implemented with an appreciation for these effects and their implications.

Today, technological innovation in the United States is not as healthy as it might be. Productivity and economic output are growing slower than in the past and slower than in most other major nations. Both unemployment and inflation are high, and our trade balance has become chronically negative. A tier of newly-industrializing countries, such as South Korea, Brazil, Venezuela, and Taiwan, are entering world markets with products formerly the province of the member countries of the Organization for Economic Cooperation and Development (OECD). Many private firms express an inability or an unwillingness to take risks on new techniques. Research and development (R&D) spending in the United States has declined as a fraction of the gross national product (GNP). Hazardous products continue to be made; work places continue to be unhealthy; and pollution continues to be a problem. While energy prices jump almost daily, new technologies that would produce more, or use less, energy are not becoming available, despite the large R&D budgets of industry and the Department of Energy. There is no need to ask whether the rate of technological innovation in the United States is slower now than in the past; in fact, this question has no unequivocal answer. However, the evidence is clear that technological innovation in the United States does need a shot in the arm if the nation is to solve many of its social and economic problems.

This chapter reviews the relationships between technological innovation and important aspects of the society, with an emphasis on the significance of technological innovation for the industrial economy. It also examines briefly the conditions that are necessary for high rates of technological innovation in firms, and in the economy as a whole. The chapter concludes with a brief discussion of the policy implications of the ideas presented. Subsequent chapters in this book develop these ideas and their implications in greater depth.

The Nature of Technological Innovation

The process of technological innovation involves the creation, design, production, first use, and diffusion of a new technological product, process, or system. While the process of technological innovation is sometimes viewed as synonymous with R&D, in fact, organized R&D is only one of several kinds of innovative activity -- in some cases R&D comes only after the fact of an innovation, or not at all. The process of technological innovation can happen rapidly, or it may require an extended period of time. The changes that occur in the costs, the performance, or the characteristics of a technology may be major or incremental, but the total of the incremental changes over a period of time can be as important as the dramatic breakthroughs in improving quality or reducing the cost of a product. The outcomes of the innovation process are unpredictable in the sense that no one can say in advance what the result would be of a decision to engage in specific kinds of innovative activity within particular firms in particular environments.

Technological innovation, as used in this book, is not the same as technological change. While the distinction is not always easy to draw in practice, the latter encompasses any change in a product or a production process, such as the adoption of an existing, but better, method by a producer. Technological innovation, on the other hand, is only said to occur when a product, process, or system embodies a new method or a new idea.

One cannot directly measure the quantity or the quality of technological innovation. Of course, engineers can recognize when a product or a process embodies a novel technical concept; yet, there is no requirement that a product or process be technically novel to be an innovation; it only needs to be the first commercially successful use.

Four indirect approaches are used to measure technological innovation. The first measures inputs to the process of technological innovation, such as the R&D budget or the number of scientists working in an area. The second measures intermediate outputs, such as the number of patents awarded, the number of technical papers published, or the number of new chemical entities synthesized. The third measures the performance of a product or process, such as its speed, durability, or cost. The fourth measures the amounts of various inputs required to produce a product, such as hours of labor, barrels of oil, or dollars worth of capital equipment. None of these measures is entirely satisfactory, and some of them introduce an element of circularity into the analyses of the relationships between technological innovation and economic performance. Nevertheless, it is extremely useful to retain the notion of technological innovation as a category of activity worthy of study and of management.

TECHNOLOGICAL INNOVATION AND THE ECONOMY

A number of macroscopic indicators provide measures of the strength and vitality of the total economy over time. Five of the most significant concepts to be measured are economic growth, productivity, inflation, employment, and the balance of trade. Each of them is related to the absolute efficiency with which the national economy uses inputs from nature to satisfy human needs and desires, as well as to the relative efficiency of doing so in comparison with other countries. Efficiency is not the only aspect of the economy reflected in the measures of these five concepts. The measures also reflect the nature of political institutions and the accidents of nature. These five concepts are strongly interrelated, and these relationships make it impossible to optimize all five at once. Yet, as the following sections will try to show, technological innovation can contribute to an improvement in all of them.

Economic Growth

The Contribution of Technological Innovation to Growth

Conventionally, the rate of economic growth is defined as the rate of increase of the GNP. Fig. 1.1 shows that the growth of the GNP in the United States in the last decade has been slower and less steady than it was during the previous two decades. Furthermore, as shown in Fig. 1.2, the U.S. GNP has grown less rapidly than that of most of the competing industrial nations over the last two decades.

The GNP is only an approximate measure of economic well-being. It includes some economic activities, such as the repair of accident damage, whose growth cannot be viewed as good, and excludes many activities that make a positive contribution, such as the noncompensated services of homemakers or the amenity values of an unpolluted landscape. Furthermore, the GNP, a measure of the dollar output of goods and services, fails to capture fully the benefits of new technology or of changes in the mix of outputs that provide better services at the same or lower prices. In fact, improvements in the quality of products at constant prices may not show up in the GNP at all.

It is reasonable to expect that economic growth is supported by technological innovation -- with better technology, more output can be produced from the same inputs of human and natural resources. During the last twenty years, a number of authors have attempted to explain the history of the rate of growth of the U.S. economy or to understand why it has differed from the growth of the economies of other countries. Their approach has usually been to assume that the rate of growth of the GNP is the sum of the rates of growth of such measurable factors as resource extraction, capital investment, the labor force, and improvements in education. Since these factors cannot account for all of the rate of growth of the GNP, some portion of the residual growth is then attributed to "technological innovation" or to "technical progress."

Robert Solow estimated the contribution of technological change to the rate of growth of the output of the nonfarm portion of the American economy to be, on average, 1.5 percent a year between 1909 and 1949. [1] He concluded that 90 percent

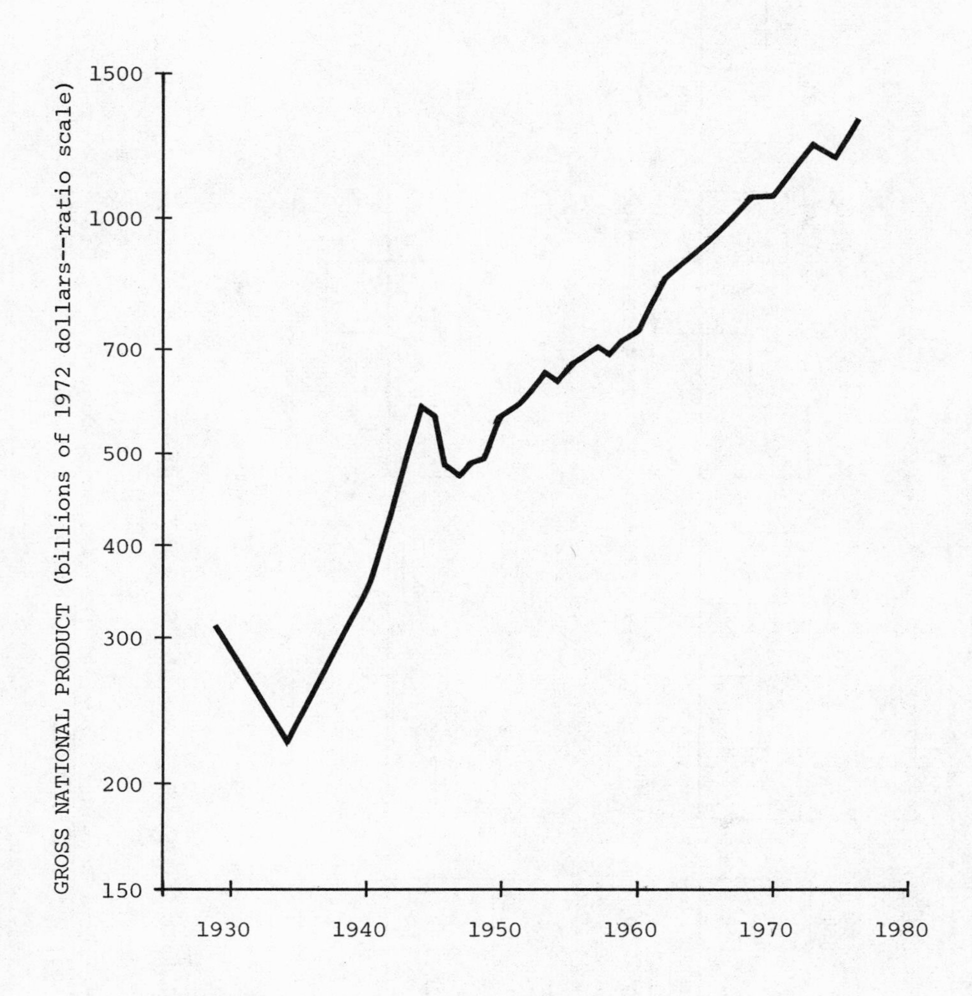

FIGURE 1.1

U.S. GROSS NATIONAL PRODUCT

Source: Economic Report of the President, January 1978 (Washington, D.C.: U.S. Government Printing Office).

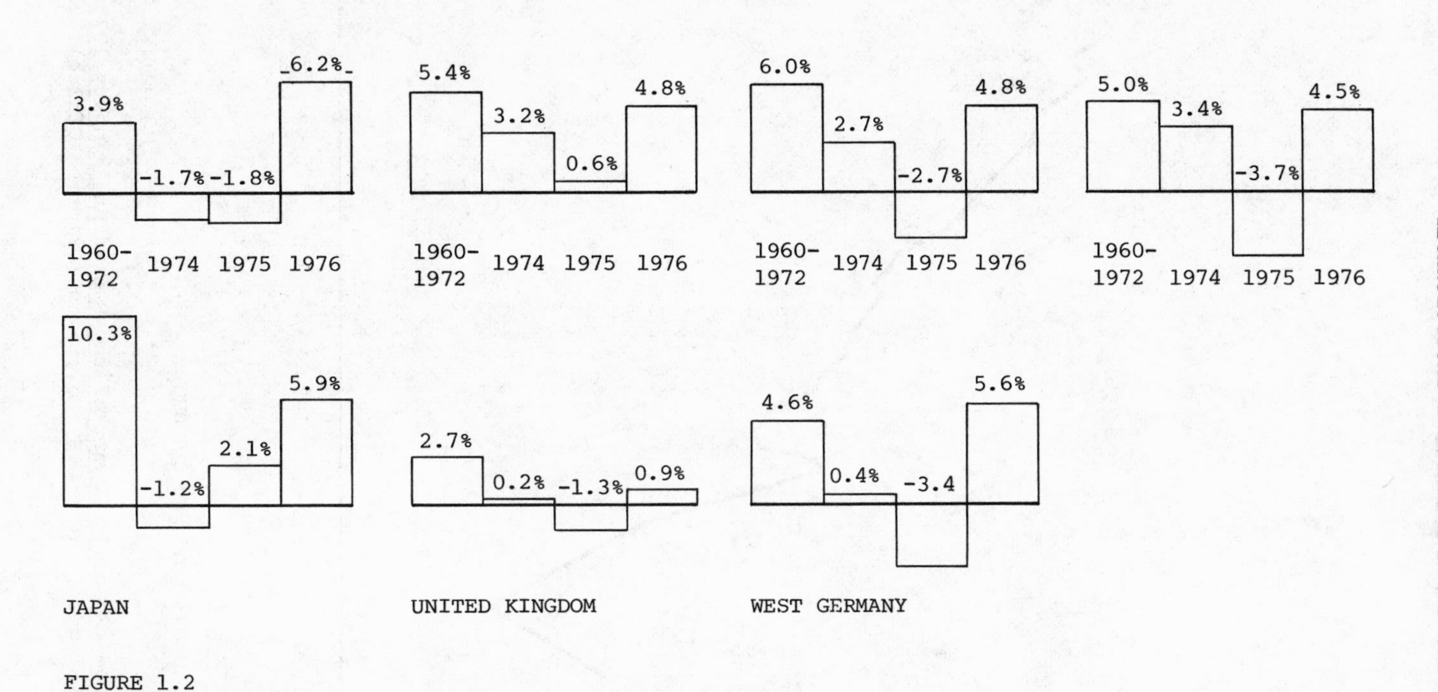

FIGURE 1.2

AVERAGE ANNUAL RATE OF CHANGE IN GNP IN SEVEN INDUSTRIAL COUNTRIES

Source: International Economic Report of the President, January 1977 (Washington, D.C.: U.S. Government Printing Office).

of the increase in output per capita was attributable to technological change, while a much smaller amount of the increase was due to higher investment in capital per worker. In other words, the country has learned how to get considerably more output from a given amount of inputs.

Dale Jorgenson and Zvi Griliches came to a considerably different conclusion. [2] They estimated the contributions of the growth in factor inputs, and of the growth in the productivity of those inputs, to the growth in total economic output for the period 1945-1965. They concluded that, of the annual rate of growth of 3.59 percent, as much as 3.47 percent could be accounted for by growth in inputs alone, and only 0.12 percent by a growth in total factor productivity. Thus, they found a very small residual to be attributed to technological innovation or other unmeasured factors. They attributed the difference between their results and studies such as Solow's to correction of the inputs for several aggregation and measurement errors made in earlier studies. Their conclusion is "not that advances in knowledge are negligible, but that the accumulation of knowledge is governed by the same economic laws as any other process of capital accumulation." [3]

Edward Denison examined the sources of growth of the GNP in nine countries during the period 1950-1962. [4] Over twenty separate determinants of growth were examined and divided into two categories: increases in inputs of the factors of labor, capital, and land; and increases in output per unit of input due to such factors as "advances in knowledge," improved allocation of resources, and economies of scale. For the period 1950-1962, advances of knowledge was inferred to be the source of 23 percent of the growth in total national income, and 34 percent of the growth in national income per person employed in the United States. (Put another way, advances in knowledge added 0.76 percent to the annual rate of growth of U.S. national income, which averaged 3.32 percent during the period.) Advances in knowledge contributed between 0.44 percent and 1.56 percent of the growth rates of the other eight countries.

Denison devoted considerable attention to the limitations of his analysis, and some of these cast considerable doubt on the quantitative results. First, advances in knowledge is actually a residual that includes all the errors in the other rates and contributions, including any factors not taken into account. Thus, it may be overestimated. Second, advances in knowledge includes both technical progress and any lag in the application of that progress, factors that cannot be estimated independently from the data. Third, since the growth of total national income does not account for the growth of activities for which output is measured by input, or for any improvements in the quality of outputs, then advances in knowledge can, in fact, include only those advances that reduce the unit costs of outputs already in existence. This factor suggests that advances in knowledge is underestimated. Finally, Denison noted that the findings for European countries are seriously confounded by the post-World War II recovery in West Germany, Italy, and other countries.

On the whole, these studies provide suggestive, but not entirely convincing, evidence that technological innovation, or at least technological change, makes an important contribution to the growth in aggregate output, or the GNP. It is then reasonable to conclude that some part of the decline in the rate of growth of the GNP is attributable to a decline in the rate of technological innovation.

Other Aspects of the Relationship Between Innovation and Growth

Even as technological innovation contributes to economic growth, a high rate of economic growth supports technological innovation. Many innovations require, for their expression or implementation, the construction of capital plant and equipment. When output is growing slowly, there is little incentive for investment in new capital; thus, the opportunity to profit from major innovation is low. When output is growing rapidly, there are many opportunities to replace obsolete plant and equipment with new and better capital, and the rate of innovation is enhanced.

Other observers have seen the relationship between economic growth and technological innovation in a somewhat different light. Concerned that economic growth, as conventionally measured, is leading to fouling of the environment and to exhaustion of the world's resources, they have suggested that in a world of "zero growth," technological innovation would become even more important than it is today, because it would be the only way to improve the quality of life within fixed limits. [5]

Still others have noted that certain technological innovations retard economic growth. For example, a government commitment to support a new technology that ultimately fails to deliver its initial promise diverts resources from other investments that might have earned better returns. Similarly, adoption of an inadequately tested new technology, which is ultimately found to be an unreasonable health hazard, may lead to a costly recall and to substitution with another technology. Thus, some degree of attention to technology assessment to determine whether new technologies might prove to be both cost-effective and socially acceptable should enhance both the rate and composition of economic growth.

Productivity

Recent Trends in U.S. Productivity

Economic society uses input factors of labor, capital, and natural resources to produce outputs of goods and services to meet its needs. In an abstract sense, technology is the formula by which the various inputs are used to produce an output. The productivity of a factor of production is defined as the ratio of the total output to the amount of that input used. An improvement in the productivity of any one of the factors of production is, therefore, a reduction in the amount of that factor required to produce a fixed amount of output, or an increase in the amount of output that can be produced using a fixed amount of the input.

To produce a product at any instant in time, it is usually possible to employ a range of each of the inputs. The actual mix of inputs used depends, in the simplest case, on their relative prices. For example, if inputs A and B are used to produce C, a firm might use more of A if A is relatively cheap than it would if A is relatively expensive. The result of the choice to use a large input of A is that the productivity of A is lower than it would be if A were more expensive, and were used more sparingly. However, if new technology is adopted over time that uses less of both A and B, the productivity of both factors increases.

An analogous situation applies at the level of an economy as a whole. A particular level of total output, or GNP, can be attained at a particular time, using many combinations of capital, hours of labor, tons of steel, acres of agricultural land, BTUs of energy, and so on. The actual choice of inputs, and, therefore, the productivity of each factor, depends, among other things, on the demands for each kind of output and on the prices of each input. As time goes on, technological change can improve the productivity of any or all of the factors.

Traditionally, economic analysts have focused their attention on the productivity of labor in manufacturing; that is, on the output of manufacturing industries per hour of labor input. Increases in labor productivity are often taken as evidence of increases in overall productivity, even though, for any particular industry, there is no reason to suspect that this relationship holds. Fig. 1.3 shows the trend in labor productivity over the last two decades in the United States, and Fig. 1.4 compares the growth in U.S. labor productivity with that in selected foreign countries. These data show that labor productivity in the United States has grown more slowly in recent years than in the past, and that labor productivity in most other industrialized countries has grown more rapidly than in the United States. [6]

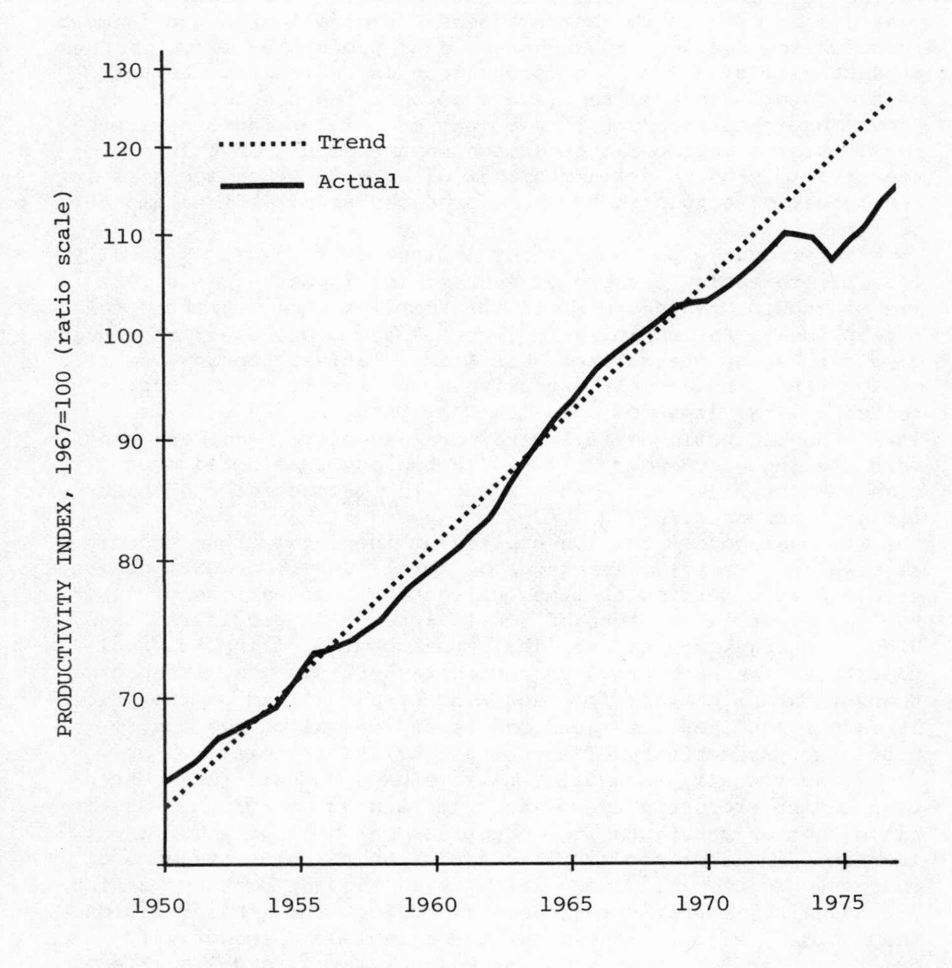

FIGURE 1.3

LABOR PRODUCTIVITY IN THE PRIVATE NONFARM BUSINESS ECONOMY

Source: Jimmy Carter, Economic Report of the President (Washington D.C.: U.S. Government Printing Office, 1978).

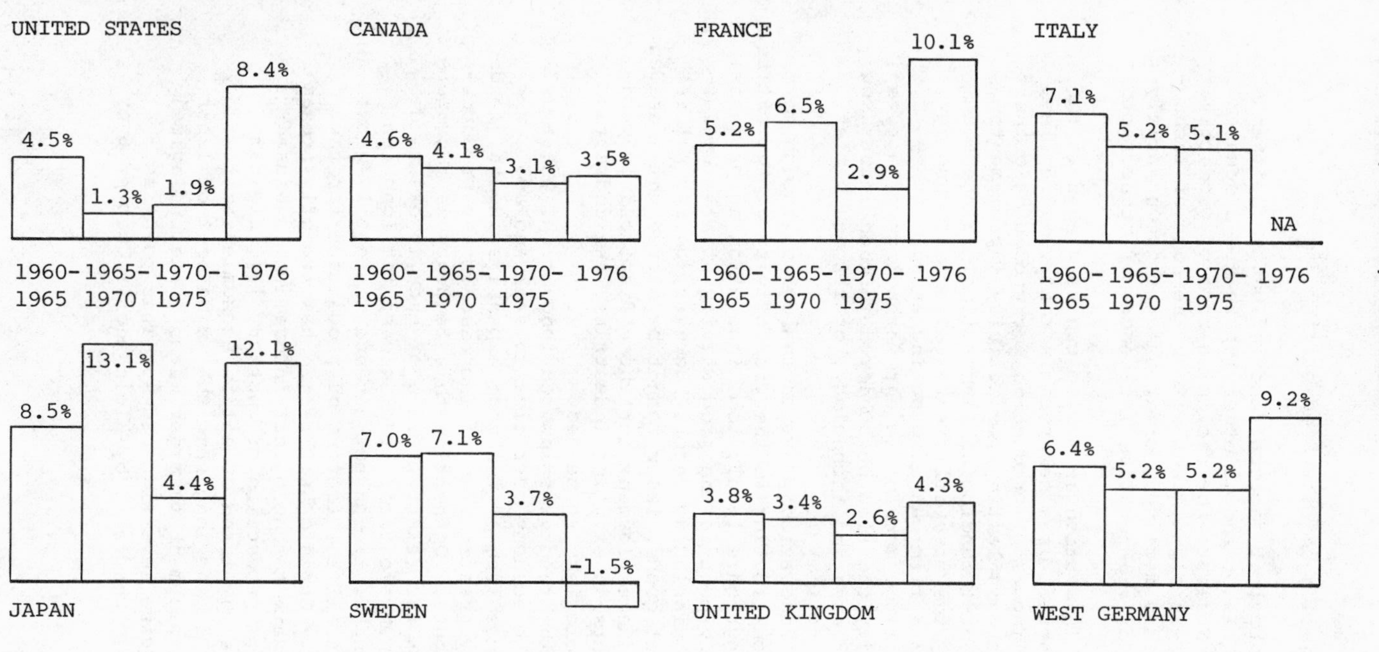

FIGURE 1.4

AVERAGE ANNUAL RATES OF CHANGE IN LABOR PRODUCTIVITY IN MANUFACTURING IN EIGHT INDUSTRIALIZED NATIONS

Source: International Economic Report of the President, January 1977 (Washington, D.C.: U.S. Government Printing Office).

Technological Innovation and the Productivity Sag

Many nontechnological factors that influence the trends in the productivity of labor and of other inputs are discussed in the next section. However, the declining rate of productivity improvement is a clear signal that either the rate of technological innovation has declined, or it has not grown rapidly enough to make up for the negative influences of other factors on productivity.

Technological innovation in a firm can be viewed as the development and adoption of a new formula that requires less of one or more of the inputs of production for producing outputs. In fact, substantial and lasting productivity improvements usually arise from major technological innovation. Since most firms do not employ the most efficient available technology, the productivity of a firm can also be increased just by adoption of an existing, more efficient technology. Improvements in productivity can also be achieved through increasing the scale of use of existing technology, or through better management and control.

Technological innovation plays an analogous, though more complex, role in contributing to the productivity of the entire economy. At the aggregate level, too, innovation in processes can enable an improvement in the productivity of all the factors of production, including labor. However, unlike natural resources or capital goods, labor cannot be stored until needed. Therefore, to provide employment for those displaced by productivity improvements in existing industries, new industries based on new technologies are needed.

According to the theory of induced innovation, there will be a tendency in the economy over time to seek and adopt new technologies that use less of the more expensive inputs, leading to a relatively greater rate of increase in the productivities of those inputs. Offsetting this tendency is a compensating tendency for the relative price of such inputs to decline as less of them are used, leading over the long run toward a more balanced, though not necessarily equal, growth in the productivity of all the factors of production for the economy as a whole. Christopher Freeman has recently reexamined whether aggregate technological innovation is biased toward one factor of production or another. [7] His data suggest that, while the trend of entire economies is toward neutral innovation in the long run, the adjustment of labor productivity and employment to major waves of technological innovation may require several decades. This point is discussed further in Chapter 3 by Klein and in Chapter 6 by Bourdon.

Other Explanations of the Productivity Sag

Several explanations, in addition to an inadequate rate of technological innovation, have been advanced to explain the current trends in productivity. Some have argued that the quality of the labor force in the United States has declined, or that people do not work as hard as they once did. Others have noted a decline in the quality of the labor force as a result of the large number of less experienced new entrants -- both women and members of minority groups. Yet another explanation is that the growth in the number of people in the work force (due to the baby boom of the 1940s and 1950s, as well as to the entry of women and minority group members) has increased the supply of available labor. According to this view, the law of supply and demand should cause a decline in the relative price of labor, accompanied by a corresponding drop in the rate of increase of labor productivity as business finds it more profitable to use less of other inputs and more labor.

Still other analysts have noted that the real prices of resources have increased in recent years, due to physical scarcity, to increased demand by other countries, and to political actions such as the formation of the oil cartel and the bauxite organization. As the prices of natural resources have increased, it is only logical that industry should have substituted the relatively less expensive commodity -- that is, labor -- for the more expensive resources. Thus, while the productivity of labor has not grown as rapidly as it has in the past, it may be that the overall output of the economy has continued to grow in the last five years as resources have become relatively more expensive, only because more labor has been used per unit of output.

A number of other factors influence the trend of measured productivity. First, the productivity statistics do not capture improvements in the quality of the outputs. Second, productivity can change just because the mix of outputs shifts in response to changes in consumer tastes or to opportunities to allocate existing resources more efficiently. Third, government regulation of environmental, health, and safety problems may cause measured productivity to decline as the costs, which were formerly not accounted for, are paid. (See Chapter 5 by Ashford, Heaton, and Priest for an elaboration of this point.) Fourth, government regulation to improve the functioning of markets (e.g., antitrust, or labeling and disclosure requirements) may enable consumers to purchase more outputs for the same total expenditure, leading to the sale of a more productive market basket of goods and services. Fifth, the long-term trends of economic activity away from manufacturing toward services, and toward the public sector, where productivity is less susceptible to technological improvement, has made it more difficult to raise the productivity of the economy as a whole.

Finally, improving productivity in some industries has become more difficult as diminishing returns to further technological development have been encountered.

Obviously, many factors have contributed to the declining rate of productivity growth, and searching for a single explanation would be fruitless. Nevertheless, it is clear that productivity improvements in the long run do not come about because people work harder -- they come about because people work smarter. And working smarter is heavily dependent on working with better technology.

Inflation

Inflation is defined as the rate of change in the overall price level. While the units of money are completely arbitrary, in the real world three major factors make inflation important. First, persons whose wealth or income are defined in terms of specific dollar amounts suffer an increasing loss if those dollar amounts are not allowed to increase with inflation. Second, as the price level in a country increases, its products become less attractive to buyers from other countries and, conversely, the products of other countries become more attractive to buyers within the country. As a result, the balance of trade deteriorates unless offset by currency adjustments. Third, a high rate of inflation tends to discourage both investment and innovative activity.

Fig. 1.5 shows the familiar fact that the rate of inflation in the United States has increased during the last decade. The classical explanations for inflation -- government deficits, expansion of the money supply, excessive wage demands by labor, excessive price demands by industry, and excessive consumer demands for goods -- have all been used as the foundation for federal policies to control inflation. All of these policies have failed.

In Chapter 3, Klein offers a more fundamental explanation for this persistent period of inflation. Briefly, he argues that the high, and increasing, rate of inflation results from a failure to develop and implement new, more productive technologies that could lower the amounts of inputs required to produce goods and services and thus reduce production costs and prices. The reasons for this failure are that American industry has become less competitive as it has matured, and that both industry and labor have sought government policies and used private contracts to protect existing products, markets, and jobs. According to his analysis, if industry were required to take more risks by the pressure of more effective competition, it would be forced to develop and adopt new technologies that are both more productive and less inflationary. Failure to do so in a competitive environment would mean eventual disappearance of firms that could not or would not take risks. Similarly, in a truly competitive environment, labor could not long achieve wages above those justified by productivity improvements. Government policy could best contribute to controlling inflation by emphasizing labor retraining and relocation, rather than by protecting existing jobs and industries when they are threatened by domestic or foreign competition.

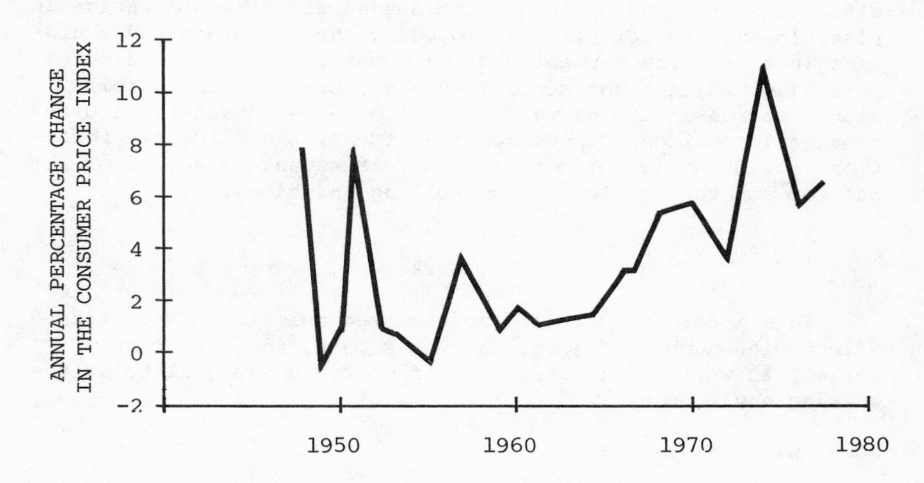

FIGURE 1.5

INFLATION IN THE UNITED STATES

Source: Economic Report of the President, January 1978 (Washington, D.C.: U.S. Government Printing Office).

The persistent high rate of inflation may be related to other phenomena involving technological innovation. First, such natural resources as petroleum, natural gas, aluminum and wheat are becoming more expensive in real terms due both to physical exhaustion and to political decisions to restrict their availability. Such cost increases are translated into inflation unless new, more productive technologies are developed that use less of these resources or of other inputs. For example, in the case of energy, evidence is growing that inflation can best be controlled by adoption of policies that would allow fuel prices to rise to market levels, rather than by policies to hold down prices through supply subsidy or regulation. [8] Such price rises can best be offset by the development and adoption of new technologies that use energy more efficiently. Otherwise, society will pay indirectly, through general taxation, for the full costs of more expensive energy, without the benefits of new, more productive, technologies that use less energy.

Finally, monetarist stabilization policies designed to control inflation in the short run by shrinking the money supply may also have contributed to the decline in productivity improvement and to the continuing high rate of inflation. As noted by James Quinn, tight money and high rates of inflation discourage the kind of investment needed for firms to engage in risky innovative activity and to build the new plant and equipment that embodies technological innovation. [9] According to this view, while tight money might be necessary for economic stabilization in the short run, the long-run impact is to discourage innovation, discourage investment, and stimulate inflation. If Quinn is correct, a fresh assessment is needed of the role of monetary policy in controlling inflation.

Work

In a modern industrial society, technological innovation affects the number of jobs, the wages paid, and the hours worked, as well as the character of work and the quality of the working environment.

Jobs, Wages and Hours

Historically, two widely divergent views have been held regarding the impact of new technology on jobs: according to the first view, innovation eliminates jobs; according to the second view, innovation creates jobs.

The heart of this dichotomy is the distinction between the local, short-run effects and the aggregate, long-run effects of innovation on jobs. In the short-run, adoption of new production technology that makes labor more productive means that fewer workers will be needed to produce the same level of output. Thus, in the short-run, jobs are lost, often with devastating impact on individual workers, families, and communities. In the long run, however, several factors offset this job loss. First, if the higher labor productivity is translated into lower prices for the product, more of the products will be purchased and more labor will be needed to produce them. Whether this production increase offsets the original job loss depends on the elasticity of demand for the product. In industries with a low elasticity of demand such as basic metals, agriculture, textiles, or railroads, demand expansion is unlikely to replace all the jobs lost by productivity increase, [10] whereas industries with a high elasticity of demand show a more-than-proportional job gain.

Second, additional wealth becomes available to society in the aggregate when a product is produced at a lower cost due to technological innovation. This wealth can be used to produce entirely new products and services whose production, in turn, creates new jobs in the economy. This mode is especially important, because production of new products usually is more labor intensive than production of older products using mature technology. Thus, a continual flow of new product technology helps to absorb jobs lost through increased productivity of processes used for older products. Third, if there is effective competition from other firms, or from foreign producers, and if new, more productive, technology is not developed, it is likely that sales of the product would be lost to competitors anyway. Thus, new technology can help protect the jobs that do remain after it is adopted.

The net impact of technological innovation on total employment remains a matter of some debate. (See Bourdon's discussion in Chapter 6.) Confusion is added by the fact that, while unemployment levels have been high for the last several years (see Fig. 1.6), total employment has grown at record rates (see Fig. 1.7) as the percentage of the population in the work force has grown to record levels. Nothing in these data suggest that technological innovation, or any other factor, has eliminated jobs over the long run. [11]

By enabling labor productivity to increase, technological innovation has contributed to a reduction in the number of hours worked and to an increase in compensation. Data on these mea-measures are shown in Figs. 1.8, 1.9, and 1.10. Since 1950, the average workweek has declined from just under 40 hours to 36 hours. Simultaneously, hourly compensation has increased, so that weekly compensation has risen from $70 to $95 per week in 1967 dollars ($50 to $170 per week in current dollars). Since about 1965, real weekly compensation has remained nearly constant in real terms, while labor's share of productivity gains has been taken in fringe benefits and in fewer hours worked.

The Character and Quality of Work

Technological innovation affects not only the quantity of work to be done, but the quality and character of that work as well. For example, much of the past innovation that has facilitated large increases in labor productivity was process innovation that enhanced the division of labor. This bureaucratization of work has tended to make individual labor less satisfying since workers have exerted less control over the nature and quality of the products to which they contribute. Under these circumstances, individual workers are less productive than they might be if they felt less alienated from their work, and this phenomenon may help to explain some of the decline in productivity growth. The reductionist nature of some new

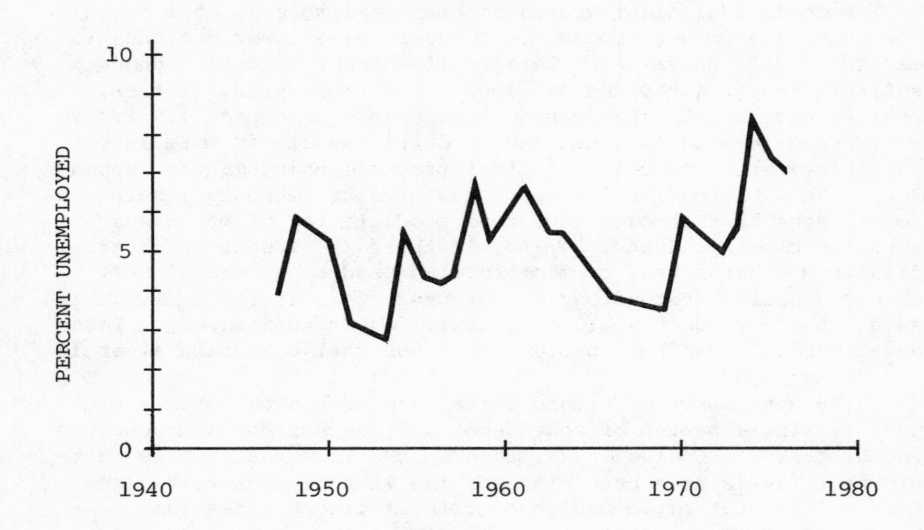

FIGURE 1.6
UNEMPLOYMENT RATE

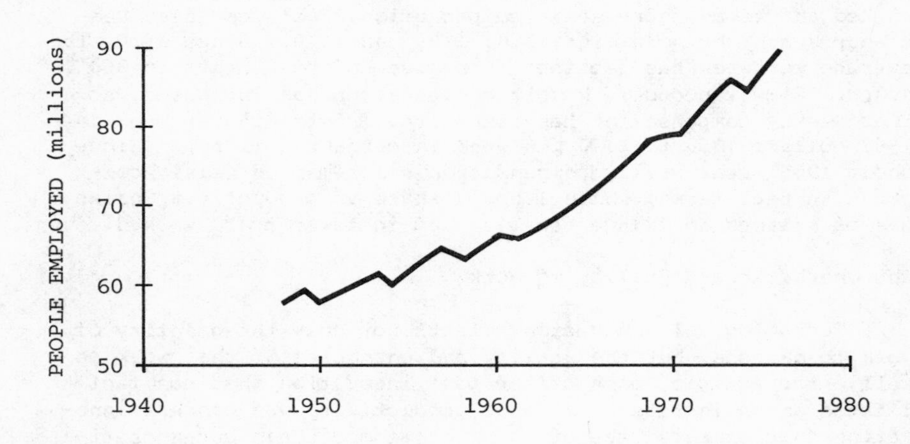

FIGURE 1.7
TOTAL NUMBER OF PEOPLE EMPLOYED

Source: Employment and Training Report of the President, 1978 (Washington, D.C.: U.S. Government Printing Office).

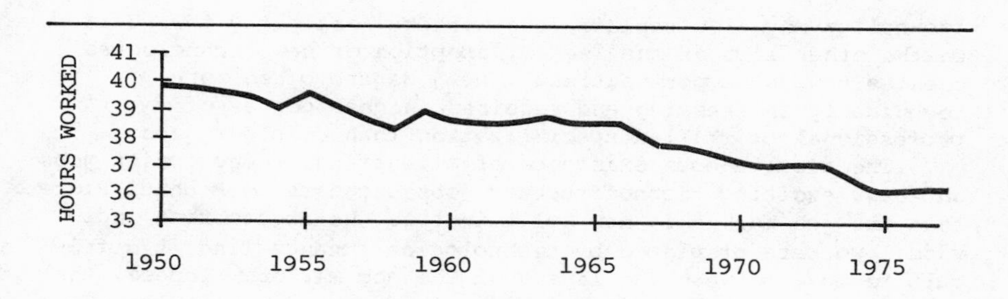

FIGURE 1.8

AVERAGE HOURS WORKED PER WEEK IN U.S. PRIVATE EMPLOYMENT

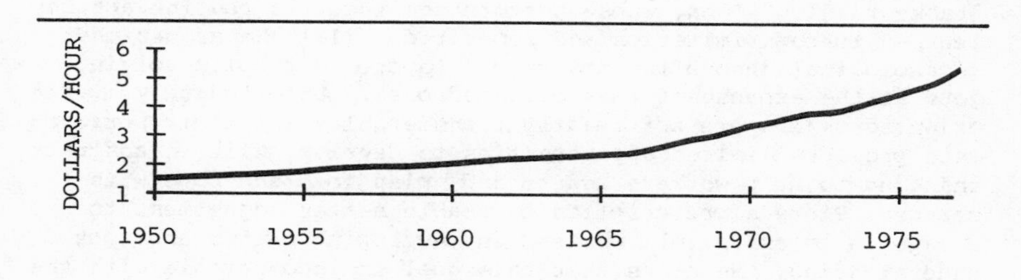

FIGURE 1.9

AVERAGE HOURLY WAGES IN THE UNITED STATES
(CURRENT DOLLARS)

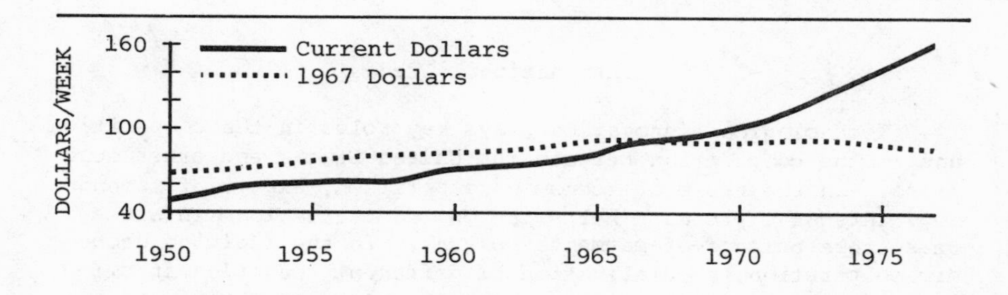

FIGURE 1.10

AVERAGE WEEKLY COMPENSATION IN THE UNITED STATES

Source: Employment and Training Report of the President, 1978 (Washington, D.C.: U.S. Government Printing Office).

technology may also explain some workers' antipathy toward it. On the other side of the ledger, adoption of new technologies creates new job opportunities. The jobs are often more technically interesting and require a higher degree of professional or skilled specialization than do older jobs.

The simultaneous existence of alienating, reductionist jobs and new, exciting technostructure jobs suggests to Michael Piore that the job market is now split in two. He argues that individual workers displaced by technological change find it difficult to move between the layers of the job market. Indeed, the United States is currently experiencing both high unemployment rates and a high level of unfilled demand for skilled employees, i.e., the phenomenon of structural unemployment. To explain this, Piore points out that an industrial society features two broad categories of jobs: (1) "problem-solving" jobs, which require original abstract thinking for their performance, and (2) "task-oriented" jobs, whose performance requires routine actions learned through imitation and repetition. [12] He argues that technological innovation has tended to create problem-solving jobs at the expense of task-oriented ones. Unfortunately, task-oriented skills are not readily transferable, and their performance provides limited opportunities to develop skill in abstract thinking to help workers understand, plan for, and cope with change. Piore's prescription to enable better adjustment to change is to assist all workers in developing better abstract understanding. He notes that this goal is incompatible with the division of labor and, therefore, that an emphasis should be placed on enriching the problem solving content, i.e., the wholeness, of work. This view is in keeping with Klein's observation in Chapter 3 that there is a trade-off between static efficiency (here, the division of labor) and dynamic efficiency (here, a labor force willing and able to cope with innovation).

International Trade

Technological innovation plays key roles in the competition and in the cooperation between the United States and other countries. In the realm of economic competition, export of technology-intensive products helps the United States to maintain a reasonable balance-of-payments posture. In the realm of economic cooperation, specialization of different countries in different areas of technology can help the world as a whole to advance rapidly, avoiding a return to protectionist foreign economic policies here or abroad. This section examines the role of technology in influencing international trade, the position of the United States in light of that role, and the role of technology transfer in the politics of trade.

The Role of Technology in International Trade

According to the classical comparative advantage theories of international trade, countries are expected to import products embodying factors of production that are relatively expensive in the importing country and to export products embodying relatively less expensive factors. In other words, a country with cheap labor and expensive capital would export labor-intensive goods and import comparatively capital-intensive products.

In testing this theory, however, Wassily Leontief discovered that it could not explain actual trade patterns. For example, exports from the United States are concentrated in product areas in which labor plays a significant role, even though U.S. labor costs are high and resource costs are low compared with most of its trading partners.

One explanation for the actual trade patterns has been found in the product cycle theory of international trade. [13] (See Chapter 4 by Graham for a review of alternative explanations.) According to this theory, U.S. exports are heavily concentrated in new technology-intensive products, even though production of such products typically involves a large input of relatively well-paid, skilled labor. However, because of these products' superior performance or unavailability elsewhere, they are successful on world markets. The second part of the theory is that the United States has been a steady source of such products both because its economy has characteristics that reward innovativeness, and because it has a large, affluent domestic market ready and willing to purchase such goods.

The product cycle theory also explains why countries that are dominant exporters of a product at one time may lose their dominance and become importers at another. Utterback points out in Chapter 2 that as a new product evolves, production processes become more rationalized and integrated, and the product's performance qualities become more standardized. Thus, production costs come to reflect less the availability of entrepreneurs, technical expertise, and skilled labor, and more the availability of unskilled labor and low-cost raw materials. In these circumstances, the advantage of technically sophisticated countries disappears, and production can move to those less-developed countries that possess an adequate infrastructure and the skills to adapt and imitate technologies developed elsewhere. Thus, as technologies and industries mature, the classical theory of comparative advantage is more able to explain patterns of trade in them. This evolution is perhaps best typified by the steel industry, whose growth center has moved from Germany and the British Isles, to the United States, to Japan, and now to such countries as South Korea, Brazil, and Venezuela.

An interesting test of the importance of new technology in international trade was recently made by Keith Pavitt and Luc Soete. [14] They found that, for many classes of products, a country's share of world exports was strongly related to a measure of its innovativeness in that industry. (The measure of a country's innovativeness in an industry was taken to be the fraction of all U.S. patents in that industry awarded to persons in that country.) Exceptions to their findings were products of the materials sectors and in nondurable consumer goods. The exceptions were tentatively explained by the fact that exports of such goods are controlled heavily by foreign firms whose innovative activity is carried out in their home countries.

Innovation and the U.S. Position in World Trade

One widely quoted piece of evidence in support of the importance of technology in international trade is the U.S. trade balance in so-called "R&D intensive" and "non-R&D-intensive" manufactured products. As shown in Fig. 1.11, the United States maintains a strong positive balance in the first kind and a highly negative balance in the latter.

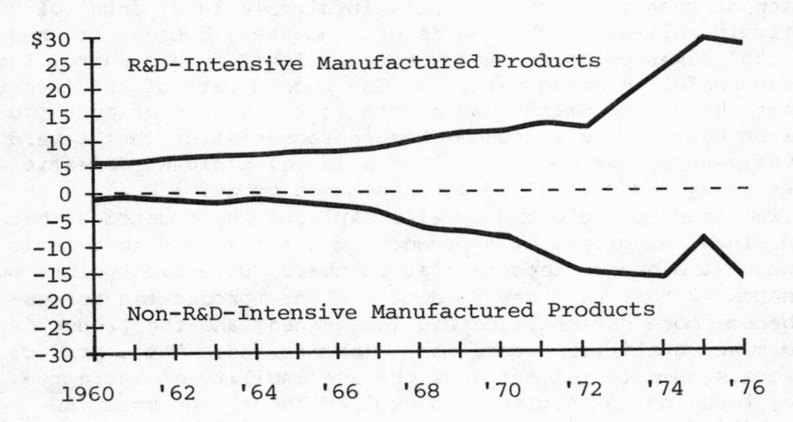

FIGURE 1.11

U.S. TRADE BALANCE (exports less imports) IN R&D-INTENSIVE AND NON-R&D-INTENSIVE MANUFACTURED PRODUCT GROUPS, 1960-76

Source: Science Indicators 1976, National Science Board, National Science Foundation (Washington, D.C., 1977).

At the end of World War II, the United States had a commanding lead over the rest of the world in many areas of technology. While the industrialized countries of Europe and Japan were struggling to rebuild, the United States advanced rapidly by capitalizing on its wartime developments and on its backlog of consumer demand. In the succeeding thirty years, concern has become widespread that the United States has lost its lead in technological innovation. [15] Solid evidence of this decline is scarce, yet corroborating evidence and the opinion of many experts supports the idea. In particular, the more rapid rates of economic growth and productivity improvement in other countries, the deteriorating U.S. balance of trade, the heavy emphasis of U.S. R&D on defense and space (see Fig. 7.6 of Chapter 7), and the declining fraction of U.S. patents granted to U.S. inventors (Fig. 1.12) suggest a loss of the U.S. lead. On the side of inputs to technological innovation, the decline of U.S. R&D spending as a fraction of the GNP in contrast to an increase in some other countries (see Fig. 1.13) also supports the conclusion that the U.S. lead in technological innovation is disappearing.

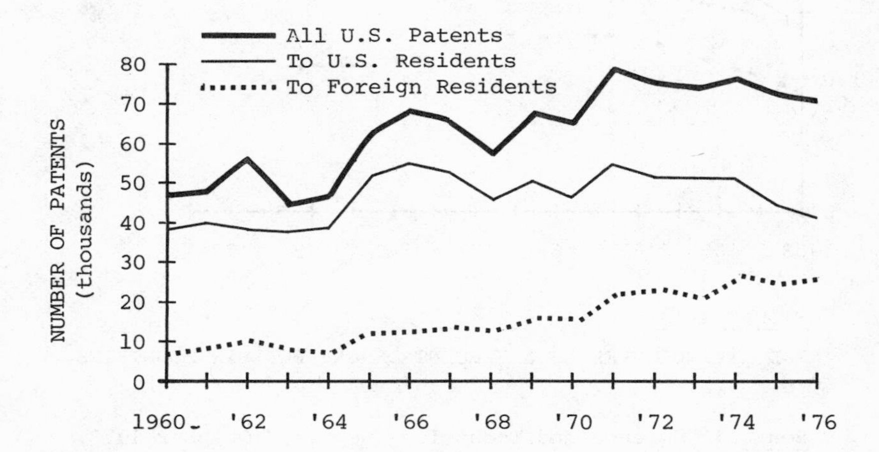

FIGURE 1.12

U.S PATENTS GRANTED TO U.S. AND FOREIGN RESIDENTS, 1960-76

Source: Science Indicators 1976, National Science Board, National Science Foundation (Washington, D.C., 1977).

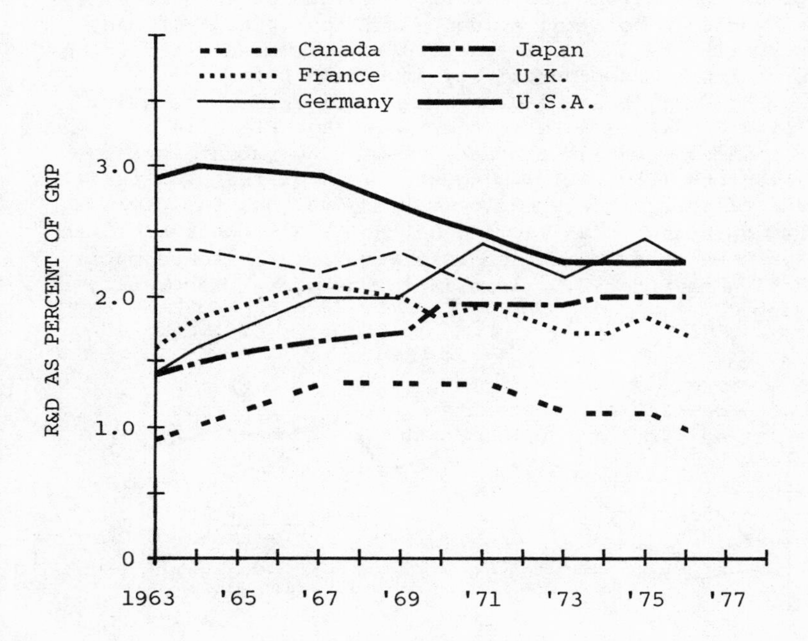

FIGURE 1.13

R&D EXPENDITURES AS A PERCENT OF GNP FOR SIX COUNTRIES

Source: "Science and Technology Report," October 1978, Committee on Science and Technology, U.S. House of Representatives, 95th Congress (Washington, D.C.: U.S. Government Printing Office).

However, there are other interpretations and evidence to the contrary. First, aided by the United States, it is reasonable that Japan, Germany, France, the United Kingdom, and Italy should have been able to recover rapidly from wartime destruction in light of their previous states of development. Thus, the relative position in technology of the United States at the end of the war would have been nearly impossible to maintain.

Second, the U.S. balance of trade in technology-intensive manufactured products has remained strong (see Fig. 1.11). Furthermore, U.S. technology-intensive agricultural exports add considerably to this balance.

Third, the patent balance decline is not unique to the United States; rather, it reflects the growing importance of technology-intensive trade worldwide. Much more attention is being paid to foreign patenting in general, [16] and there is preliminary evidence that most major countries are experiencing a declining fraction of patent awards to nationals. Beyond this fact, the current area of technological growth for the United States is semiconductor-based electronics, and there is evidence that innovations in this area are not being patented at all. It is too easy to "invent around" a patent in this field, and the field is changing too fast for a patent to provide protection worth its cost. Instead, to protect their innovations, companies depend on trade secrets, on maintaining physical secrecy by encapsulating entire devices in epoxy resin, and on a rapid rate of product obsolescence.

Fourth, the widely noted decline in the U.S. support for R&D may not be significant for international trade. Much of the decline has been in military and space R&D, rather than in basic research or in industrial R&D for producer and consumer markets. While there are commercial spin-offs from military and space research, it tends to be spent heavily on testing expensive prototypes and building expensive facilities. These expenditures probably contribute less to the trade balance than do other kinds of R&D.

Fifth, a different picture emerges of the U.S. relative R&D intensiveness when it is compared to value-added in manufacturing rather than to the GNP. Table 1.1 shows that the United States is considerably more R&D intensive than all countries, except Japan, in terms of enterprise-funded R&D, and that it is nearly twice as intensive as all countries, including Japan, in terms of total funding of R&D by both government and industry.

What emerges, then, from the data on the U.S. lead in technological innovation is a mixed picture: some indicators suggest a problem and some possible causes, others do not. What can be most surely said is that if there has been a decline in the relative position of the United States, one cannot find a cause for the decline in the R&D or patent statistics.

TABLE 1.1

THE RATIO OF R&D EXPENDITURES TO VALUE ADDED IN MANUFACTURING IN TWELVE COUNTRIES (by percent)

	Enterprise-funded R&D					Total R&D				
	1963-1964	1967	1969	1971	1973	1963-1964	1967	1969	1971	1973
Belgium	2.3	2.1	n/a	2.1	2.4	2.4	2.3	n/a	2.4	2.6
Canada	1.1	1.4	1.3	1.2	1.2	1.3	1.7	1.6	1.5	1.6
Denmark	n/a	n/a	n/a	1.9	1.9	n/a	1.4	n/a	2.0	2.0
France	1.4	n/a	1.6	1.8	1.9	2.2	3.3	2.8	2.7	2.8
Germany	2.0	2.2	2.3	2.6	2.3	2.1	2.6	2.6	3.0	2.9
Italy	1.3	1.5	1.5	1.9	1.3	1.3	1.6	1.6	2.1	1.6
Japan	2.2	2.2	2.5	2.8	3.0	2.2	2.2	2.5	2.8	3.7
Netherlands	n/a	n/a	4.1	n/a	n/a	n/a	4.7	4.3	4.3	n/a
Norway	0.9	1.3	1.3	1.4	1.0	1.0	1.5	1.5	1.7	2.1
Sweden	2.4	2.6	2.4	2.8	2.8	3.3	3.3	2.8	3.3	3.5
U.K.	n/a	n/a	n/a	n/a	2.0	n/a	n/a	n/a	n/a	3.5
U.S.	2.7	2.7	3.2	3.3	3.1	6.3	5.8	5.9	5.6	5.0

Source: Sumiyo Okubo, Rolf Piekarz, Eleanor Thomas, "International Comparison of Enterprise-Funded R&D in Manufacturing" (Paper presented at the Engineering Foundation Conference, Easton, Md., October 1977). Reprinted in "Science and Technology Report," Ocotber 1978, Committee on Science and Technology, U.S. House of Representatives, 95th Congress (Washington, D.C.: U.S. Government Printing Office).

Technology Transfer and the Politics of Trade

Historically, nations have often attempted to alter their trade patterns by establishing tariffs, quotas, tax preferences, and other barriers to imports or stimuli for exports. As they have grown to appreciate the significance of new technology in determining export opportunities, countries have adopted explicit policies related to the transfer of technology. Such policies are designed either to accelerate or to retard the flow of technology among nations.

The entire field of technology transfer has become politicized, [17] and decisions about the flow of technology across national borders have become a matter of public policy. Simultaneously, the development of rapid communications and transportation among most nations of the world, and the adoption of the English language as the standard language of world scientific and technological communications, have made it increasingly difficult for individual countries to keep technology at home. This is true whether technology is patented, treated as a trade secret, or licensed. Countries are also limited by the growth of the multinational corporations, which view international technology transfer as an internal business decision.

Technology can be transferred across national borders in several ways. Abstract understanding and know-how can move through scientific and technical publication, the patent literature, education, training, and technical assistance. In the short run, these methods of technology transfer are relatively ineffective in building a base for competition, because countries need time to absorb the transferred information and to convert it into operating technical systems. For developing countries, however, these methods may offer the best long-run strategy for building a domestic ability to adapt technologies to meet local needs and opportunities, as well as to innovate.

A second method of technology transfer is the actual sale or movement of productive facilities, along with the persons trained to operate them. This kind of technology transfer is more immediately effective in establishing a competitor to the originating country's firms than the flow of information alone. Other methods of international technology transfer include the direct sale of technology-intensive products; licensing of domestic technology to foreign firms; direct investment by foreign nationals in domestic firms; and the participation of design, engineering, and construction firms in foreign project development. Each of these methods has been the focus of proposals to limit or to accelerate technology transfer from the United States, especially to the less-developed countries.

An important implication of the product cycle theory of trade is that the ability to imitate (to adopt and adapt) is nearly as effective as the ability to innovate in terms of maintaining a favorable trade position. The experience of Japan in the postwar period bears this out, and it appears that the Peoples' Republic of China is about to embark on the same course.

It has been said that U.S. firms are less successful today in adopting foreign technology than firms in Japan, for example. There are several explanations for this observation. Americans face more serious cultural and linguistic barriers to adoption of foreign technologies than do citizens of other countries, who have found knowing English and understanding American culture to be a necessity. Furthermore, the myth of "Yankee ingenuity" has probably strengthened the reluctance to accept technology not invented in the United States -- the "not invented here" syndrome. Finally, the number of American students who study business, engineering, and the like overseas is miniscule when compared with the number of foreign students studying in these fields in American universities. [18] In an increasingly competitive world, the United States needs to pay greater attention to the potential advantages of technology adaptation and to methods for encouraging it.

Control of technology transfer is only one way to intervene politically in trade; it may not be as effective as strengthening the domestic capability to innovate. The first rank of countries in terms of innovation (e.g., the United Kingdom, Sweden, Germany, Japan) are recognizing that domestic policy can strengthen trade by emphasizing measures to enhance technological innovation. (See Chapter 8 by Hollomon for a review of programs for this purpose in several countries.) The less developed countries have begun to recognize that they have limited ability to short-circuit the product cycle through such strategies as rapid industrialization or urbanization. Instead, they have begun to turn to policies that build on their labor and resource advantages, that strengthen basic infrastructure, and that build an indigenous capability to adapt foreign technology to fit local needs and capabilities.

TECHNOLOGICAL INNOVATION AND THE QUALITY OF LIFE

The preceding section is concerned largely with the relationships between technological innovation and measures of the health of the economy, such as economic growth, employment, productivity, inflation, and the balance of trade. Of course, the economy is not an end in itself. The ultimate goal of each of these economic activities is improving the quality of human life. Of equal importance to the quality of human life are other activities that are affected by technological innovation and that are captured imperfectly or not at all in the economic framework discussed so far. This section examines the implications of innovation for some of the other aspects of the quality of life.

Supporting the Quality of Life

New technology plays an important role in improving the human condition. Few people disagree with the assertion that the quality of life has been directly enhanced by development of technologies for sanitation, disease prevention and treatment, communication, transportation, housing, clothing, food production, and so on. In a very real sense, intelligent development and use of tools (technology) is what has distinguished humankind from the rest of nature.

Threats to the Quality of Life

While its positive contributions are manifold, the rapid rate of technological innovation in the last century has also created many serious new threats to the quality of life. Rapid population growth, threats to the productivity of the natural environment, hazards to human health and safety, the enhanced power of terrorism, and highly destructive military weapons are all examples of problems traceable directly to the development of new technology. Paradoxically, technological innovation in other areas is essential to managing these problems.

Development of new technologies has helped to understand and control the negative effects of innovation. For example, developments in chemical instrumentation and processes have contributed to our ability to identify, understand, and ameliorate the health hazards posed by the release of toxic chemicals in the environment. Similarly, manipulation of large data bases on computers using the tools of epidemiology and econometrics has enabled people to identify and characterize other kinds of environmental and social problems. Satellite reconnaissance helps monitor world-wide military developments. Rapid communication and transportation have made it possible for scientists, engineers, citizens and policy makers to convene for addressing new problems when they arise. In each of these examples, some new technology has contributed to the creation of problems, while other new technology has contributed to attempts to resolve them. It is clear that society needs and values rapid development of new, low-cost technologies designed to identify, measure, and control its problems.

However, coping with the negative effects of technological change requires new institutions for social control as well as new technologies of control. Thus, in the last decade, the United States has featured rapid growth of government regulatory programs intended to manage the effects of new technology, especially in the areas of the environment, health, safety, and consumer information. The need for much of the social welfare legislation can also be traced to the impacts of technological change on individuals, families, and communities. Some have argued that the regulatory efforts cost too much [19] and that they inhibit innovation, thus contributing to inflation and to a slowing of economic growth. To see whether these claims have any merit, it is necessary to view the connection between economic well-being and government regulation of the side effects of technological innovation in the short, medium, and long runs. (See Chapter 5 by Ashford, Heaton, and Priest for a detailed assessment of this issue.)

In the short run, the primary effect of government regulation is to redistribute the costs and benefits of production and consumption. Generally, production costs and consumer prices go up, and measured productivity and measured economic output and growth decline. At the same time, other costs in the society go down, such as accident repair, amenity loss, health care, and loss of life. Thus, if regulations are well-designed, their immediate effect is likely to be no net change, or even an increase in total welfare -- often with redistribution of that welfare from "haves" to "have-nots." [19] As Allen Kneese and Charles Schultze have noted, the higher prices that result from including the social costs of production in the market costs are not inflationary, they represent only a change in accounting conventions. [20]

In the "medium run," government regulation of the effects of technological innovation may pose a barrier to some further innovation, while stimulating other innovation. With some luck in regulatory design, innovation will be directed away from problem areas (e.g., highly flammable fabrics) and toward more benign ones (e.g., safe fire retardants or fire-resistant fabrics). There is also the possibility that regulation dampens risk taking -- that it discourages the would-be innovator by reducing the chances that a new product will prove acceptable to both the market and the regulators. However, there is also the possibility that regulation disrupts entrenched and mature organizations and that by doing so it serves as a catalyst for change. (See the discussions of innovation in mature organizations by Utterback in Chapter 2, Klein in Chapter 3, and Ashford, Heaton, and Priest in Chapter 5.) At the current level of knowledge, no one can claim to be able to account for (let alone predict) what the net effect of any particular regulatory program has been (or will be) on technological innovation over the medium term. [21]

Over still longer periods of time, the real question is not whether government regulation of the negative effects of technological innovation is bad for the economy, but whether there could even _be_ an economy in the absence of at least some regulation. With the power and complexity typical of many new technologies (consider thalidomide, carcinogens in drinking water, microwaves, computerized personal records, and the sawed-off shotgun), failure to manage their adoption and use puts the survival not only of the economy, but also of the society, in jeopardy. Thus, the real issue with regard to government regulation and technological innovation is the careful design of regulation that recognizes the explicit and implicit trade-offs being made and that provides constructive stimuli for, and minimizes destructive barriers to, innovation.

The Social Limits to Growth

This chapter has argued that a healthy rate of technological innovation can support growth in output, improvements in productivity, control of inflation, provision of employment opportunities, and an acceptable balance of trade.

This optimistic view of the prospects for the United States and for the world must be tempered, however, by the realization that there may be both social and physical limits to growth in the world. On the one hand, technology helps to overcome the physical limits to growth posed by limited resources and by the limited carrying capacity of the environment. On the other hand, the physical limits to growth noted by Thomas Malthus, and brought to the forefront by Jay Forrester [22] and Donella Meadows, et al. [23] may one day no longer yield to the advances in productivity derived from technological innovation noted by Harold Barnett and Chandler Morse, [24] William Nordhaus, [25] and H.S. Cole, et al. [26] The laws of thermodynamics, genetics, and atmospheric chemistry do set limits, however distant.

Perhaps of more immediate concern are the social limits to economic growth, defined by Fred Hirsch in terms of the scarcity of positional goods. [27] He noted that once basic needs are met, part of the satisfaction of a high rate of growth relative to one's neighbors is that it gives one access to goods they cannot afford. Such goods, which he called "positional goods," are valued more for the fact that few have them than for their utility. However, as all people, or all countries, compete and grow together, the value of such positional goods is degraded by widespread use. [28] Thus, Hirsch argues, diminishing returns to the advantages of growth based on acquiring positional goods will naturally limit the attractiveness of further growth.

Hirsch's theory has great relevance for a world in which countries emphasize technological innovation and growth as a strategy for getting or staying ahead of the others. As each country attempts to improve its position, all can advance together, since international trade and welfare need not be a zero-sum game. As this happens, diminishing returns from satisfaction of discretionary demand may begin to emerge in the wealthy countries, and the desire to grow will diminish. The developed countries may then turn their full energies to sharing their growing output with other, less wealthy nations, and to seeking ways to overcome the physical limitations to growth by developing ever-more productive technology.

THE CONDITIONS FOR TECHNOLOGICAL INNOVATION

In this chapter, I have argued, successfully I hope, that a high rate of technological innovation, if directed wisely, can contribute to both the economic strength of the United States, and the quality of life of its people. If these arguments are true, it is then important to ask how the rate and direction of technological change might be enhanced. The remaining chapters of this book will attempt to answer this question in depth. Here, it is only possible to begin the discussion by considering the factors that contribute to technological innovation in individual firms, as emphasized by Utterback in Chapter 2, and in the economy as a whole, as emphasized by Klein in Chapter 3.

At the level of the firm, several factors influence the level and success of innovative activity: a flexible organizational structure; a high diversity of staff experience; an adequate financial condition; a good recognition and understanding of market needs; a good recognition and understanding of competitive and other environmental pressures; a willingness, ability, and need to take risks; and a set of technological possibilities. [29] Innovation is facilitated if a firm is experiencing, or can anticipate, a rapid growth in demand for its products, and if the workers, owners and managers can expect to earn a financial reward for their efforts.

An economy that is innovative on the whole is able to respond to a changing environment, whether challenged by problems or opportunities. Klein has called this a "dynamic economy." [30] According to Klein, such an economy is optimized not for the purpose of making the best choice among a set of available options, but for generating new options as needs and opportunities arise. Such an economy has at least four characteristics:

1) effective competition among firms, such that each not only can take risks but also can impose risks on the others (thus, all firms must take risks from time-to-time to survive);

2) opportunity for the entry of new firms in established areas or in new areas of business;

3) few institutional barriers to taking risks (such as private contracts or public policies that protect firms or jobs from competition), or standards of design or performance that are barriers to new technologies or approaches; and

4) mechanisms in firms and in the society to generate and to integrate new knowledge and new scientific understanding.

Earlier sections of this chapter, as well as the following chapters, present evidence suggesting that the United States economy is not as dynamic today as it needs to be. Similar conclusions have been reached regarding Japan, Sweden, Italy, the United Kingdom, France, and the OECD countries. [31] Several reasons can be identified for the inadequate dynamic performance of the United States.

First, it appears that the willingness and ability of firms, institutions, and individuals to take risks have declined. One reason is that static criteria of economic efficiency, such as present value models and cost/benefit analysis, have been accepted as normative criteria for decision making in both the private and public sectors, rather than as useful descriptive models of average behavior. This has led to an excessive focus on near-term performance and has weakened the ability of the economy to adapt to substantial changes in the environment, such as postwar recovery of trading competitors or energy price increases. The use of such methods of analysis can be traced, at least in part, to the widespread adoption of "scientific management" techniques in industry and government. In addition, the structure of U.S. industry has evolved in such a way as to dampen the enthusiasm of business for risk taking. This can be attributed to the tendency of many industries to become less competitive over time as they take advantage of scale economies and maturing technology, as well as to the movement of American business from the entrepreneurial to the managerial model.

Second, formation of new firms has become considerably more difficult in recent years. In the United States, publicly funded new ventures fell from 204 in 1969 to 0 in the first half of 1975. [32]

Third, both domestic and foreign policies have been used by many interests in the society to stabilize their environments and to insulate them from risk taking. For example, both U.S. industry and labor have sought and obtained legislation and executive actions to protect existing firms and jobs from foreign competition. Thus, they have been able to partially insulate themselves from pressures to take risks in response to the evolution of the world economy. In this evolution, the position of some U.S. industries has weakened as a result of the recovery of European nations and Japan from the devastation of World War II, and as a result of the emergence of a number of third world countries as potent world economic actors, such as the oil exporting nations and the new producers of iron and steel. While use of the power of government to protect existing jobs and firms in current locations may be in the best interests of the protected parties, it makes the society as a whole less adaptable to the changing world.

Fourth, both industry and Federal funding for basic research have declined in real terms since 1967, as shown in Fig. 1.14. (However, there was an encouraging upturn in 1977 and 1978.) This followed a doubling of total basic research funds in the previous seven years. While not all technological innovation requires basic research, reduced growth of the storehouse of new fundamental knowledge contributes to a reduced capability of the United States to adjust to new circumstances.

POLICY IMPLICATIONS

In order to maintain a strong economy and to ensure a high quality of life, the government should pursue policies to maintain a high rate of technological innovation consonant with reaching other social goals. A high rate is encouraged in turn by policies that support a dynamic economy; those policies include support for basic research, encouragement of new entrants, maintenance of effective competition, improvements in education and retraining assistance, and avoidance of measures to protect sluggish industries. An acceptable direction of innovation can be encouraged by well-designed programs of technology assessment and government regulation of the negative side effects of the use of new technology.

The design and implementation of these kinds of policies is not easy or simple. The authors of the remainder of this book are helping to begin the policy debate on America's technological future in the third millenium.

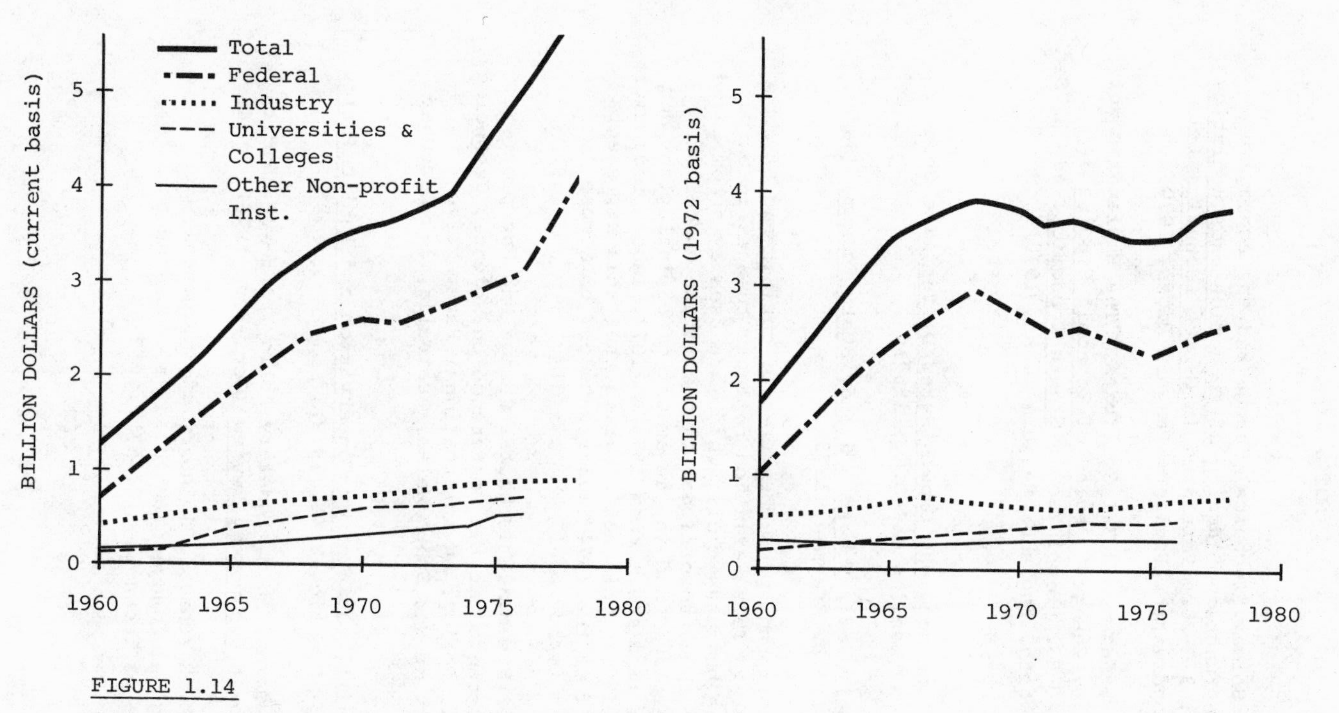

FIGURE 1.14

FUNDING FOR BASIC RESEARCH IN THE UNITED STATES

Source: Science Indicators 1976, National Science Board, National Science Foundation (Washington, D.C., 1977), and "Science and Technology Report," October 1978, Committee on Science and Technology, U.S. House of Representatives, 95th Congress (Washington, D.C.: U.S. Government Printing Office).

NOTES

1. Robert M. Solow, "Technical Change and the Aggregate Production Function," Review of Economics and Statistics 39 (1957): 312-20. Reprinted in Growth Economics, ed. A. Sen (Harmondsworth, England: Penguin Books, 1970).

2. Dale W. Jorgenson and Zvi Griliches, "The Explanation of Productivity Change," Review of Economic Studies 34 (1967): 249-83. Reprinted in Growth Economics, ed. A. Sen (Harmondsworth, England: Penguin Books, 1970).

3. Ibid.

4. Edward F. Denison, Why Growth Rates Differ, The Brookings Institution (Washington, D.C., 1967).

5. Hermann Daly, ed., Toward A Steady State Economy (San Francisco: W.H. Freeman, 1973).

6. In the short run, productivities of capital and of labor vary widely. This variation reflects changes in output during the business cycle, which are usually accompanied by a less than proportional change in the number of persons employed and in the amount of capital plant and equipment in place. From 1975 to 1977, labor productivity grew rapidly. It is not clear whether this represents recovery from the 1974-1975 recession, or a more persistent improvement.

7. Christoher Freeman, "Technical Change and Unemployment," (Paper presented at the Conference on Science, Technology, and Public Policy: An International Perspective, University of New South Wales, Australia, 1-2 December 1977).

8. John Sawhill, ed., Energy Conservation and Public Policy (New York: Prentice-Hall, 1979).

9. James B. Quinn, "U.S. Monetary Policy: A Heavy Hand in Technology," Technology Review (Oct.-Nov. 1976): 39-46.

10. William D. Nordhaus, "A Theory of Endogeneous Technological Change," (Ph.D. diss., Masschusetts Institute of Technology, 1967), p. 266.

11. The data on total employment and the unemployment rate have been affected in recent years by three growing phenomena: (1) requirements that people register as unemployed when they are not actually seeking work, in order to qualify for certain social welfare benefits; (2) unreported employment of illegal immigrants; and (3) "underground" employment of persons receiving assistance, or of persons seeking to avoid paying taxes. These phenomena may lead to an overstatement of the actual unemployment rate and to an understatement of the size of the work force.

12. Michael J. Piore, "Labor's Role in Technological Change," in *Technical Innovation and Economic Development*, National Technical Information Service CONF-760491 (April 1976), p. 57-70.

13. Raymond Vernon, ed., *The Technology Factor in International Trade*, National Bureau of Economic Research (New York: Columbia University Press, 1970); and Louis T. Wells, ed., *The Product Life Cycle and International Trade*, Division of Research, Graduate School of Business Administration, Harvard University, Boston, Mass., 1972.

14. Keith Pavitt and Luc Soete, "Innovative Activities and Export Shares: Some International Comparisons," Draft of Chapter 3 in "Technical Innovation and British Economic Performance," manuscript, August 1978.

15. Betsy Ancker-Johnson and David B. Chang, "U.S. Technology Policy -- A Draft Study," Office of the Assistant Secretary for Science and Technology, U.S. Department of Commerce (National Technical Information Service PB-263806, March 1977). *Technological Innovation and Economic Development: Has the U.S. Lost the Initiative?* Proceedings of a Symposium on Technological Innovation (National Technical Information Service CONF-760491, 1976): Stephen Feinman and William Feuntevilla, "Indicators of International Trends in Technological Innovation," (Final Report under National Science Foundation Contract NSF-C889, Gellman Research Associates, Jenkintown, Penn., April 1976); and *The Effects of International Technology Transfers on the U.S. Economy, Papers and Proceedings of a Colloquium*, National Science Foundation (Washington, D.C.: U.S. Government Printing Office, July 1974).

16. Dennis Schiffel and Carole Kitti, "Rates of Invention: International Patent Comparisons," *Research Policy* 7 (1978): 324-40.

17. *Science and Technology Policy Outlook*, Organization for Economic Cooperation and Development (Paris, July 1978).

18. Robert P. Morgan et al., "The Role of U.S. Universities in Science and Technology for Development: Mechanisms and Policy Options" (Final Report under National Science Foundation Grant No. INT78-0892, Department of Technology and Human Affairs, Washington University, St. Louis, Mo., September 1, 1978; scheduled for publication by Pergamon Press).

19. More likely, any particular regulation represents undercorrection or overcorrection of the market failure to which it is directed. Society has much to learn about regulating effectively.

20. Allen V. Kneese and Charles L. Schultze, *Pollution, Prices and Public Policy*, The Brookings Institution (Washington, D.C., 1975), p. 105.

21. Christopher T. Hill, ed., *Federal Regulation and Chemical Innovation*, Symposium Series No. 109, American Chemical Society (Washington, D.C., 1979).

22. Jay W. Forrester, *World Dynamics* (Cambridge, Mass.: Wright-Allen Press, 1971).

23. Donella H. Meadows, *The Limits to Growth* (New York: Universe Books, 1972).

24. Harold Barnett and Chandler Morris, *Scarcity and Growth* (Baltimore, Md.: Johns Hopkins University Press, 1962).

25. William D. Nordhaus, "World Dynamics --Measurement Without Data," *Economics Journal* 83 (1973): 1156-83.

26. H.S.D. Cole, C. Freeman, M. Jahodu, and K.L.R. Pavitt,; *Models of Doom* (New York: Universe Books, 1973).

27. Fred Hirsch, *Social Limits to Growth* (Cambridge, Mass.: Harvard University Press, 1976).

28. For example, when the first person in town buys a snowmobile, he enjoys unprecedented access to the wonders of the winter forest. When everyone has a snowmobile, the forest is filled with infernal noise, the wonders disappear, and everyone is poorer by the price of a snowmobile.

29. Gerald Zaltman, Robert Duncan, and Jonny Holbek, Innovations and Organizations (New York: John Wiley and Sons, 1973).

30. Burton Klein, Dynamic Economics (Cambridge, Mass.: Harvard University Press, 1977).

31. Umberto Columbo, "Strategies for Europe. Proposals for Science and Technology Policies: Industrial Innovation in Europe," Omega 5 (1977): 511-27; The Conditions for Success in Technological Innovation, Organization for Economic Cooperation and Development (Paris, 1971); and Keith Pavitt, "Governmental Support for Industrial Research and Development in France," Minerva 14 (Autumn 1976): 330-54.

32. Science Indicators 1976, National Science Board (Washington, D.C.: U.S. Government Printing Office, 1977), p. 105.

2 The Dynamics of Product and Process Innovation in Industry

James M. Utterback

INTRODUCTION

Innovation has been linked to rising productivity, to growth in employment, and to an improved quality of living. It has also been linked to increasing economic growth and to strong positions in export markets and trade. Conversely, a lack of innovation in the face of rising competitive challenges may contribute to inflation, to unemployment and dislocation of labor, to stagnation of growth, and to the rising importation of more attractive or lower priced goods. But understanding this is of little importance unless one understands how innovation occurs and how to influence it.

The central theme of this analysis is that the conditions necessary for rapid innovation are much different from those required for high levels of output and efficiency in production. The pattern of change observed within an organization will often shift from innovative and flexible to standardized and inflexible under demands for higher levels of output and productivity. Different creative responses from productive

---

Note: The author is especially indebted to William J. Abernathy. Our collaboration over the past four years has led to many of the ideas and findings expressed here. Many others were originated by him and are explored in the context of the auto industry in his recent book The Productivity Dilemma (Baltimore: Johns Hopkins University Press, 1978). This chapter is based on work supported by the National Science Foundation, Division of Policy Research and Analysis under Grant No. PRA76-82054 to the Center for Policy Alternatives at the Massachusetts Institute of Technology.

units facing different competitive and technological challenges may be expected, and this in turn suggests a way of viewing and analyzing the possible policy options for encouraging innovation. [1]

Many alternative definitions and conceptions have been used for various purposes in studies of innovation. One perspective sees innovation as a creative act synonomous with invention; while another sees innovation as a thing, i.e., a piece of hardware and possibly its design and production; and still another views innovation as a choice to use a thing, including possibly the ways which it is used and its diffusion. The first of these definitions focuses on the originality and newness of the innovation; the second, on its tangible form and use in the market or production process; and the third, on marketing approaches to different classes of users. In order to encompass these varied perspectives, innovation has been defined here as a process involving the creation, development, use, and diffusion of a new product or process.

The innovation process involves both organizations and the environment in which they operate. Innovation usually requires deliberate, concerted action by an organization following a strategy for competition, survival, and growth. In order to know how to manage and influence the innovation process, the appropriate unit of analysis is the firm itself.

Forces both inside and outside the industrial firm influence the process. Outside forces include users' needs, changing prices of inputs, competitive stresses, and government stimuli and regulations. The inside forces include the firm's resource allocations; the product and process technologies themselves; the people, organization, and communication patterns involved in producing innovations, and the technical resources and strengths of the firm.

Major innovations come to fruition in ten to thirty years, so cross sectional and survey data are not terribly helpful in understanding how innovation occurs. [2] This analysis, therefore, is built on historical studies of innovations in their organizational, technical, and economic settings. Such data are necessarily incomplete, but at the same time they yield a rich variety of insights.

Innovations vary greatly, and some of the differences among them appear to correspond to markedly different patterns in the process through which they arise. In particular, it is important to distinguish between product and process changes, and between innovations which require change in many facets of the firm and those which require only modest change. [3] This chapter considers product and process innovation in turn, and discusses their origin and development; their diffusion and use; their economic impacts; and, finally, the policies which might stimulate or constrain each type of innovation. [4]

A DYNAMIC MODEL OF PRODUCT AND PROCESS CHANGE

One way of viewing different types of innovations and their relationships is to think of them as successive steps in the development of a line of business. The business starts through the origination of one or more major product innovations. These are usually stimulated by users' needs through frequent interaction with users of the innovation. Exploration of the product's potentials in different applications follows. Rising production volume may lead to the need for innovation in the production process. Demands for greater sophistication, uniformity, and lower cost in the product create an ongoing demand for development and improvement of both product and process. This means that product design and process design become more and more closely interdependent as a line of business develops. A shift from radical to evolutionary product innovation will usually occur as a result of this interdependence. This shift is accompanied by heightened price competition and increased emphasis on process innovation. Thus, small-scale units that are flexible and highly reliant on manual labor and craft skills and that use general-purpose equipment will develop into units that rely on automated, equipment-intensive, high-volume processes. Changes in innovative pattern, production process and scale, and kind of production capacity will all occur together in a consistent predictable way. [5]

These relationships are summarized in Fig. 2.1. The rate of major product change is shown to be high at first and gradually diminishing as major process innovation increases. Both product and process change subsequently become incremental in a situation marked by production of standardized products in high volume. Competitive emphasis is first on functional product performance, later on product variation and finally on cost reduction. Innovation is at first stimulated by information on users' needs and even by users' technical inputs. As the product line and process develop, opportunities created by expanding internal technical capability increasingly provide the stimulus for innovation. Later, pressure to reduce cost and improve quality are expected to be the major stimuli for change. The initial product line is diverse, often being mainly custom designs. Innovative emphasis will begin to shift when it includes at least one product design stable enough to have significant production volume. The line of business will consist mostly of undifferentiated, standard products when it is fully developed.

Production begins in a flexible and inefficient form, and major changes are easily accommodated. As volume expands, processes become more rigid, with changes occuring in major steps. Ultimately the production process assumes an efficient, capital-intensive, and rigid form, and the cost of change is consequently high. General-purpose equipment, requiring highly skilled labor, will be used at first. Later, some subprocesses will be automated, creating "islands of automation" linked by manual processes. Special purpose equipment which is mostly automatic, with labor tasks consisting mainly of monitoring and control, will be the hallmark of highly developed productive units. Early on, materials inputs are limited to those generally available. Later, specialized materials may be demanded from some suppliers. If specialized materials are demanded, but not available, vertical integration to provide them will be extensive.

As the line of business develops, location will also shift. Early plants will be small-scale and near users and sources of technology. Ultimately, plants will be large-scale, highly specific to particular products, and located to minimize materials, labor and transportation costs. In sum, small-scale units that are flexible and highly reliant on manual labor and craft skills using general-purpose equipment will develop into units that rely on automated, equipment-intensive, high-volume processes, which are highly productive but correspondingly less flexible. In this setting, major product or process innovations will tend to be viewed as disruptive and will tend to originate through invasion of the line of business by new entrants.

As a unit moves toward large-scale production, the goals of its innovations change from meeting ill-defined and uncertain targets to meeting well-articulated design objectives. In the early stages, there are many product performance requirements which frequently cannot be stated quantitatively. Their relative importance or ranking may be quite unstable. It is precisely under conditions where performance requirements are ambiguous that users are most likely to produce major product innovation and where manufacturers are least likely to do so. One way of viewing regulatory constraints in the later stages of a product's evolution, such as those governing auto emissions or safety, is that they add new performance dimensions to be resolved by the engineer -- and so may lead to more innovative design improvements. They are also likely to open market opportunities for innovative change of the kind characteristic of entrepreneurial firms in such areas as instruments, components, and process equipment.

The stimulus for innovation changes as a unit matures. Initially, market needs are uncertain, and the relevant technologies are as yet little explored. Uncertainty about markets and appropriate targets is reduced as the unit develops, and larger research and development investments are justified. At some point, before the increasing specialization of the unit makes the cost of implementing technological innovations prohibitively high and before increasing price competition erodes profits with which to fund large indirect expenses, the benefits of research and development efforts reach a maximum. Then, technological opportunities for improvements and additions to

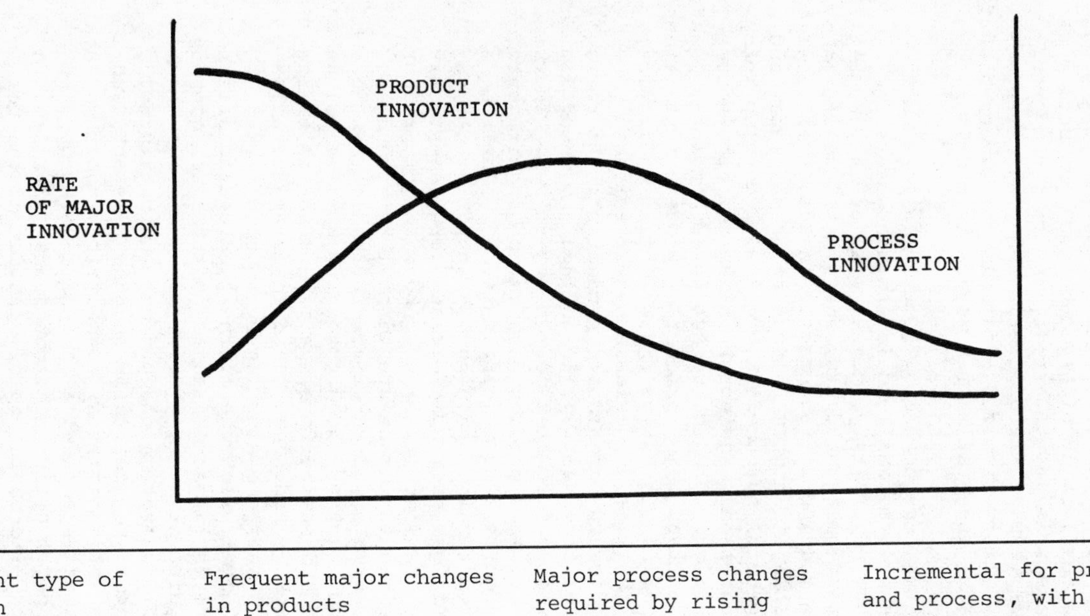

Predominant type of innovation	Frequent major changes in products	Major process changes required by rising volume	Incremental for product and process, with cumulative improvement in productivity and quality
Competitive emphasis on	Functional product performance	Product variation	Cost reduction
Innovation stimulated by	Information on users' needs and users' technical inputs	Opportunities created by expanding internal technical capability	Pressure to reduce cost and improve quality

Product line	Diverse, often including custom designs	Includes at least one product design stable enough to have significant production volume	Mostly undifferentiated standard products
Production processes	Flexible and inefficient; major changes easily accomodated	Becoming more rigid, with changes occuring in major steps	Efficient, capital-intensive, and rigid; cost of change is high
Equipment	General-purpose, requiring highly skilled labor	Some subprocesses automated, creating "islands of automation"	Special-purpose, mostly automatic with labor tasks mainly monitoring and control
Materials	Inputs are limited to generally-available materials	Specialized materials may be demanded from some suppliers	Specialized materials will be demanded; if not available, vertical integration will be extensive
Plant	Small-scale, located near user or source of technology	General-purpose with specialized sections	Large-scale, highly specific to particular products

FIGURE 2.1

A MODEL FOR THE DYNAMICS OF PROCESS INNOVATION IN INDUSTRY

Source: Willian J. Abernathy and James M. Utterback, "Patterns of Industrial Innovation," Technology Review 80 (June/July 1978): 2-9.

existing product lines become clear. A strong commitment to research and development is characteristic of productive units in the middle stages of development. Such units invest heavily in formal research and engineering departments, with emphasis on process innovation and product differentiation through functional improvements.

Although data on research and development expenditures are not readily available on the basis of productive units, divisions, or lines of business, an informal review of the activities of corporations with large investments in research and development shows that they tend to support business lines that fall neither at one extreme nor the other but are in the technologically active middle range. Such productive units tend to be large and to have a large share of their markets. [6]

Units in different stages of evolution respond to differing stimuli, or respond differently to the same stimuli, and therefore, undertake different types of innovation. This idea can readily be extended to question of barriers to innovation, and probably to patterns of success and failure in innovation for units in different situations. New entrepreneurial firms tend to view as barriers any factors that impede market aggregation, while firms with stable products and markets tend to rank uncertainty over government regulations or vulnerability of existing investments as more important disruptive factors. [7]

MAJOR PRODUCT INNOVATIONS

Origin and Development

Any innovation is necessarily a combination of a users' need and a technological means to meet that need. Often knowledge of the need to be met or the technology with which to meet the need, or even both, lie outside the organization which actually produces and puts a major new product into use. It is the combination of the need and the technology which is new. Major product innovation may require adaptation of the technology, or combinations of several pieces of technology, in a novel way. Innovation is not necessarily the creation of the new technology per se. Innovation is often more an entrepreneurial act which arises in a totally unplanned for and unexpected manner. Many latent, and even many clearly articulated but unmet, needs exist, as do many technical solutions seeking uses. This adds emphasis to the importance of synthesis as a necessary first step in innovation. [8]

Since innovation is synthesis, and since most knowledge of users' needs and most knowledge of technological resources must necessarily lie outside any particular organization, it is not surprising to find that communication with sources of technology and need information outside the organization is strongly correlated with innovation, particularly if the discussion is informal and allows different ideas to be modified, tested, and explored. [9] Further, the diversity of communication is strongly correlated with the occurrence of major product innovations.

In creating the idea for an innovation, the need input often takes precedence over the technology input. [10] Recognition of the need, in turn, often stimulates the entrepreneur to search for technical resources and information to meet the need. Therefore, users are frequently the source of technology information as well as need information, and users often design prototypes and solutions for their own use, which may later be developed by them or by other manufacturers and suppliers. [11] A consequence of this search pattern is that there are substantial time lags between the initial generation of technological and scientific potentials and their use. In this sense, basic research enters the innovation process through education and development of trained people, who later apply in meeting users' needs what they learned years earlier. In the short run, a policy of stimulating markets and users' needs is crucial if major product innovations are to occur, but in the longer term, investment in basic research and education is also crucial if the process is not to atrophy.

Diffusion and Use

The initial uses of major product innovations tend to be in small, often vacant, market niches in which the superiority of the new product, in one or two ways, allows it to command a temporary monopoly, high prices, and high profit margins per unit. For example, ice was first manufactured for refrigeration in the inland South where harvested ice was prohibitively expensive. Mechanical refrigeration was first used on ships for exporting meat and later in food processing plants. [12] Rayon was first produced and used as a uniform filament for incandescent lamps, and later as a high performance tire cord. [13] Radio telegraphy was first used for ship-to-shore communications and later for broadcast. [14] The jet engine and many other innovations were first used for military purposes. A major product innovation does not initially compete directly with the technology that it may ultimately replace, and it may even complement older technology by augmenting it in important ways. A major product innovation may initially be crude, expensive, fragile, and unreliable, and so diffusion starts very slowly while it is constrained by these various problems. For example, ice was costly to manufacture and early plants were dangerous to run, but it later became an economic replacement for harvested ice even in the North. [15] Rayon was difficult to dye but was uniform and could be produced with high tensile strength.

Research and experience with early applications led to ways to dye and weave it into fabrics. [16] Early transistors were expensive and had poor temperature stability and frequency response, but they were light, rugged, and had low power requirements. Thus, they were ideal for missile guidance and for hearing aids. [17] As such problems are overcome, diffusion of an innovation becomes more rapid.

The important use of a major product innovation is almost never the one initially envisioned. For example, the use of radio in broadcasting far surpassed its use for point to point communication. Similarly, the use of the computer in information processing surpassed expectations for its limited use in complex calculation. Gelatin based photographic film failed in the established professional market, but opened an amateur market of far greater importance under the Kodak name. [18] With time, experiments by users create a new understanding on their part, and on the part of the producer, of the capabilities required. Ways may be discovered to incorporate required capabilities within the product design, and this may lead to further understanding of potentials of the product. As the new product is modified, its use expands through the capture of successive market niches and segments. Competition with more traditional technology will become more and more direct, though the older technology for which the new is a substitute may be protected in many applications for a long time. [19] As volume grows and competition becomes more intense, prices will fall, and the diffusion of the innovation will become yet more rapid until it may reach a point of becoming, essentially, a standardized product.

Synthesis of a need with a means to meet it is a prerequisite for all innovations. Greater emphasis falls on the definition of needs and or uncertainty about the usefulness and potential uses of an innovation in the early stages of its development. As requirements that must be met by the innovation are better and better understood through experiments with early versions and uses, the focus of innovation will shift toward advancing the firm's technical resources to meet those requirements. Finally, when a product becomes standardized, incremental innovations aimed simply at improvements can proceed with little user interaction to meet a few well understood and accepted requirements. [20] Major product innovations essentially open new possibilities for performance and further development.

Economic Impacts

Not only do major product innovations enter vacant market niches, but they also are often introduced by new entrants in these markets. These may be either a new enterprise based on technology or an established firm bringing its technological expertise into a new area of the economy. Edison with incandescent lighting, Marconi with wireless telegraphy, Eastman Kodak with roll film, Polaroid with instant photography, Xerox with plain paper copying, Digital Equipment with mini-computers, Texas Instruments and Fairchild with integrated circuits, Advent with projection television, and Wang with word processing systems are but a few of the more familiar among hundreds of entrepreneurial firms entering markets with major product innovations. Each revolution in electronics technology has been marked with a wave of new entrants. [21] New fields such as biotechno- logy seem to be following a similar pattern. IBM with electric typewriters, General Motors with the diesel electric locomotive, Boeing with commercial jet aircraft, and Texas Instruments with the electronic watch are representative of larger firms moving into established product areas from a different technological base to create major product changes. [22] New applications of technology tend to have a greater return on investment and sales and to produce a more rapid growth in sales and profits, than do continued investments in more fully developed market areas. [23]

In order to understand the productivity impacts of major new product innovations, it is critical to understand that what is a product for one firm may be the process equipment, components for assembly, or materials used by other firms. Before cloth is made into clothing or furnishings for final consumption, fiber producers provide yarn to weavers who in turn sell the material they produce to fabric finishers. Equipment manufacturers serve all of these segments. Stronger fibers may allow increases in spinning and weaving speeds and may enable productive changes to be made in equipment design. [24] Conversely, demands for fire retardant fabrics may require changes in chemicals and fibers, and in the earlier production steps in different firms. Thus, what is a major product innovation for one firm may alter the production possibilities of other firms. When one examines the relationships between various producer and user firms it quickly becomes apparent that, in this broader sense, most innovation is not product innovation at all, but contributes to productivity directly through the linkage of different firms in the physical flow of production to final demand. [25]

At first, general-purpose equipment and highly skilled labor may be used on a job shop or pilot plant basis for producing the new product in small volumes. In such a case, the labor involved is expensive, and the process is highly labor intensive. This explains the greater than usual rates of increase in employment and the use of highly paid skills in rapidly evolving product areas. [26]

This also begins to explain why the United States tends to be most export competitive in the very industries in which labor is most intensively used and well paid, such as aircraft and aircraft engines, computers, industrial electronics, and heavy machinery, while imports tend to be greatest in those areas involving greater amounts of capital investment and less labor paid at lower wages. Innovative products are far more competitive in export markets than are other products, especially standardized products and commodities. [27] An implication of this view is that the competitive position of an industry is not likely to remain constant for long periods. When its products become standardized, and when innovative and entrepreneurial skills available in the United States become less critical, then its dominance in trade is weakened as well. [28] This issue is discussed fully by Graham in Chapter 4.

MAJOR PROCESS INNOVATIONS

Origin and Development

The manufacturing firm itself demands high performance as a user of process equipment. Process innovation is in large part originated and developed by the manufacturing firms which initially put into use the innovations in equipment and systems. For example, Ford pioneered in the continuous production of plate glass to meet its heavy demand for car windows and windshields. [29] Pilkington, which assisted Ford later, perfected a different type of continuous process for plate glass production which involves floating the molten glass over a pool of molten tin. [30] General Electric developed a continuous process for drawing copper wire from a melt as opposed to cold drawing. Firms in the steel industry have developed the basic oxygen furnace, continuous casting, and direct reduction of iron ore. [31] Producers of semiconductor chips have played a dominant role in the development of process equipment to meet their own needs. [32] These examples should not be surprising since users initiate and often do initial development work to meet their own needs. Process innovation tends to be stimulated by rising demand and production volume, by rising costs of various factors of production, or by increasing cost or regulatory constraints. [33] Opportunities created by an expanding internal technical capability within the productive unit will play an important role, and process change may be required by a growing sophistication of the product itself. Process change in turn may enable or require further innovation in products, resulting in an increasingly tighter linkage between product change and process change. Process innovation tends to follow product innovation by several years, but the lag between major product change and major process change tends to be shorter as a product becomes standardized. [34]

Major process innovation begins when significant volume achieved in one or a few product lines encourages standardization. Some tasks are automated and emphasis is placed on a systematic flow of work. Machines may be installed where operations are required, leading to faster throughput and perhaps less continuous use of some pieces of equipment. Levels of automation will vary widely, with "islands of automation" being linked by manual operations. Steps taken to expand capacity will most frequently involve breaking bottlenecks. [35]

Major process change often involves adding together two or more "islands of automation" into one larger step. This may require a completely different approach and technology than any of the earlier steps. [36] The Pilkington process entirely eliminated the need for grinding and polishing cast plate glass. Continuous casting combined the steps of ingot casting and the blooming or slabbing mill. Ford's automated engine production at their Cleveland plant, machinery for automated manufacture of light bulbs, Kodak's early process for continuous film production, and several advances in the production of rayon and open-end spinning of cotton provide other examples. [37] One of the consequences of such changes is the simplification of the production process, since many of the earlier operations are combined in a few pieces of equipment. A larger and larger initial investment will be required to enter a line of business as more major innovations in the production process occur.

Diffusion and Use

A manufacturer will frequently call on an equipment supplier or engineering firm in the scale-up and introduction of major process changes. Diffusion may then occur through the subsequent installation of the change in other productive units by the equipment supplier, or it may occur through others' imitation and independent development of variations of the original idea. Diffusion of process change depends on the size of the investment required relative to the size of the productive unit, and on the relative advantage of adopting the innovation. The smaller the investment required and the greater the advantage of adopting the innovation in terms of productivity increase, product quality, and product uniformity, the more rapidly will its diffusion occur. Conversely, as a greater number of competitors adopt the innovation, the disadvantage of not adopting will become greater and greater; therefore, diffusion will also accelerate as more productive units in an industry adopt an innovation. [38]

Because major process innovation often results from adding several steps in a process together, it tends to require larger productive units. [39] Thus, industry structure will change as a result of the process change, resulting in larger units and higher entry costs, and often in fewer units. [39] Manufacturers may demand greater uniformity and consistency of materials from their suppliers. The new process may make new locations economically attractive if it shifts the attractiveness of various raw materials, energy sources, transportation modes, and labor skills. [40]

Economic Impacts

Major process innovations tend to result in large and immediate step increases in productivity that are often followed by more gradual increases resulting from further refinement of the production process. When these steps are examined in detail for many past innovations, it becomes apparent that no amount of automation of the old process could have produced a similar increase in productivity.

Though many observers emphasize new product innovation, process innovations may have equal or even greater commercial importance. A high rate of productivity improvement is associated with process improvement. The cost of incandescent light bulbs, for example, has fallen more than 80 percent since their introduction. [41] Airline operating costs were cut by half through the development and improvement of the DC-3. [42] Semiconductor prices have been falling by 20 to 30 percent with each doubling of cumulative production. [43] The Model T Ford was reduced in price by roughly 70 percent over the course of its production. [44] Similar dramatic reductions have been achieved in the costs of computer core memory and television picture tubes. [45]

Because process innovations tend to reduce production costs, greater gains tend to accrue to holders of larger market shares. This may explain why new entrants tend to stress product rather than process change, and why waves of entry in a business appear to occur around new generations of product technology rather than process technology. Major process innovations on the other hand tend to originate with firms already included in the set of competitors in an area of business. [46] An exception to this general rule occurs when major process change so dramatically shifts material requirements or productivity that optimal plant sizes decrease. This may then result in expanded entry in the business, as in the cases of current processes for manufacturing ammonia and polyvinylchloride and the early processes for ice manufacture using steam. [47]

Government has influenced process change in many ways, including purchase and lease to manufacturers of specialized equipment, assisting in the development of production process equipment and in the diffusion of its use, and guaranteeing production volume for national defense purposes as in the case of transistors. The government has also created volume markets for products for its own use, largely from defense and aerospace manufacturers as was true, for example, in the cases of aircraft engines and integrated circuits. [48] All of these actions have reduced risk and raised the advantage of major process innovations.

INCREMENTAL IMPROVEMENTS IN PRODUCTS AND PROCESSES

Origin and Development

Incremental innovations are extensions of existing technology that improve product performance, cost, or quality step by step. Incremental improvements in products and processes are not as dramatic as major product innovations or major process innovations, nor are they as fully covered in the literature on technological change. But they may be the most important type of innovation because of their short run economic and competitive impact. Incremental innovations are stimulated by demands for reduced cost and improved quality as products become less differentiated and, therefore, more directly in competition with one another. While each step improvement in either cost of production or quality may be small, such changes are clearly focused and may add up over a period of time to very consequential gains.

Incremental innovations often occur imperceptibly through adjustments in operating procedures and materials, and through slight variations in the production process, which may reduce design margins as operating characteristics of the process become known. Incremental innovation also occurs imperceptibly through simplification of a product and redesign of parts of the product. [49] As volume of production grows, the firm may demand special design of a greater proportion of the components it uses. Substitution of less costly or more prefinished materials for those in earlier designs may also occur. [50]

Incremental innovations may be originated by suppliers of equipment and components who can aggregate rewards to be gained from the improvements which, while small for any one user, would be large for a group of users taken together. [51] Incremental innovations also often originate from inside a larger manufacturing firm. They may derive largely from the experience of people within the firm and may be pursued informally -- rather than being defined or recognized as formal projects, formal allocations of resources, or formal research and engineering efforts. [52]

Incremental innovations appear to be especially important in efficient and capital-intensive production processes involving large-scale plants devoted to the manufacture of one or a few standardized products. In this setting, innovation is constrained by past investments in both product design and production process technology. Continuing small improvements in both product and process can maintain the production unit's competitive vitality and can lead to relatively swift and certain gains, which, because of large production volume, can have an immediate and important impact.

When an organization is focused on volume production of a standardized product, major change in either the product or the process may be disruptive and difficult to accomplish. Organization structure may also be highly focused, with emphasis on goals and rules for controlling costs and flows of materials and product. Major change in technology in this setting also changes the control system and the basis of authority within the organization, thus compounding the problem of introducing change. Conversely, incremental changes generally reinforce the existing structure and do not have such a disruptive effect.

Use, Diffusion, and Economic Impacts

Incremental innovations, almost by definition, do not tend to be published, patented, sold separately, or even formally identified within the innovating firm. However, such innovations are recognized clearly enough to be transferred rapidly from unit to unit or from plant to plant within a larger firm -- probably mainly through personal communication and transfer of individuals. The data suggest that use and diffusion of incremental innovations tends to occur primarily within the originating firm. This may well explain why the "experience curve," a regular relationship between cumulative production volume and declines in unit costs of production, is often an organization-based rather than an industry-based phenomenon. [53]

Incremental innovations tend to be cumulative and focused, each one adding a bit to the gains achieved in earlier steps. The sum of performance gains or cost reductions from such innovations usually turn out to be greater than the initial gains made through more radical innovation in product or in process. For example, more than half the decline in costs over a period of years in the production of rayon were traced to incremental changes. The same finding has held in studies of products as diverse as light bulbs, liquid propelled rocket engines, automobiles, and computer core memories. [54] As mentioned earlier, major innovations in a production process may lead to a large initial step gain in productivity, but incremental improvements of the process which it stimulates may advance it even further. At the same time, it is difficult for units producing standardized products in high volume and stressing incremental innovations to respond to the challenges of major change. Thus, they become vulnerable to invasion of their markets by other products or alternative solutions. For example, the major manufacturers of mechanical typewriters did not introduce the electric typewriter. [55] Few major manufacturers of mechanical calculators now manufacture electronic calculators. [56] And few manufacturers of vacuum tubes were successful in making the shift to transistors. [57] Unless productive units that reach the stage of high volume standardization are dealing with a long-lived product, they may eventually be forced out of business by an invading technology.

INNOVATIVE CHALLENGES AND DEFENSIVE RESPONSES

While a major innovation may be viewed as an opportunity by its originator, it may be viewed as disruptive and destructive of established investments by competitors. Periods of continuity and predictable change in innovative patterns, as explained above, appear to begin and end in periods of sharp discontinuity. Discontinuities seem to be associated with radical shifts in product and process technologies. [58] A radical innovation is one which can create new businesses and transform or destroy existing businesses. Substantial portions of the capital stock will essentially be swept away and replaced. A crisis can result for established firms when they are faced with a new competitor whose products serve the same function as theirs, but from a different technological base. Among the more familiar cases are the electronic calculator replacing electromechanical calculators, transistors replacing vacuum tubes, jet engines replacing piston driven engines in aircraft, and diesel electric locomotives replacing steam locomotives for rail transport. [59]

Dramatic change in process technology can also create new businesses and undermine existing investments. For example, Pilkington's float glass process for making plate glass removed the need for laborious grinding and polishing in finishing plate glass and resulted in a highly automatic and integrated process with superior economics. Another example is the introduction by a plant engineering firm of a process for making ammonia based on centrifugal compressors to replace reciprocating compressors. This not only quickly replaced the existing plants of many manufacturers, but also encouraged the entry of a number of new firms in the production of ammonia and encouraged the production of the product in new locations. [60]

Direct regulation of products and processes may result in higher barriers to entry and more entrenched market positions for existing products. It may also result in greater conservatism in design, focusing primarily on improvement of the existing technology. But regulatory impacts may also produce revolutionary changes, at least from the viewpoint of the productive unit. For example, restriction of chlorinated hydrocarbons as dielectric fluids for use in electrical transformers has resulted in the entry of new competitors with alternatives for this market. [61]

Drastic change in the cost of inputs may also result in rapid shifts in technology. Of course, the most familiar example is the recent increase in oil prices which has made past investments predicated on declining prices unattractive. But shifts in the availability of other materials, for example, the loss of natural rubber supplies during World War II, have resulted in major new businesses in the past and may well also in the future. [62]

All of these challenges may be viewed as invasions of an existing market. Usually an invasion begins with the origination of a major innovation outside the recognized set of competing units in an industry. Small new ventures, or larger firms entering a new business, appear to introduce a disproportionate share of the innovations that create major threats and, conversely, opportunities.

Established firms often respond to an invasion of their product line by new technology with redoubled creative effort and investment in what is well known. The new technology may be viewed as expensive and relatively crude at first, leading to the belief that it will find only limited application. But even though it may be crude, the new technology may have great performance advantages in certain submarkets and may gain ground by first competing in these more limited markets. Use of the new technology expands by means of its capture of a series of submarkets. [63] As the market expands the new technology may also have much greater potential for improvement and cost reduction than does the existing technology. Thus, price cutting by established units as a defense may be ineffective.

The new technology often opens new applications an captures most of the resulting expansion of the market. During an invasion, the defensive efforts of established firms may cause the old technology to reach much higher levels of performance and sophistication than those previously attained. But this usually ultimately proves to be a futile response, resulting in loss of market share and exit from the business. Only in exceptional cases is the creative response of the old technology so vigorous as to drive back or nearly overturn the new technology. The once promising aluminum engine block was driven back by thinwall casting techniques that greatly reduced the weight of traditional iron blocks. Under pressure from electric incandescent lamps, the efficiency of gas lighting was increased five-fold, using a set of glowing filaments called a gas mantle. [64] Ice boxes tripled in efficiency to no avail as home electric refrigerators began to capture a major share of that market. [65] Today solvent-based coatings are improving dramatically under competition from newer products cured by ultraviolet light.

The emphasis in competition will shift to product change and away from cost and quality, while, at the same time, prices may drop with extraordinary rapidity, and many new options and performance dimensions may be available to users. The total market may expand as a direct consequence of the invading innovation. [66] This usually only postpones the abrupt decline of the established technology and lends false strength to arguments against withdrawal from the old and rapid investment in the new on the part of established actors in the business. At this point, the situation in the business has dramatically changed. The technological base, raw materials, labor skills, and support functions necessary in the business may be completely different. Firms which were not competitors a few years before will be major factors in the business and may even dominate it. Firms which were strong earlier may be markedly weaker and may exit from the business. In Chapter 3, Klein claims that this sort of rivalry leads to the most dynamic periods of growth and employment in the economy at large.

SUMMARY OF POLICY OPTIONS ARISING FROM FINDINGS ON INDUSTRIAL INNOVATION

A dynamic interpretation of technological innovation implies that government actions will have different effects on innovation in different stages of the evolution of a productive unit. Radical innovation disrupts existing competitive relationships, which means that government must encourage rivalry between existing firms and from new firms in order to encourage some important innovations. A dynamic interpretation of technological innovation implies that different resources are required and different conditions must be met for different innovations to occur. The ideas presented in this chapter are based on a model in which characteristics of a productive unit, its products, and its production processes are related in a consistent pattern.

A variety of programs enhance innovation and the competitive viability of firms. No one government policy is the key to effective stimulation of change. Lack of any one critical factor may be a barrier to innovation. Timing, interaction with other programs, and the details of implementation are often crucial. Provision of support for research and technical development alone, for example, is clearly not sufficient, nor is stimulating demand for innovative products sufficient if a base of technical knowledge and trained people does not exist. The dynamic interpretation given here means that to encourage major innovations, government must use policies which stimulate demand and competition, policies which stimulate investment and market entry, policies which support the technical base and education, and policies which limit and ameliorate the social costs of change in an integrated way. No one policy alone, however helpful it may be when other parts of the environment for change are present, will ensure the desired result.

Different government actions will encourage innovation differently in the start-up, growth and stable phases of the evolution of a manufacturing organization and of its product and process technology. During the early stage, which features major product innovation, the most effective government actions are to provide an adequate resource base of knowledge and skilled technical and crafts people, and to facilitate the formation of new innovative firms and entries in new markets by established firms, both through market stimuli and encouragement of saving and investment. Such actions serve to reduce the market, technological, and financial uncertainties surrounding major product innovations. As an organization grows and its focus shifts to major process innovation, the range of government actions that increase the rewards to successful innovation expands to include encouragement of the exchange of technical information, standards setting, and investment in plant and equipment. When the organization's conditions shift to those surrounding stable products and market shares and corresponding incremental innovation, government becomes concerned with maintaining competition and rivalry, limiting social costs involved in the use of products and processes, and minimizing and ameliorating the dislocations which occur when change renders products and productive units no longer viable in their existing form. Neither increasing rewards for change nor reducing risk of change would be expected to produce major innovations from stable firms. Rather, as Klein argues in Chapter 3, the only way to successfully stimulate major innovation in stable firms is to increase their risks of losing existing business.

A degree of consistency is required among policies adopted to encourage innovation. Policies that will be helpful for stimulating innovation in one area may deter innovation and reinforce the status quo in another. There is no simple relationship between policies and effects. There will be winners and losers as a result of any action taken. A dynamic interpretation of the innovative process makes some of these effects predictable. Inconsistent policies can negate attempts to innovate. A dynamic interpretation of the innovation process implies that demands for innovation are inconsistent with high demands for productivity improvement or standardization. Conversely, demands for greater competition and rivalry may prevent the industrial consolidation needed to produce a stable product at high levels of output and efficiency.

Creating new competitive conditions ensures that better alternatives and ideas have a chance to become established in the marketplace, whether their source is an invading firm or an entrenched actor in the business. The important point is to avoid policies which simply reinforce the existing structure and set of products and possibilities, and to create policies which allow invasion by new products and new competitors.

NOTES

1. Firms vary greatly in size, diversity of product lines, resources, and in their attempts to create different kinds of innovations. To deal with this diversity, the convention of looking at what might be termed a productive unit or simple firm -- that is, a part of an organization which produces a related group of products, and its associated production technology -- will be adopted. In the case of a diversified or multidivisional company, this would be one of its divisions, often geographically and usually managerially separate from other parts of the organization. These units often exhibit consistent patterns of innovation, with some stressing new products and product performance, others stressing major advances in production technology, and yet others constantly improving product quality, costs, and productivity. While these patterns are not completely exclusive, it will be shown later that the way in which innovation occurs, the forces which stimulate innovation, and the types of policies which influence it, will vary directly with these different sorts of innovations and emphases.

2. By analogy this might be extended to the case of services (although indeed there is very little information available on service innovations) whether they be major new services offered to the public or to industry, completely different ways of providing a service, or improvements in service quality and cost.

3. James M. Utterback, "Innovation in Industry and the Diffusion of Technology," Science 183 (February 1974): 620-26.

4. W.J. Abernathy, and J.M. Utterback, "Patterns of Industrial Innovation," Technology Review 80 (June/July 1978): 41-47. J.M. Utterback and W.J. Abernathy, "A Dynamic Model of Product and Process Innovation," Omega 3 (1975): 639-656.

5. W.J. Abernathy and P. L. Townsend, "Technology, Productivity and Process Change," *Technological Forecasting and Social Change* 7 (1975): 379-96.

6. Abernathy and Utterback, "Patterns of Industrial Innovation."

7. S. Myers and E. Sweezy, "Why Innovations Fail," *Technology Review* 80 (March/April 1978): 40-46.

8. Sumner Myers, and Donald G. Marquis, "Successful Industrial Innovations: A Study of Factors Underlying Innovation in Selected Firms," Report to the National Science Foundation, No. 69-17 (Washington, D.C.: Government Printing Office, 1969).

9. Thomas J. Allen, *Managing of the Flow of Technology*, (Cambridge, Mass.: MIT Press, 1977). J.M. Utterback,"The Process of Innovation in Instrument Firms" (Ph.D. diss., Sloan School of Management, Massachusetts Institute of Technology, 1969).

10. James M. Utterback, "Innovation in Industry and the Diffusion of Technology."

11. Eric A. von Hippel, "The Dominant Role of Users in the Scientific Instrument Innovation Process," *Research Policy* 5 (July 1976): 212-39.

12. Oscar Edward Anderson, Jr. *Refrigeration in America: A History of a New Technology and Its Impact* (Princeton, N.J.: Princeton University Press, 1953).

13. Arthur A. Bright, Jr., *The Electric-Lamp Industry: Technological Change and Economic Development from 1800 to 1947* (New York: Macmillan Co., 1949). Henry A. Gemery, "Productivity Growth, Process Change, and Technical Change in the United States Glass Industry: 1899-1935" (Ph.D. diss., University of Pennsylvania, 1967). T.C. Nolet and J.M. Utterback, "Product and Process Change in Non-Assembled Product Industries" (Working Paper-78-12 of the MIT Center for Policy Alternatives, Cambridge, Mass., 1969).

14. W.R. Maclauren, *Invention and Innovation in the Radio Industry* (New York: Macmillan Co., 1949).

15. Anderson, *Refrigeration in America*.

16. Nolet and Utterback "Product and Process Change in Non-Assembled Product Industries."

17. James M. Utterback, and Albert E. Murray, <u>The Influence of Defense Procurement and Sponsorship of Research and Development on the Development of the Civilian Electronics Industry</u>, MIT Center for Policy Alternatives (Cambridge, Mass.: June 1977).

18. Reese V. Jenkins, <u>Images and Enterprise: Technology and the Photographic Industry, 1839 to 1925</u> (Baltimore: Johns Hopkins University Press, 1975).

19. Arnold C. Cooper and Dan Schendel, "Strategic Responses to Technological Threats," <u>Business Horizons</u> 19 (February 1976): 61-69.

20. Abernathy and Utterback, "Patterns of Industrial Innovation."

21. Ernest Braun and Stuart MacDonald, <u>Revolution in Miniature: The History and Impact of Semiconductor Electronics</u> (Cambridge: Cambridge University Press, 1978). A.M. Golding, "The Semiconductor Industry in Britain and the United States: A Case Study in Innovation, Growth and the Diffusion of Technology" (Ph.D. diss., University of Sussex, 1971). John E. Tilton, <u>International Diffusion of Technology: The Case of Semiconductors</u>, The Brookings Institution, (Washington, D.C., 1971).

22. George N. Engler, "The Typewriter Industry: The Impact of a Significant Technological Innovation" (Ph.D. diss. University of California, Los Angeles, 1965). Richard S. Rosenbloom, "Technological Innovation in Firms and Industries: An Assessment of the State of the Art," in M. Kranzberg et al., "Technological Innovation: A Critical Review of Current Knowledge" (Report from Georgia Institute of Technology to the National Science Foundation, Washington, D.C., 1974).

23. E. B. Roberts, "Entrepreneurship and Technology," in <u>The Human Factor in the Transfer of Technology</u>, eds. Gruber and Marquis (Cambridge, Mass.: MIT Press, 1969). Edward B. Roberts and Alan L. Freeman, "Strategies for Improving Research Utilization," <u>Technology Review</u> 80 (March/April 1978): 32-40.

24. Roy Rothwell, <u>Technological Change in Textile Machinery: Manpower Implications in the User and Producer Industries</u> (Prepared for NSF/BMFT Seminar "Public/Private Cooperation for Technological Innovation," (Geneva, June 1977).

25. Abernathy, and Townsend, "Technology, Productivity and Process Change."

26. W.J. Abernathy "Production Process Structure and Technological Change," Decision Science 7 (October 1976): 607-19.

27. Wells, L. T., ed., The Product Life Cycle and International Trade, Division of Research, Graduate School of Business Administration, Harvard University (Boston, Mass., 1972).

28. C. Freeman, J. E. Harlow, and J. K. Fuller, "Research and Development in Electronic Capital Goods," National Institute Economic Review 34 (November, 1965).

29. R.W. Douglas and S. Frank, A History of Glass Making (Henley-on-Thames: Fowles, 1972).

30. Nolet and Utterback, "Product and Process Change in Non-Assembled Product Industries."

31. Linsu Kim, "The Evolution of Technological Innovation in the Iron and Steel Industry" (Working Paper-78-10 of the MIT Center for Policy Alternatives, Cambridge, Mass., 1969).

32. Eric A. von Hippel, "The Dominant Role of the User in Semiconductor and Electronic Subassembly Process Innovation," IEEE Transactions on Engineering Management, vol. EM-24, no. 2 (May 1977): 60-71.

33. John L. Enos, Petroleum Progess and Profits: A History of Process Innovation (Cambridge, Mass.: MIT Press, 1962).

34. W.J. Abernathy and Kenneth Wayne, "Limits of the Learning Curve," Harvard Business Review 52 (Sept./Oct. 1974): 109-19.

35. James R. Bright, Automation and Management, Division of Research, Graduate School of Business Administration, Harvard University (Boston, Mass., 1958).

36. Nolet and Utterback, "Product and Process Change in Non-Assembled Product Industries."

37. Abernathy and Wayne, "Limits of the Learning Curve"; Bright, Automation and Management; Jenkins, Images and Enterprise; Nolet and Utterback, "Product and Process Change in Non-Assembled Product Industries"; Rothwell, Technological Change in Textile Machinery.

38. Utterback, "Innovation in Industry and the Diffusion of Technology."

39. D.C. Mueller and John E. Tilton, "R&D Cost as a Barrier to Entry," *Canadian Journal of Economics* 2 (November 1969): 570-79; Engler, *The Typewriter Industry*; Richard H. Fabris, "A Study of Product Innovation in the Automobile Industry During the Period 1919-1962" (Ph.D. diss., University of Illinois, 1966); Jenkins, *Images and Enterprise*; William A. Reynolds, "Innovation in the U.S. Carpet Industry, 1947-1963" (Ph.D. Diss., Columbia University, 1967); and Rothwell, *Technological Change in Textile Machinery*.

40. John S. Hekman, "The Product Cycle and New England Textile" *Quarterly Journal of Economics*, in press.

41. Bright *The Electric-Lamp Industry*; Gemery, "Productivity Growth, Process Change, and Technical Change in the United States Glass Industry."

42. R. Miller and D. Sawers, *The Technical Development of Modern Aviation* (New York: Praeger, 1970).

43. David L. Bodde, "Riding the Experience Curve," *Technology Review*, (March/April 1976): 53-9.

44. Abernathy and Townsend, "Technology, Productivity and Process Change."

45. Kenneth E. Knight, "A Study of Technological Innovation -- The Evolution of Digital Computers" (Ph.D. diss., Carnegie Institute of Technology, Pittsburgh, Penn., November, 1963); Abernathy and Utterback, "Patterns of Innovation in Industry." For other examples, see R. Clarke, "Innovation in Liquid Propellant Rocket Technology" (Ph.D. Diss., Stanford University, 1968), and Samuel Hollander, *The Sources of Increased Efficiency: A Study of DuPont Rayon Plants* (Cambridge, Mass.: MIT Press, 1965).

46. For examples, see Robert D. Buzzell and Robert E. M. Nourse, *Product Innovation in Food Processing, 1954-1964*, Division of Research, Graduate School of Business, Harvard University (Boston, Mass., 1967); Mueller and Tilton, "R&D Cost as a Barrier to Entry"; and Arthur D. Little, Inc., *Patterns and Problems of Technical Innovation in American Industry: Report to the National Science Foundation*, PB 181573, U.S. Department of Commerce, Office of Technical Services (Washington, D.C.: Government Printing Office, September 1963).

47. Edward Greenberg, Christopher T. Hill, and David J. Newburger, Regulation, Market Prices, and Process Innovation: The Case of the Ammonia Industry (Boulder, Colo.: Westview Press, 1979). Anderson, Refrigeration in America.

48. Utterback and Murray, The Influence of Defense Procurement and Sponsorship of Research and Development.

49. Badiul A. Majumdar, Innovations, Product Developments and Technology Transfers: An Empirical Study of Dynamic Competitive Advantage, The Case of Electronic Calculators (Ph.D. diss., Case Western Reserve University, 1977).

50. Abernathy, "Production Process Structure and Technological Change."

51. Christopher Freeman, "Chemical Process Plant: Innovation and the World Market, National Institute Economic Review 45 (1968): 29-51.

52. For examples, see Clarke, Innovation in Liquid Propellant Rocket Technology; Enos, Petroleum Progess and Profits; Hollander, A Study of DuPont Rayon Plants; and Jordan P. Yale, "Innovation: The Controlling Factor in the Life Cycle of the Synthetic Fiber Industry" (Ph.D. diss., New York University, 1965).

53. Unit production costs are supposed to go down by a constant percentage with each doubling of cumulative production of a product. This relationship was first observed in learning and in the performance of repetitive tasks, thus its name. In fact it depends greatly on product simplification, development and adjustment of the production process and the organization and management, as well as the quality, stability, and experience, of the work force. This falling function of cost and volume has been shown to exist in a wide variety of settings, although the data plotted are usually prices used as a surrogate measure of costs. See also Bodde, "Riding the Experience Curve;" and Alan R. Fusfeld, "The Technological Progress Function: A New Technique for Forecasting," Technological Forecasting and Social Change, 1 (1970): 301-12.

54. See sources in note 52.

55. Engler, "The Typewriter Industry."

56. Majumdar, "An Empirical Study of Dynamic Competitive Advantage."

57. Braunand and MacDonald, *The History and Impact of Semiconductor Electronics*; Golding, *The Semiconductor Industry in Britain and the United States*; Tilton, *International Diffusion of Technology*.

58. This observation has led Burton H. Klein, *Dynamic Economics*, (Cambridge, Mass.: Harvard University Press, 1977) and Jenkins, *Images and Enterprise*, to pose the idea that market structure depends on and follows from innovation, rather than the more widely accepted idea that market structure influences decisions to innovate. Perhaps both are true, one holding in periods of invasion and discontinuity, the other in periods of relative stability in the set of competitors.

59. James M. Utterback, "Business Invasion by Innovation" (Working Paper-78-13 of the MIT Center for Policy Alternatives, Cambridge, Mass., 1978).

60. Greenberg, Hill, and Newburger, *Regulation, Market Prices, and Process Innovation*.

61. Adam B. Jaffe, *Regulation of Chemicals: Product and Process Technology as a Determinant of the Compliance Response* (M.S. thesis, Massachusetts Institute of Technology, 1977).

62. Robert A. Solo, "The Development and Economics of the American Synthetic Rubber Industry," (Ph.D. diss., Cornell University, 1952).

63. Cooper and Schendel, "Strategic Responses to Technological Threats."

64. Bright, *The Electric-Lamp Industry.*

65. Anderson, *Refrigeration in America*.

66. Cooper and Schendel, "Strategic Responses to Technological Threats."

3 The Slowdown in Productivity Advances: A Dynamic Explanation

Burton H. Klein

INTRODUCTION

How is the slowdown in the rate of U.S. productivity advance, shown in Fig. 3.1, to be explained? A number of reasons have been given for the decline in the rate of productivity gain since the mid-1960s, including the impact of safety and environmental regulations, the rising cost of investment relative to depreciation allowances, and the increasing degree of constraint imposed on productivity gains by labor unions. While these factors, discussed by Hill in Chapter 1 and Bourdon in Chapter 6, contributed to the severity of the decline, it is my conviction that it would have occurred anyway. To understand this assertion, it is necessary to understand how productivity advances occur.

Individual gains in productivity can come about in a variety of ways. For example, since the physical efficiency of machines is determined by the best available fuels and materials, one way of improving productivity consists of improving the existing machines or discovering new ones. Another way of improving productivity is to overcome scaling problems, such as those that had to be solved in reducing the size of turbine engines before they could be used in airplanes or in scaling up the size of coal burning electric power plants to make them more efficient.

Almost invariably, important productivity gains arise from significant advances in technology, and advances in technology that lead to productivity gains are almost always followed by organizational changes. Conversely, organizational change can trigger advances in technology by permitting entrepreneurs to view improvements from a new perspective. For example, when the automobile industry organized its assembly lines as a continuous production process, rather than by type of machine,

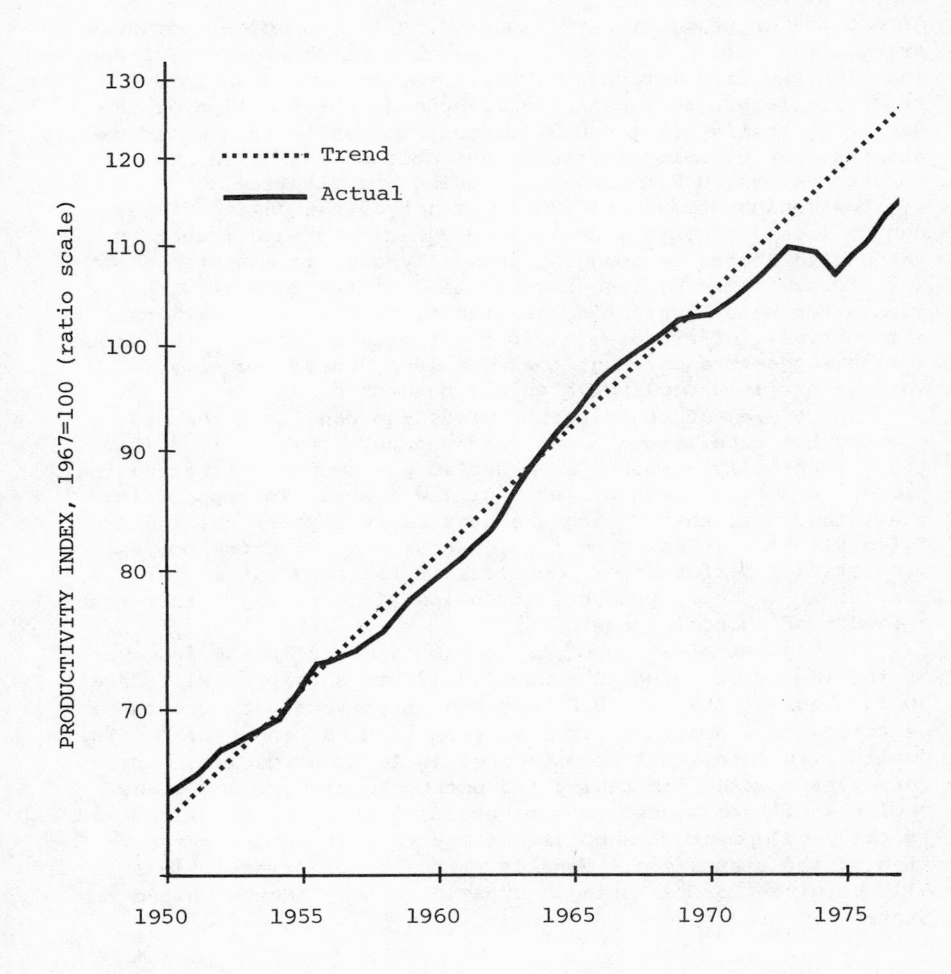

FIGURE 3.1

LABOR PRODUCTIVITY IN THE PRIVATE NONFARM BUSINESS ECONOMY

Source: Economic Report of the President, 1978 (Washington D.C.: U.S. Government Printing Office).

entirely new possibilities, including automatic production lines, became apparent.

Gains in productivity triggered by technological advances or by organizational changes always involve some degree of dealing with new circumstances. Therefore, the slowdown in U.S. productivity growth can be looked upon as a reflection of the declining ability of the U.S. economy to deal with new circumstances. By becoming less and less able to deal with new circumstances, the U.S. economy is losing its vitality.

Worsening productivity performance of the United States can be looked upon as a decline in dynamic behavior; that is, the United States is changing from a dynamic to a static economy. "Dynamic" is defined here as the ability to deal with new circumstances by generating new technical and organizational alternatives. Therefore, the productivity decline in the United States suggests a movement toward a less dynamic economy in which there is a decline in entrepreneurship.

There are, of course, economists who deny that the U.S. economy has experienced a profound change. They believe that the productivity slowdown is a cyclic phenomenon reflecting the slower economic growth of the last few years. To support this view, they note that during the past several years the industries with the lowest growth rates have exhibited the poorest productivity performance. However, if the argument in this chapter is correct, productivity gains are a cause, rather than a result of economic growth. [1]

Through examination of cases and evidence in the framework of the theory of "dynamic economy," [2] this chapter will develop an argument that the U.S. decline in productivity growth can be traced to a declining rate of technological innovation. This decline, in turn, will be explained in terms of declining propensities toward risk taking and entrepreneurship, and these declines will be traced to the declining entry of new firms and to the development of a political system that favors preservation of the status quo. Finally, a number of prescriptions will be given for restoring a dynamic economy in the United States.

The Nature of a Dynamic Economy

In a static economy, the manager takes both tastes and alternatives as givens; in a dynamic economy, the entrepreneur constantly searches for ways to satisfy a latent demand or to satisfy an existing demand with better or less expensive alternatives. That is to say, static economics is a theory of how to choose among a set of alternatives whose characteristics are known with some degree of certainty, while dynamic economics is a theory of how to create entirely new alternatives.

Incentives play a more significant role in a dynamic economy than in a static economy. Dynamic economics contains not only positive incentives in the form of the quest for higher profits (the "invisible hand" argument of Adam Smith), but also negative incentives in the form of the possibility of a rival dislodging a firm from a well-established market (a concept hereafter referred to as the operation of "hidden foot"). The difference between the invisible hand and the invisible foot can be illustrated by considering an industry in which the historical record shows that there is a 50-50 chance of a firm being dislodged from an important market. In such an industry both the hidden hand and the hidden foot play important roles. The more that entrepreneurs are able to guess correctly about promising new markets, the larger will be the profits. On the other hand, in this industry the penalties involved in ignoring the new advances of ones' competitors are very severe. A firm seeking to make its profits as large as possible by making a product only slightly different from its competitors' products faces almost certain bankruptcy.

Another fundamental difference between static and dynamic economics is that static economics is completely deterministic and luck plays no role. Dynamic economics is nondeterministic not only in the sense that luck plays a significant role in setting outcomes, but also because it acknowledges a positive relationship between necessity and luck. Necessity and luck are related because there is nothing like the hidden foot to motivate entrepreneurs to ask searching questions and to seek new alternatives. Since chance plays such an important role in a dynamic economy, the entrepreneur cannot hope to know the probability of bringing about a significant advance in technology or productivity. Such calculations can only be made if the advance is relatively trivial. In a dynamic economy, the principal difference between an entrepreneur and a manager is not that the entrepreneur is more reckless, but that his advantage lies in posing searching questions, while the manager's consists of answering well-defined questions.

The Relationship Between A Dynamic Economy and Productivity Advance

The High Price of Industrial Peace

During the Victorian period, the blame for the slowdown in productivity advance in Britain was placed mainly on the belief that the world was running out of ideas. The chances for technological change were believed to be small. Since the Victorians regarded themselves as God's chosen people, there could be no other explanation. However, a decline in dynamic behavior does not indicate a drying up of ideas; rather, it is a sign that as technologies mature the associated organizations tend to become less dynamic -- to feature managers rather than entrepreneurs. Furthermore, a decline in dynamic behavior in an economy reflects the fact that the growing costs of entry of new firms in established fields causes the rate of formation of new firms to decline as well.

In a static economy, easy entry of new firms in an established field performs the function of eliminating excess profits of established firms. In a dynamic economy, the advantages of new firms are that they have a relative advantage in generating ideas and that they increase the propensity to take risks in an entire industry; that is, when new firms enter an industry, existing firms must consider ideas they would otherwise dismiss. Thus, unless there is the entry of new firms, real challenges from other industries, or challenges from foreign firms, it can be predicted that as the industry matures, the diversity of ideas generated will become narrower and narrower until finally an equilibrium is reached in which there are only trivial differences between products, and in which rivalry among firms is replaced by market sharing. A system of mutual deterrence develops, and very tight regulations are imposed on entrepreneurs lest they disturb the peace! One of the major questions facing the United States is whether stagflation is too high a price to pay for this kind of industrial peace.

The Role of Tight Cost Constraints in Stimulating Productivity

A tight cost constraint is one which forces an entrepreneur to develop less costly alternatives than those now available. A tight cost (or tight quality) constraint is simply a way of forcing oneself to ask whether a better job can be done. To answer this question, the firm must search for ways to bring about major reductions in costs, which usually requires that it spend money on R&D.

Not all firms will be equally successful in keeping costs within tight constraints. In fact, a wide range of outcomes of R&D projects can be hypothesized: the larger the roles that both good and bad luck play, the wider will be the variance of outcomes. However, if one firm turns out to be successful, it will have found a way to bring about a savings in capital or labor or both; in other words, it will have found a way to bring about an increase in "total factor productivity." In principle all firms could be equally successful in observing tight cost constraints -- with the result that market shares would remain constant. But statistically speaking, such outcomes are very unlikely.

There are other ways to reduce costs, such as cutting workers' wages or cutting salaries of managerial officials. However, Keynes was not the first to discover that it is hard to cut wages. As early as the 1860s, entrepreneurs in the British steel industry regarded improvements in productivity as a better way to meet competition than reducing wages.

A number of studies have shown R&D expenditures and productivity gains to be highly correlated. [3] However, it would be wrong to conclude that the more spent on R&D, the more rapid will be the rate of productivity gain. A high correlation can be expected only as long as all the firms in an industry impose the same cost constraint on themselves. If, as rivalry declines, firms impose a smaller degree of constraint on themselves, their R&D expenditures may or may not increase. R&D expenditure is a good predictor of productivity advances only so long as the degree of cost constraint and, more basically, the degree of rivalry remain constant.

The Role of Feedback in a Dynamic Economy

In dynamic economic theory, two concepts play central roles: unambiguous feedback and ambiguous feedback. Unambiguous feedback is measured in terms of changes in market shares for various categories of similar products (e.g., intermediate automobiles, long-range commercial aircraft). If, for example, during the past five years, a 40 or 50 percent gain in market share has been typically associated with the introduction of a new product, then it is quite certain that the hidden foot is providing other firms in the industry with a genuine incentive to take risks. Consequently, it can be predicted that in such a situation searching questions will be asked, and a high degree of constraint will be imposed on cost or quality. It also can be predicted that as the competitors' market share gains are successfully challenged, less searching questions will be asked and a lower degree of constraint will be imposed.

On the other hand, ambiguous feedback consists of all the hints and clues relevant for the generation of better alternatives to existing products and processes. These hints can be gleaned from the firm's own activities to generate better alternatives, from those of its competitors, and by asking why particular "experiments" were not more successful. (Although an R&D project may be written off as unsuccessful, it may contain some good ideas relevant to generating a more successful alternative.) An entrepreneur can also obtain important clues to better alternatives by observing the technology of another industry, from studying basic scientific research, or from realizing that some ingenious ideas embodied in one product can be used for an entirely different application in another. In short, ambiguous feedback consists of the good luck the entrepreneur might enjoy if he asks the right questions.

The relationship between the two types of feedback is simply this: the higher the degree of unambiguous feedback from the market, the tighter will be the cost constraints, the greater will be the incentive to ask searching questions, and, therefore, the greater will be the need for ambiguous feedback as hints and clues to answering the searching questions.

HOW MAJOR PRODUCTIVITY ADVANCES OCCUR

This section draws on the history of four major advances in the productivity of U.S. industry to illustrate how the theory of dynamic economics works in real situations. It uses significant technological and organizational changes in the automobile, aircraft, and semiconductor industries to illustrate the importance of entrepreneurship, new entrants, risk taking, the hidden foot, tight cost constraints, and both ambiguous and unambiguous feedback in contributing to productivity improvement. This section draws heavily on material from the author's book, Dynamic Economics, and on research now in progress. [4] Following the examples, the dynamic theory of technological innovation and productivity advance is developed in considerable detail to show the conditions necessary for rapid improvements in productivity.

Examples of Major Productivity Advances

The Model T Ford

The Model T Ford represented the fulfillment of Henry Ford's dream: a practical car for farmers. Having been raised on a farm, he believed that farmers constituted an important segment of the market for new automobiles, though other car manufacturers tailored their products for city dwellers. For this city market, two distinctly different kinds of cars had been developed: the inexpensive but nondurable runabouts, ranging in price from $495 for a Sears & Roebuck runabout to $750 for a Century Steam Car, and the durable, but more expensive, automobiles ranging in price from $1,000 to well over $5,000. In 1908, Buick, which produced a medium-priced car for $1,000, was the top automobile producer of the nation, enjoying 25 percent of the market, as compared with 10 percent for Ford.

However, the inexpensive runabouts were not rugged enough for farmers and the more rugged cars were too expensive. Therefore, the constraints imposed upon the Model T design by Ford were durability and a price close to that of the inexpensive runabouts. These constraints helped make possible two important discoveries. The first discovery came from Henry Ford's observation that a French racing car constructed of vanadium steel was lightweight and durable. While other automobile manufacturers had observed the same racing car, Ford's cost constraint made this observation seem more relevant, and the Model T became the first passenger car to be built with vanadium steel.

Though the first Model Ts were stronger than other low-priced cars, they did not meet the second constraint of being inexpensive. When introduced, the Model T cost $850, and in 1909, when the price went up to $950, Ford's share of the market began to decline. Ford then adopted a price of $600 as his goal and initiated a wide search for ideas to rationalize and speed up production. The second discovery emerged from this search: moving production lines. This idea is generally credited to Clarence W. Avery, who had seen the concept employed in a meat packing plant. He rationalized that if moving production lines could be used for disassembling carcasses, why not for assembling automobiles!

Was it luck that Avery happened to visit a meat packing plant? Of course it was. Luck plays an important role in all discoveries. On the other hand, were it not for the need to meet Ford's tight cost constraints, there is little chance that the relevance of the process used in meat packing would have been appreciated.

Thus, Ford Motor Company created a loophole in the law of supply and demand -- it developed an entirely new option. If it is assumed that the farmers' needs for such an automobile were known, it is difficult to explain why other automobile makers did not develop a similar car unless one appreciates the importance of the factors in dynamic theory. (Of course, Henry Ford did not repeal the law of supply and demand; after the Model T was developed, Ford had to face the fact that his sales dropped as his price increased).

While no one can say to what extent Henry Ford accepted the risk of developing a new kind of automobile because of fear of his rivals, it is nonetheless true that at the time at least six automobile manufacturers who had enjoyed more than 10 percent of the market had already gone out of business. In that kind of environment, there was clearly a big risk in developing an automobile that would be only marginally different from those of his competitors; in short, it was an environment in which it paid to be different.

The story of Ford and the Model T helps to clarify one of the big debates in economics: the extent to which a large market was responsible for the rapid economic development of the United States. There is no question that a large market provides a country with an important potential advantage. But with a invisible hand alone, there is no assurance that this potential will be rapidly exploited. Nevertheless, a large and diverse market provides one important advantage. It helps to ensure that entrepreneurs will come from different backgrounds, and that they will, therefore, think differently about demand conditions. In other words, the advantage of size is a statistical advantage. In European countries, the probability was smaller that an automobile entrepreneur would come from a farm, and would, therefore, think like a farmer.

The DC-2 and DC-3 Commercial Airplanes

During the 1930s, more than one-half of total airline company costs were made up of airliner operating costs and depreciation. Shortly after the passage of the Air Mail Act of 1934, which drastically curtailed subsidies to the airlines, the airline companies began to press for more economical airliners. Trans-World Airlines asked several firms, including Douglas, to enter a design competition for an airplane whose performance would be equivalant to that of the recently developed Boeing 247, but whose costs would be significantly lower. Although a two-engine airplane was generally regarded as a simpler and more economical concept than the trimotor airplanes (developed during the late 1920s), because of the safety regulations of the Federal Aviation Administration (FAA), the plane was required to fly on a single engine. With that requirement, neither of Douglas' competitors at that time (Sikorsky and General Aviation) were willing to risk development of a two-engine design.

Douglas went ahead with their development of the DC-2, which turned out to be 30 percent overweight. Without the ability of the people in charge to quickly reorient the program to take advantage of the recently developed variable-pitch propellor and of a more powerful engine, Douglas could not have met the FAA performance requirement either. Thus, luck also played an important role here. (Later, it will be seen that conducting R&D programs to take advantage of good luck, and to minimize the consequences of bad luck, constitutes the essence of dynamic efficiency.)

It is true that, as of the time, there was no airplane in direct competition with the DC-2 (only somewhat later did Lockheed develop a close competitor). However, prior to the time of the DC-1 and the DC-2 (the production version of the DC-1), Douglas had never developed a commercial airplane. So, as far as the commercial market was concerned, Douglas was a newcomer. If it were to become an important factor in the commercial business, it obviously had to take risks. (Trying to get into the commercial market by merely developing another version of the trimotor airplane would have undoubtedly involved even greater risk). In his Wilbur Wright Memorial Lecture, Arthur Raymond, who was on the team that developed the DC-1 and the DC-2, and later was in charge of the DC-3 program, recalled the questions that a new entrant to a market asks:

> What is my competitor doing? Is he so entrenched that it will be extremely difficult, if not impossible, to make headway against him? On the other hand, has he been established in the field so long that he is perhaps growing complacent? Should he have brought out a new model some time ago and failed to do so, thus giving me an opening? [5]

Although the DC-2 operating costs were 25 percent below those of the trimotor airplanes, and although Lockheed's L-10 (similar in design to the DC-2) resulted in an even greater savings, this was not enough. Even with more economical airliners, the major airlines were still operating at a loss. The next step came when William Littlewood, Chief Engineer for American Airlines, proposed that larger engines be utilized to develop an airplane that could carry twenty-four, instead of fourteen, passengers. As a consequence, the DC-3 was developed as a stretched version of the DC-2. The development of the DC-3 brought the operating costs down by 50 percent.

The development of commercial aircraft differed from that of automobiles in one important respect. Aircraft development was supported by research undertaken directly by the government in the laboratories of the National Advisory Committee for Aeronautics [NACA] and in the universities. For example, Arthur L. Klein of the California Institute of Technology contributed to the development of the DC-2 by discovering a way to reduce drag that was previously unknown to designers of low-wing monoplanes.

Development of the Boeing 707

Boeing's experience in developing the B-47 jet bomber gave it an advantage when it came to developing commercial jets. But bombers and commercial airplanes are developed with entirely different cost constraints. For military airplanes the primary constraint is the performance requirement, and cost is a secondary consideration, while for commercial airliners cost is also an important restraint. Boeing had learned this the hard way when it developed the Stratocruiser airplane, a modificiation of the B-29 bomber, for commercial use. Although it might have been successful if it had been delivered to the airlines before the DC-6, it was not. Consequently, the relatively high operating costs, as compared with the DC-6, prevented Boeing from selling enough Stratocruisers to recoup its initial investment, and the company lost about $50 million.

After this experience, Boeing officials were determined to go about the development of a commercial jet in a different manner. The principal question about the economics of the airplane arose from the fact that FAA certification requirements for commercial jet aircraft were not yet established. Moreover, the airplane's stalling speed directly affected its economics. The higher the stalling speed, the lower would be the certifiable weight and the poorer would be the economic characteristics of the airplane, since the payload would be smaller and longer runways would be required. Further, relatively small errors in estimating the stalling speed could result in relatively large errors in estimating costs. Therefore, an experimental airplane was required to provide more accurate measurements.

Boeing adopted a cost constraint that required a commercial jet to have operating costs no greater than piston-engine commercial aircraft. With company funds, Boeing undertook the development of an experimental airplane that would serve as a prototype for both a commercial jet and an Air Force tanker. As it turned out, the 707 had lower operating costs than either the DC-7 or the Lockheed turboprop airliner, though relatively few people in the industry anticipated that the jet would have this advantage.

The Batch Process for Producing Transistors

Originally, transistors were regarded as substitutes for vacuum tubes. Although a half-dozen large receiving tube companies went into the transistor business, by the late 1950s, Texas Instruments occupied the same position in the semiconductor industry as Buick did in the auto industry when it was challenged by Ford.

The story of the challenge to Texas Instrument's market leadership began in the Shockley Laboratories, established as a wholly owned subsidiary of Beckman Instruments in the mid 1950s. (Along with John Bardeen and Walter Brattain, William Shockley received the Nobel Prize for the discovery of the transistor.) Shockley Laboratories was staffed mainly by a group of young scientists and engineers recruited from universities. However, shortly after the company was founded, a serious disagreement arose between Shockley and his staff. According to reliable sources, the disagreement arose not over whether the transistor would have uses other than as a replacement for the vacuum tube, but, rather, how visionary it paid to be given the current status of the technology. While Shockley was thinking in terms of ideas which might pay off ten or twenty years later, eight of the people he had hired believed that they had an idea that would pay off in about six months.

As is the case with many important discoveries, the dissidents' concept was relatively simple. The manner by which transistors were being produced, they reasoned, was analagous to producing postage stamps one at a time. Why not produce sheets of many stamps and then cut the sheets into individual stamps? Their motivation, however, was not so much to reduce the cost as it was to improve the performance of transistors without increasing the cost. They reasoned that if transistors with an order of magnitude improvement in performance could be produced at no more than what transistors then cost, there would be additional markets for semiconductors -- most immediately in the computer industry.

Not able to persuade Shockley, the young entrepreneurs managed to convince Fairchild Camera and Instrument to finance a new firm to develop the Planar process. Fairchild in return received an option to buy the firm, which it later exercised.

This, however, is not the end of the story; the young entrepreneurs continued to search for better alternatives. If many transistors can be produced in one operation, why take them apart and reassemble them to create complete circuits? Why not make many transistors and a complete circuit in one operation? This thinking eventually led to the drastic decline in the price of integrated circuits by more than 90 percent between 1963 and 1968, a decline in which Fairchild played a primary role.

It is no accident that the evolution of semiconductors took this particular form. Had entrepreneurs continued to think in terms of producing semiconductors one at a time, the idea of producing an inexpensive integrated circuit probably would not have occurred to them. But the tight quality and cost constraints not only led to one cost reducing discovery (the Planar transistor which made batch production possible), it also helped speed up the entire evolution of the technology. The same process has occurred in many other industries.

The development of transistors was but another case similar to that of the Model T Ford. By assuming that the demand for products of superior quality and constant price would be high, the people who developed the batch process were able to discover a loophole in the law of supply and demand. (If the market potential, and the means of achieving it, were well known, why did not many firms do likewise?) Again, it was fortunate that not all the people in the industry had made the common estimates of a low probability of such enhanced demand. If all firms had acted upon the assumption that the semiconductor was merely a replacement for vacuum tubes, the real potentialities of the technology would never have become known.

Conditions for Rapid Improvements in Productivity

Movements Away from the Center of a Distribution of Hypotheses

All of the examples presented exhibit the two principal characteristics of the dynamic process -- a process that results in steady reductions in cost, steady improvements in quality, or both. First, the alternatives do not become more and more alike, i.e., there is no movement toward the center of a distribution of all the possibilities. On the contrary, highly discontinuous changes are observed that change the shape of the distribution. Second, the changes that occur are irreversible changes that change history. For example, after the development of the Model T, the world no longer looked the same to automobile makers or to farmers; their subjective probability distributions of the alternative possibilities had changed.

Of course, not all the newly developed products and processes represent either significant movements away from the center of the distribution of possibilities, or major changes in the shape of the distribution. In fact, in a famous article written in 1929, Harold Hotelling lamented that he observed not only trivial differences between products, but also trivial differences between political parties. [6] On the basis of his observations, he developed a model to explain such convergence toward the center of a distribution: a model that has had a profound effect on both economics and political science. The reason for calling the Hotelling model to the reader's attention is to highlight the fact that dynamic processes involving feedback are entirely different from those that occur in his model, in which there is no feedback. For the dynamic model, the presence of feedback provides a mechanism for avoiding Hotelling's inevitable convergence by creating entirely new alternatives.

Alternatives move rapidly away from the center of a distribution when an industry generates a large amount of feedback. When large changes in market shares hinge on success in developing a new product, entrepreneurs are likely to impose tight cost or quality constraints upon themselves that result in the generation of entirely new alternatives.

The general picture associated with a series of cost reducing discoveries is shown in Fig. 3.2. At any given time, there are both relatively inexpensive and expensive alternatives: hence, the probability distribution. But an entrepreneur does not know where on the distribution a particular idea will fall. An idea that might seem to lie on the extreme left of the diagram when development is started can end on the extreme right side by the time development is completed.

How widely the entrepreneur searches will depend on the amount of feedback generated. When changes in market shares are small, he will search until an alternative is found that corresponds with his belief with respect to the middle of the distribution; then he will stop (point A on Fig. 3.2). This search pattern is associated with an industry that has an invisible hand but no hidden foot. However, a small degree of rivalry will result in a bit broader search and a modest degree of movement away from the center of the distribution (point B on Fig. 3.2).

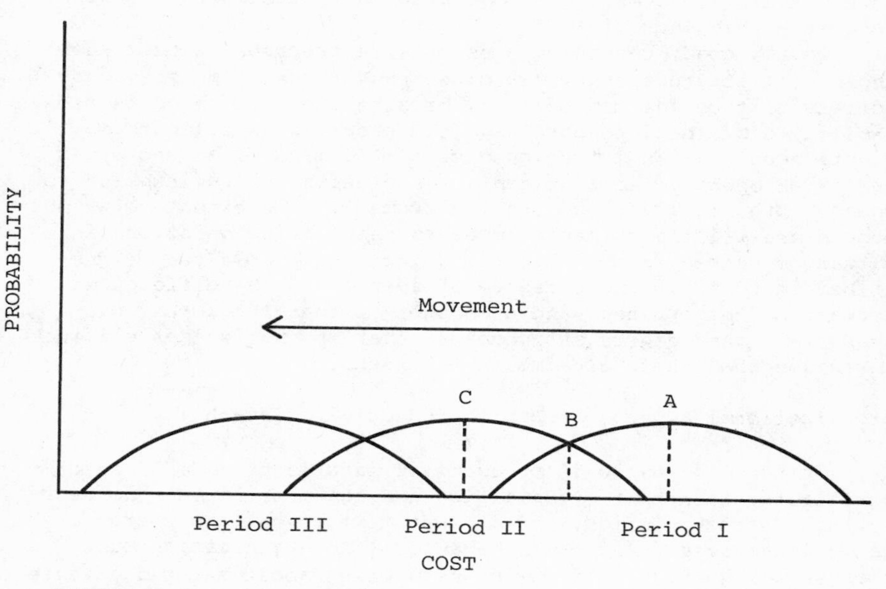

FIGURE 3.2

MOVEMENTS AWAY FROM THE CENTER OF A PROBABILITY DISTRIBUTION OF THE COSTS OF ALTERNATIVES

If feedback from large changes in market shares is high, firms will set their costs constraints beyond the tail of the distribution. Not all firms in the industry will be equally successful in meeting the constraint. If one is successful, the distributions will move to the left as shown on Fig. 3.2, with the most successful project appearing at the middle of the next distribution. The word "middle" is used because the entrepreneur knows that, in the next generation, some alternatives will turn out to be less successful than the most successful alternative developed during the previous period.

If an entrepreneur hopes to make history, he must not assume that he knows the probability that a particular hypothesis will turn out to be successful. Naturally, he will want to make some assumptions about the expected demand for his alternative (although if he hopes to make as much money as possible he must never assume he has complete knowledge of the elasticity of demand). [7] Furthermore, he must ask himself some penetrating questions -- for example, why can't an alternative be developed to have a 30 or 40 percent advantage over existing alternatives? But he must never assume that he knows how to bring about a discontinuous advance; for if he thinks he knows the answers (even in probablistic terms), his most precious potential asset will be lost -- his luck. [8]

To ask good, searching questions, entrepreneurs must adopt the mental attitude that, "at close range none of my previous beliefs or theories can be true, because they are based on probability distributions obtained from previous experiments or experiences." This is the essence of the mind of an entrepreneur -- an open system that can interact with its environment to change both its ideas and its environment. The extent to which people are willing to ask themselves penetrating questions is a matter of degree rather than kind; that is, people can be described in terms of their degree of openness. The difference between an entrepreneur and a manager is that the former possesses a higher degree of openness; that is, he is less willing to assume that there are immutable truths.

Organizational Aspects of Rapid Productivity Growth

If rapid advances in productivity are to be made in an organization, then that organization and its members must possess certain characteristics, and it must feature certain characteristic incentives for dynamic behavior. An organization must possess a high degree of openness to bring about a rapid advance in productivity. This means it must be able to engage in unconstrained interactions necessary to multiply the possibilities of good luck. Not only must it be staffed with imaginative and inquisitive people, it must be highly interactive internally and with the external world. It should be so interactive that the precise authorship of particular discoveries is always in dispute, because the more interactive an organization, the greater will be the possibility for encountering good luck. An entrepreneur may be good at asking questions, but if he always interacts with the same people, the probability of gaining new insights is most unlikely. As for outside interactions, universities are very important potential sources of new ideas. It is no accident that such interactions were at their peak when the most rapid progress was being made in the development of the commercial airplane, new pharmaceuticals, and computers.

The degree of openness of an organization might be measured in several ways. For example, data could be collected from an R&D department to find out who calls whom over several months. Then, using this data, predictions could be made of the pattern of communication for the next several months. If the predictions prove to be accurate, then the department probably possesses a low degree of openness. The more inaccurate the predictions, the greater the degree of openness in the department.

Organizations cannot long remain dynamic if they do not interact with their customers to understand their <u>real</u> problems. In the Minnesota Mining and Manufacturing Company, the key organizational "invention," which played a very important role in the company's success, consisted of having the salesmen, who previously only interacted with distributors, call upon the firms actually using the company's newly developed sandpaper. They acquired firsthand information about users' problems, and, subsequently, a research and development organization was set up to see what might be done to make the product more useful. This resulted in the development of completely new products.

The openness of organizations is determined by the degree of feedback to which they must respond and by their internal incentives. In dynamic firms, the correlation between seniority, position in the hierarchy, and salary is low. As an extreme case, during the 1950s, one chemical company paid twelve highly creative chemists more money than the chairman of the board. In static organizations, not only does salary depend on administrative position, but there is often a system of fringe benefits which makes it more and more expensive to leave the company. In other words, instead of providing positive incentives for risk taking, the incentives are better designed to ensure obedience.

Dynamic Efficiency and Productivity Advance

Dynamic efficiency is defined as maximizing the probability of recognizing good luck, while minimizing the consequences of bad luck. As used here, dynamic efficiency is similar to Harvey Leibenstein's concept of "X-efficiency"; that is, it is the efficiency associated with producing a given output with fewer inputs. [9] However, while Leibenstein agrees that competition plays an indispensable role in promoting dynamic efficiency, he does not discuss what kind of behavior is involved in promoting dynamic efficiency, nor how it differs from that involved in promoting static efficiency.

The key idea in static efficiency is to make good use of existing knowledge. As Adam Smith long ago pointed out in his famous discussion of a pin mill, the organizational principle in static efficiency is specialization; that is, larger tasks are broken down into smaller and smaller tasks. The organizational corollary associated with static efficiency is not a firm with a high degree of openness; on the contrary, the pursuit of static efficiency inevitably drives organizations to a lower and lower degree of openness.

By contrast, dynamic efficiency involves undertaking research and development and production activities to maximize the likelihood of good luck and minimize the consequences of bad luck. The first requirement for a high degree of dynamic efficiency, therefore, is an open and highly interactive organization to multiply the possibilities of good luck. Second, it is imperative to design projects so knowledge can be obtained early in development programs at a relatively small cost. Creative engineers are fond of making the point: "It is not what you do not know that can kill you -- it is what you think you know." Therefore, it is necessary to explore new concepts as quickly and inexpensively as possible. The third requirement for a high degree of dynamic efficiency is the easy incorporation of new knowledge. For example, not only did a variable-pitch propellor and a new engine save the DC-3 when it could not meet its performance requirements, but also its design made the changing of engines relatively easy. Performance of the DC-3 might have been increased somewhat by closely optimizing the design for a particular engine, but to do that would have made the incorporation of new technological ingredients more difficult. Because saving time often means production activities are started before all uncertainties are eliminated, dynamic efficiency requires that processes be designed to make the incorporation of product changes relatively inexpensive. When introducing new airplanes or pharmaceuticals, firms do not try to minimize production costs, but, rather, to minimize the cost of changes.

That rational people behave differently in a world of high uncertainty was recognized long ago by Keynes when he pointed out that one reason for holding money was to be better prepared to deal with "strong uncertainties"; that is, by holding money, one is better prepared to take quick advantage of good luck and to minimize the possibility of bad luck. But apparently he did not recognize that the same argument holds with respect to improving the efficiency of capital. In a world in which knowledge changes, the entrepreneur must be sufficiently flexible in planning so that it will be relatively easy to incorporate either new knowledge or newly recognized luck into his plans.

The Trade-off Between Static and Dynamic Efficiency

There is a trade-off between static and dynamic efficiency, since organizations cannot be optimized simultaneously for both low and high degrees of openness. A highly informal organization that is ideal for responding quickly to new conditions is not ideal for managing well-defined tasks. For example, from an automobile manufacturer's perspective, firms in the semiconductor industry appear chaotic. Conversely, in the static organizations described by Richard M. Cyert and James G. March in their Behavioral Theory of the Firm [10], the primary tasks are subdivided into smaller and smaller subtasks, the pattern of communications is almost entirely up and down echelons, and search is very narrow -- ironically such an organization could never have developed the computer used in Cyert's and March's simulation models.

As a consequence of the trade-off between static and dynamic efficiency, the cost of experimentation skyrockets as organizations become more complex and more production-oriented.

It was for this reason that the Ford Motor Company contracted with the Honda Corporation to set up a small shop at Ford where Japanese engineers could work on the development of a stratified charge engine. As a Ford engineer explained it to me: "While Honda still employs the same type of organization we used to develop the Model T, at Ford today our organization is too complex to undertake such a task inexpensively." Apparently, engineers at the Ford Motor Company are not convinced of Galbraith's argument that making major advances requires massive organizations. [11]

The trade-off between static and dynamic efficiency also can be observed in the declining ability of American industry to cope with foreign competition. This trend cannot be blamed upon any U.S. disadvantage in terms of scale economies or vertical integration, that is, on a disadvantage in static efficiency. In fact, in terms of static efficiency, the United States is still well ahead of the other industrialized countries. The problem, rather, is that organizations optimized for a high degree of static efficiency are incapable of highly flexible responses. The exploitation of scale economies and vertical integration requires massive organizations that have a minimal ability to respond to new circumstances. By contrast, from the standpoint of dynamic efficiency, small is indeed beautiful.

The Role of New Entrants into an Industry

There is an abundance of evidence that entrants into an industry -- whether the meat packing, automobile, aircraft engine, radio, or the semiconductor industries -- have played an indispensable role in maintaining the dynamism of those industries. Because new firms have great incentive to be different, they create rivalry which increases the incentive for the other firms in the industry to take risks. For example, the principal cause of the railroad cartels breaking down again and again during the nineteenth century was the entrance of new firms into the industry. [12] Contrary to common wisdom, railroad companies desired regulation not because the government was more efficient in running a cartel, but because regulation was needed to put an end to the entry of disobedient new firms.

I am not the first to point out the role of new firms in providing an industry with a greater incentive to engage in risk taking. The same point is made in F.M. Scherer's book, <u>Industrial Market Structure and Economic Performance</u> [13].

While the incentives of new firms differ from those of established firms, new firms have an even more profound advantage because they can exhibit a greater degree of dynamic efficiency. To exploit the discoveries it has already made, an established firm is almost inevitably driven to a high degree of organization; that is, it becomes less interactive and, as such, possesses a lower degree of openness.

Once a firm is driven to a higher degree of organization, each advance seems more significant than before. For example, when executives of the Ford Motor Company describe the Mustang as a fundamental breakthrough, they are describing an organization whose perceptions of an advance have been greatly modified since the introduction of the Model T. In other words, when organizations become more structured, the tolerance of their members for ambiguity is likely to decline to the point that small differences appear to be large ones.

There are exceptions to the rule that as a firm exploits its earlier discoveries its openness declines. For example, each new generation of communications satellites contains some very significant advances. The people who bring about these advances might describe their jobs as discovering loopholes in the laws of nature. However, to be successful, a new communications satellite must not only embody some marvelous new ideas, it must also be highly reliable. Consequently, sooner or later everyone on the team must stop being an entrepreneur and do his best to make the satellite perform reliably. Unlike other industries, however, once the satellites are in orbit, there is no way to bring about further improvements. Consequently, the teams are disbanded and once again managers become entrepreneurs.

New firms, on the other hand, tend to be rather informally organized. Typically, their employees have not known each other very long, and internal competition mainly takes the form of competition in ideas. Only later does position in the organization's hierarchy become the chief form of competition. Moreover, successful new firms pay a good deal of attention to understanding the real problems of their customers. It is only in well-established firms, in which the R&D process has become highly routine, that inventiveness for its own sake (unrelated to customer need) becomes the order of the day. Thus, this kind of environment is ideal for luck to play an important role.

New firms have other advantages. Their cost of experimentation is likely to be a good deal lower. Newly established firms are not as concerned as well-established firms about making discoveries that might make highly specialized facilities or staff obsolete.

Toward a Dynamic Model

To begin thinking about the formulation of a dynamic model, it is necessary to understand two concepts: Type I and Type II uncertainties.

Type I Uncertainty

Type I uncertainty is the uncertainty associated with unambiguous feedback. Type I uncertainty implies risk for a firm, and since a firm cannot buy insurance to protect it from this risk, it provides an incentive to deal with risk.

If firms are insured against competitive risks by forming a perfect cartel in which changes in market shares occur within very narrow limits, there are only trivial differences among old and new products. In this situation firms can make nearly perfect predictions about each other's products and there is a negligible degree of Type I uncertainty, because there is a negligible amount of ambiguous feedback.

As firms take greater risks, changes in market shares will become greater, more feedback is generated, and Type I uncertainty increases.

A high degree of uncertainty, however, does not necessarily imply real surprises. If every three or four years some firm in an industry generates a significant discovery, the firms in the industry may not be able to predict each other's new products, but they can predict that some surprise will occur. Such surprises are not necessarily startling from the point of view of insiders who are accustomed to highly discontinuous changes. If an industry is unaccustomed to dealing with the unexpected, even a relatively small deviation from the "normal" will be regarded as surprising. For example, when the rates fell by as much as 20 percent during the successive breakdowns of the railroad cartels in the 1880s, the railroad companies described this as a "price war." On the other hand, in the commercial aircraft industry during the 1930s, or in the semiconductor industry -- where a 20 percent advance was regarded as an incremental improvement until fairly recently -- if a firm's rivals had behaved like the railroads, it would not have been regarded as war: it would have been regarded as peace!

The main point to keep in mind is that when the hidden foot is doing a good job of keeping Type I uncertainty high, it operates just as automatically as the invisible hand. For example, as Fig. 3.3 shows, progress in reducing the operating and depreciation costs of commercial airliners was so smooth that it almost seems to have been preordained. In addition, the figure shows that, during the period between 1925 and 1940, when progress was the most rapid, it was also the most unpredictable. The fact that both good and bad luck played important roles in the outcomes during that period is illustrated by the fact that the difference between the highest and lowest costs of the alternatives developed each year averaged over 50 percent. After World War II, there was a definite slowdown in the rate of progress. Government regulation of airline rates, established in the late 1930s, took the pressure off the aircraft developers to reduce costs. Also, during the 1930s,

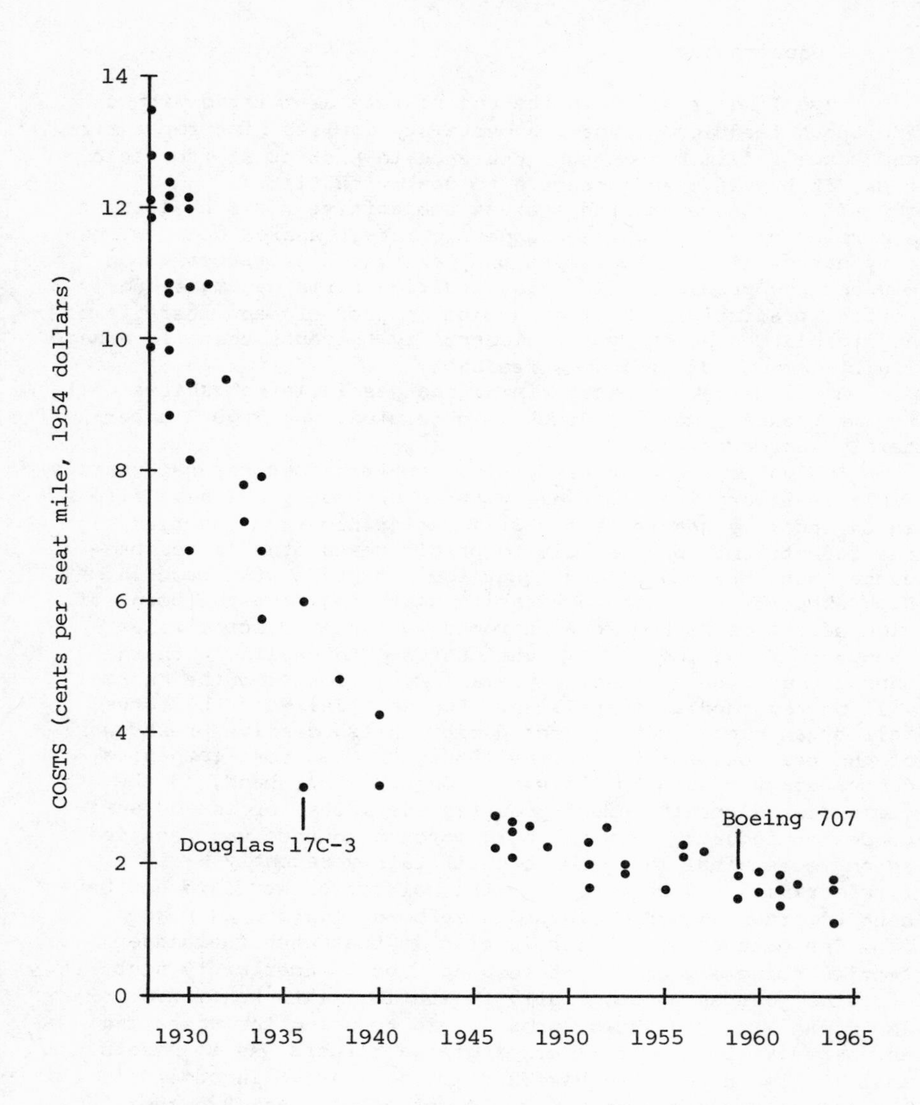

FIGURE 3.3

AIRLINER OPERATING COSTS

Source: Almarin Phillips, Technology and Market Structure (Lexington, Mass.: Heath Lexington Books, 1971), pp. 50-51.

entry of new firms into the industry came to a standstill. Today, one of the most important reasons for deregulating the airline industry is that aircraft companies will find themselves under more pressure to reduce costs as a consequence of a greater degree of rate competition.

One way to measure Type I uncertainty is in terms of the probability of a good or bad surprise. Table 3.1 shows the fate of 26 major automobile firms during the period 1903 to 1921, and 31 semiconductor firms during the period 1957 to 1975. Only one automobile company and one semiconductor company stayed in the top five for the entire period. In the automobile industry the downfall was more severe; all of the unpleasant surprises involved either dropping to 10th place or going out of business, while a typical downfall in the semiconductor industry involved something like a 50 percent decline in a firm's share of the market. Of the automobile firms which made it into the top ranks, or which remained in the top ranks in 1921, only one was in business as early as 1903, two entered the field in 1915, and two in 1916. In the semiconductor industry, two newcomers played much the same role.

It should be apparent from these data that the successful princes in modern industry are not the firms which started out lucky, spent more on R&D, and increased the probability of further success. Rather, like many of Machiavelli's princes, many of the modern princes met their downfall because they could not adapt to new circumstances.

Competition from other industries or from foreign firms can play an important role in generating feedback for an industry and, hence, in making Type I uncertainty high. Moreover, firms can face a high degree of Type I uncertainty for reasons other than competition. In particular, uncertainty with respect to inputs can play the same role. For example, some petroleum companies have much poorer oil reserves than others, and for this reason, they are more dynamic than the other petroleum companies. Or, to take a more extreme example, because of input uncertainty, Japanese steel firms have a far greater incentive to engage in dynamic behavior than do American firms.

Type II Uncertainty

While Type I uncertainty provides firms with challenges, Type II uncertainty provides opportunities. To clarify the concept of Type II uncertainty, it will be useful to distinguish between three states of the world. The first is a completely predictable world in which chance plays no role. For example, in general equilibrium economics, the entrepreneur is assumed to know the final equilibrium price before any trading takes place. In such a completely deterministic world, entrepreneurs have no opportunity to be lucky.

TABLE 3.1

FATE OF MAJOR FIRMS IN TWO INDUSTRIES

26 Automobile Firms, 1903-1924

Never fell from top 5	3 firms
Fell from top 5	6 firms
Rose into top 5	2 firms
Remained minor entities	15 firms

Source: Burton Klein, Dynamic Economics (Cambridge, Mass.: Harvard University Press), p. 100.

31 Semiconductor Firms, 1957-1975

Never fell from top 5	3 firms
Fell from top 5	5 firms
Rose into top 5	1 firm
Remained minor entities	22 firms

Source: "The Semiconductor Industry: A Survey of Structure, Conduct and Performance," January 1977, p. 23.

The second is a world in which entrepreneurs can accurately predict each other's actions in probabilistic terms. Chance plays a role, but a completely prescribed one in the world, because it does not acknowledge the probability of events not included in an initial subjective probability distribution.

Third is the world of strong uncertainties -- a world in which the uncertainties cannot be measured in terms of probability distribution, because quite unexpected events do occur. The stronger the uncertainties in this world, the greater the roles likely to be played by both good and bad luck.

Type II uncertainty is measured by the variance of outcomes of R&D projects. A high variance outcome indicates that entrepreneurs recognize the possibility of opportunities. A low variance outcome implies that entrepreneurs have stopped asking sharp questions. For example, the discussion of aircraft operating costs indicated that during the period of rapid progress, the difference between the lowest and highest cost alternative in any year averaged about 50 percent. This outcome implies a high degree of Type II uncertainty. On the other hand, a 10 percent difference would imply a low degree of uncertainty.

The central proposition of dynamic theory is this: the greater the degree of Type I uncertainty, in other words, the greater the degree of feedback -- the greater the degree to which potential applications will be exploited.

Responses to Changes in Type I Uncertainty

The Different Effects of Increases and Decreases

The responses to an increase in the degree of Type I uncertainty are of a very different character from the responses to a decline. When Type I uncertainty increases, at some point an industry will respond to the increase in feedback by (1) creating more favorable internal incentives for risk taking, and (2) acquiring a greater degree of openness. If the rate of progress is measured in terms of cost reductions, a more rapid rate of progress should occur. On the other hand, when Type I uncertainty declines just the opposite occurs, with entrepreneurs tending to pose less searching questions and with luck playing a less important role in R&D outcomes. Therefore, the rate of progress is slower. Performance then follows an S-shaped curve with a period of rapid progress in moving alternatives away from the center of a distribution (fast history) followed by a period of slower progress (slow history).

Suppose Type I uncertainty declines and fast history turns into slow history. How can cause and effect be determined? Are firms merely responding to a decline in their opportunities or are the opportunities declining because the degree of openness is becoming smaller and smaller? There are two fundamental reasons for assuming that the basic explanation is to be found in a declining degree of openness brought about by the failure of new firms to enter the industry. The first is an empirical reason. Assume that the industry in question is experiencing slow history (that it is on the relatively flat upper portion of an S-shaped curve of performance) and a discovery comes along to accelerate the rate of progress. If the rate of progress had been limited by a lack of opportunities, it seems reasonable to expect that major firms in the industry would have played a major role in the revitalizing discovery. But this is usually not the case. Of fifty cases for which evidence was found, all of the revitalizing discoveries came either from a new firm in the industry (e.g., the Polaroid camera), a firm from another industry (e.g., diesel locomotives, synthetic fibers), or from a university laboratory (e.g., computerized machine tools). [14]

The second and more basic reason to believe that a decline in opportunities is not the fundamental cause of slowdowns is that the phenomenon of diminishing returns does not apply to ideas. Obviously, if a technology is defined narrowly enough (say, railroad transportation is defined as only using steam engines), it can be predicted that sooner or later the rate of progress will diminish. However, if the definition of a technology is continuously broadened, there need be no decline in the rate of progress. Insiders will not necessarily agree with this point of view. On the contrary, they can be counted upon to insist that progress has slowed because bringing about major advances has become enormously more difficult. In one sense they are absolutely right -- a more constrained organization will have greater difficulty in bringing about such advances.

Responses to Increases in Type I Uncertainty

An increase in Type I uncertainty in an industry is equivalent to an increase in unambiguous feedback. That is, the market shares in an industry undergo more rapid change. How do firms respond in such a situation?

In dynamic economics, response is measured in terms of outputs rather than inputs. At one extreme, there is the response that is so rapid and so effective that the firm in question experiences no serious decline in market share. Conversely, there is the response that is so delayed and so weak that a decade must pass before the decline is somewhat arrested. However, it can be assumed that the response is nonlinear because the more entrenched the internal technostructure and the more cohesive the internal alliances, the greater the amount of feedback that will be required to elicit any response.

Statistically, there is a good reason to assume that the speed and effectiveness of a firm's response will be highly correlated, inasmuch as both are dependent on its internal characteristics. Much empirical work remains to be done to develop reliable dynamic response curves for firms. However, on the basis of fragmentary information, I have made some rough guesses of the likely differences between dynamic response curves for firms from several industries. These are shown in Fig. 3.4.

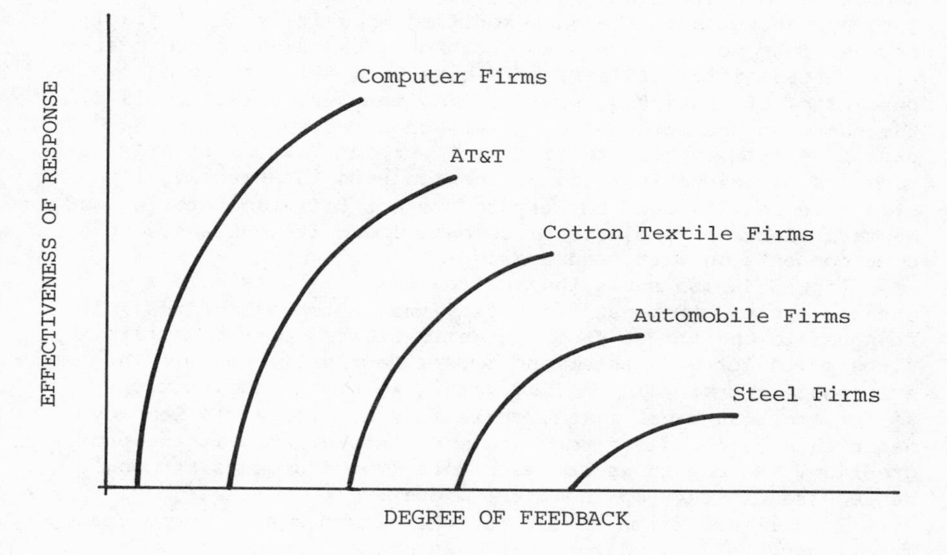

FIGURE 3.4

DYNAMIC RESPONSES TO ADDITIONAL FEEDBACK OR TYPE I UNCERTAINTY

AT&T provides an interesting case from Fig. 3.4. Most people think of AT&T as a monopoly and of Bell Laboratories as an organization which for many years has remained highly dynamic without the need of a hidden foot. This commonly held belief about AT&T is wrong. [15] Informed people are in agreement that Bell Laboratories was not nearly as dynamic an organization before World War II as it was after. It is not difficult to explain the rise in its dynamic behavior. In addition to the minor challenges with which AT&T dealt before World War II, after World War II it encountered three major ones: microwave long-range communications, communication satellites, and common carrier private communications systems. In short, there can be no doubt whatsoever that the hidden foot, often in the form of attacks by relatively new firms, played a major role in explaining why Bell Laboratories became a more dynamic organization after World War II.

One principal difference between AT&T and General Motors is that by having a relatively unstructured organization in the form of Bell Laboratories, as well as a more structured organization in the form of Western Electric, AT&T has a substantial ability to deliver a quick technological response to additional feedback. On the other hand, by being regulated by the Federal Communications Commission (FCC), it also has an advantage in getting the rules of the game modified when its technological prowess does not suffice. For both of these reasons, competitive inroads into AT&T's markets have been held to around 5 percent of its business, while during the period 1955 to 1975, the share of the automobile market accounted for by foreign cars rose from about 5 to around 20 percent. Thus, if AT&T made its decisions in terms of the long-run interests of its stockholders, it would not try to protect Bell Laboratories and Western Electric from outside competition. It would welcome some competition with open arms!

Fig. 3.4 also shows that cotton textile firms have a greater dynamic response capability than automobile firms. In response to the threat from synthetic fibers, cotton textile firms moved forward faster and more effectively than did the automobile industry in dealing with its foreign competition. As for American steel firms, while they are certainly less dynamic than automobile firms, how much less dynamic is an open question. No one knows how much more Type I uncertainty would be required to wake up the steel industry.

Established firms differ in their responses to increased Type I uncertainty. Often the ideas needed for a firm to respond to a sharp and persistent increase in Type I uncertainty already exist. They are ideas that entrepreneurs far down in the hierarchy have been trying to sell to the vice-presidents. Or, if ideas are lacking, the entrepreneurs will typically begin to search outside the organization, while the vice-presidents are still busily engaged in convincing each other that the decline in market share is only temporary. However, sooner or later one of the more adventurous vice-presidents will begin communicating with the entrepreneurs, and, after an organizational struggle, new coalitions will be formed. In short, when the internal power structure is not so completely cemented together that it is totally insensitive to increases in Type I uncertainty, the predicted organizational response is a greater diversity of internal and external interactions. Moreover, while internal incentives may or may not be changed, it can be assumed that those who are successful in taking risks will be rewarded.

Responses to Decreases in Type I Uncertainty

Suppose there has been a persistent decline in Type I uncertainty in an industry. Such a decline might occur because an industry was able to reduce uncertainty with respect to its inputs or because new firms are not entering the industry at the rate they once were. In order to better describe the response, a concept will now be introduced that plays a central role in dynamic economics: the distinction between <u>microstability</u> and <u>macrostability</u>.

An industry with a high degree of microstability is a highly predictable industry: market shares remain approximately constant, there is a relatively small degree of variance in the outcomes of R&D projects, and the pattern of communications within the associated firms is highly predictable. Both the Swiss chocolate and the Scottish textile industries have a high degree of microstability. Conversely, macrostability is a dynamic concept of stability. It is concerned with the ability of an industry as a whole to make smooth adjustments to new circumstances. The degree of macrostability is measured by the rate at which the performance of a technology improves, whether measured in terms of reductions in costs or improvement in quality.

The U.S. semiconductor industry is a good example of an industry with a high degree of macrostability. Up until about 1970, the rate of technological progress in this industry followed the rapidly rising portion of the S-shaped curve shown in Fig. 3.5. The vertical lines on the curve represent such major discoveries as the silicon junction transistor, the tunnel diode, the Planar transistor, and the integrated circuit. Viewed as isolated events these discoveries were very unpredictable, but the fact that the entire process was both rapid and smooth indicates an industry in which there was a good deal of feedback.

As a technology matures, scale economies and vertical integration became more important. Consequently, if a new firm wishes to enter the industry to compete with existing firms, it is likely to find the cost of entry very high. Thus, potential entrepreneurs must be very clever not only in generating new ideas, but also in persuading somebody to provide risk capital. Thus, as a technology matures, the probability of new firms entering an industry declines substantially -- and as entry declines, so will the degree of macrostability and the rate of productivity improvement.

When entry into an industry slackens, the degree of macrostability as measured by the rate of progress will decline, as shown in Fig. 3.5. As long as new firms enter an industry, it can be assumed that from a micropoint of view, outcomes will be highly unpredictable, whether measured in terms of the outcomes of R&D projects or changes in the market shares. On the other hand, it is fairly certain that as long as this situation exists, the continuation of rapid progress will be highly predictable. Let U_1 be the uncertainty with respect to the continuation of the steeply ascending portion of the curve shown in Fig. 3.5 (Type I uncertainty); and let U_2 be the degree of uncertainty with respect to micro-outcomes (Type II uncertainty). Then, as in shown in Fig. 3.6, a low degree of uncertainty with respect to the continuance of the fast history trend implies a high degree of uncertainty with respect to microoutcomes (A). However, as more and more certainty is introduced into microoutcomes (due to the failure of new firms to enter the industry), continuance of the previous rate of progress becomes more and more uncertain as it goes from A to B and from B to C. That is to say, it can be predicted that sooner or later the rate of progress will decline.

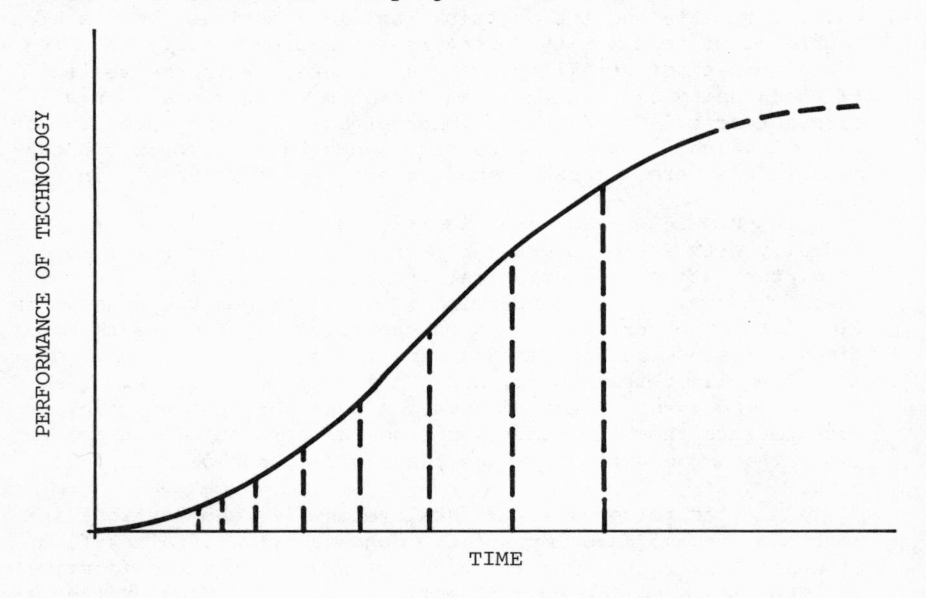

FIGURE 3.5

CHANGING PERFORMANCE OF A TECHNOLOGY

The two most important differences between increases and declines in Type I uncertainty are the following. First, increases involve highly discontinuous responses, while declines do not. As entrepreneurs are phased out of an operation, managers acquire an ever more important role. Second, when Type I uncertainty declines, changes in incentives and organizational characteristics are much easier to measure. Indeed, even without any measurement, it is not necessary to be a social anthropologist to discern what is happening.

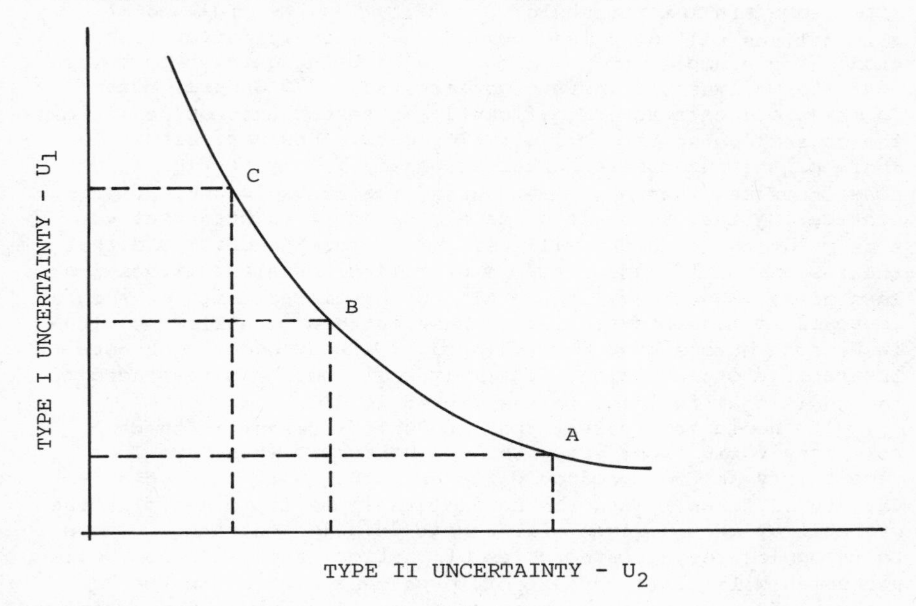

FIGURE 3.6

THE RELATIONSHIP BETWEEN TYPE I AND TYPE II UNCERTAINTIES

Suppose that no new firms enter the industry in question for a very long time, so that Type I uncertainty is very low. What kind of equilibrium will be ultimately reached, and what is the nature of the process when a dynamic equilibrium collapses into a static equilibrium? Consider the automobile industry during the 1950s when neither foreign competition nor other forces provided much Type I uncertainty. At that time, competition took place mainly in terms of style and, at any time, only a very narrow range of styles were successful. That is, only small movements away from the center of the distribution of commodity attributors was acceptable. While it is an open question whether automobile makers created a taste for style change or merely responded to such tastes, the fact of the matter is that firms that failed to recognize a change in style, i.e., a movement to a new center of the distribution, were seriously penalized. For example, Chrysler, which was very slow to adopt long tailfins, found that from 1952 to 1954 its sales had declined from 20 percent to 12 percent of the market.

It must be emphasized again that movement toward the center of a distribution occurs in an industry in which there is a negligible amount of Type I uncertainty. Consequently, after some minimum threshold in rivalry has been reached, alternatives will be driven away from the center of distribution. For example, as a consequence of being quicker to borrow ideas formerly found in foreign cars, the 1979 General Motors intermediate cars are significantly different automobiles from the corresponding Ford and Chrysler cars. However, since the share of the automobile market accounted for by foreign firms rose from less than 5 percent during the 1950s to over 20 percent during the 1960s, it certainly can be predicted that at some point the industry will respond to more feedback and that the response will take the form of making the alternatives look less alike. Moreover, it can also be predicted that for such a response to have occurred, the inner pattern of alliances within General Motors must have changed. Without becoming a more interactive organization, it simply could not have responded to the additional feedback in the manner it did.

It should be apparent that product differentiation can take many forms, some very subtle. Moreover, when Type I uncertainty is low, product differentiation will take on a variety of forms. When the name of the game is to minimize the probability that a rival will engage in price cutting, it pays to recognize the idiosyncrasies of particular consumers. Scale economies will limit the extent to which a product can be tailored for a particular consumer, but firms try to go as far as possible in recognizing individual differences in tastes. For example, in an excellent article on the ready-to-eat breakfast food industry (an industry no new firm has entered for some years), Richard Schmalensee has noted that variations include differences in sweetness, protein content, shape, grain base, and crunchiness. [16] According to his argument, firms in the industry went about as far as they could in recognizing these differences -- with the consequence that price cutting has rarely occurred. However, while the firms in the breakfast food industry were good at recognizing differences in tastes, they were apparently not nearly so expert in detecting changes in tastes. When natural foods became popular in the early 1970s, quite a number of new firms entered the market; only after that did the well-established firms respond to the change in demand conditions.

Even when demand conditions are fairly well known, established firms are often slow to recognize changes in tastes. Consider, for example, the long time lapse before the hamburger chain restaurants began to provide more variety in their menus.

When firms no longer have to deal with feedback, they lose their ability to engage in anything but routine tasks. They have entered a stable Hotelling equilibrium. Calcification does not occur immediately; it often takes five to ten years.

As it occurs the obvious response of firms is to protect themselves from feedback. Consequently, they engage in product differentiation and advertising. However, while product differentiation ensures firms against small price cuts on the part of their rivals, it does not protect against the possibility of a large cut. Under what circumstances are large cuts likely to occur? Firms occasionally cut prices and profits to drive others out of business. More probably, a firm would be willing to reduce its prices if it has brought about a significant reduction in its costs. But calcified firms are not likely to be able to reduce their costs significantly. Hence, a Hotelling equilibrium is very stable, since it minimizes the probability of either small or large price cuts. What makes it stable is that a firm is not likely to start a war it cannot finish. Thus, such an equilibrium can be described as a system of stable mutual deterrance. [17]

However, it is important to keep in mind that, while such an equilibrium is stable from the point of view of managers, it is not from the point of view of the dynamic stability of the economy. Firms in a Hotelling equilibrium possess an almost zero degree of macrostability.

The Limitations of Dynamic Theory

Although it is my conviction that only by taking feedback into account can economics become a predictive science, a few words are in order about the limitations of dynamic theory.

First, individual discoveries cannot be predicted. Dynamic economics is not concerned with making microscopic predictions. Rather, it is concerned with making macroscopic predictions -- predictions of the rate of progress as a function of the degree of Type I uncertainty.

Second, once the rate of progress in a particular industry has leveled off, it is not easy to predict the dynamic response to increases in the degree of feedback. Empirical work relating increases in the degree of feedback to the dynamic response will enable us to make better predictions. However, since a few key people can often make a significant difference in the timing of the response, it may never be possible to make really good predictions.

Third, dynamic theory does not permit comparisons across industries; that is, it cannot be inferred that progress is higher in one industry than in another because of differences in the degree of feedback. The reason is that in some industries it is easier to broaden the definition of the technology. Some examples of this are: jet engines broadening the definition of aircraft technology, Bessemer steel broadening the definition of steel technology, and semiconductors broadening the definition of computer technology.

As the definition of railroad technology has been broadened, the rate of advance in total factor productivity has been very respectable. As Fig. 3.7 shows, the average rate of productivity increase in the railroad industry was not as great as in transportation as a whole, but it was greater than for manufacturing in general. Although railroad technology is very old, there is no evidence of a retardation in its rate of productivity increase.

Were it not for railroad rate regulation, progress in reducing costs would have been faster. In general, economic regulation stabilizes market shares, removes feedback from industries, and weakens the incentive to adopt new ideas quickly. Consequently, the rate of diffusion of new advances in the railroad industry has been very slow. According to estimates made by Edwin Mansfield, once one railroad adopted an invention it took a long time before the industry as a whole adopted it: about fifteen years for the diesel locomotive, twenty-five years for the Mikado locomotive, twenty years for the four-wheel trailing truck locomotive, twenty-five years for centralized traffic control, and thirty years for car retarders. [18] Since the adoption rate depends in part on the profitability of inventions and the size of the investment required, it is necessary to hold these factors constant when making comparisons with other industries. By making calculations on that basis, Mansfield found that, compared with the railroad industry, the time between 10 and 90 percent adoption was about six years shorter in the steel industry, two years shorter in the coal industry, and fourteen years shorter in the brewing industry. As Mansfield points out, the differences between the railroad and coal industries may not be statistically significant; but none of these can be regarded as dynamic industries. [19]

Yet, even the railroads seem to be moderately responsive to increases in the degree of unambiguous feedback. From 1950 to 1960, intercity freight traffic from railroads declined by about 25 percent, due mainly to competition from trucks. During the period from 1960 to 1970, the average rate of increase of total factor productivity doubled that of the previous decade (see Fig. 3.7). About one-half the gain came from closing down less-profitable lines and half from speeding up the rate of technology diffusion.

This example shows that, even though comparisons cannot be made across industries, there are ways of detecting whether an industry is making good use of its potential. Moreover, it is always possible to recognize when feedback begins to play a significantly more important role in an industry.

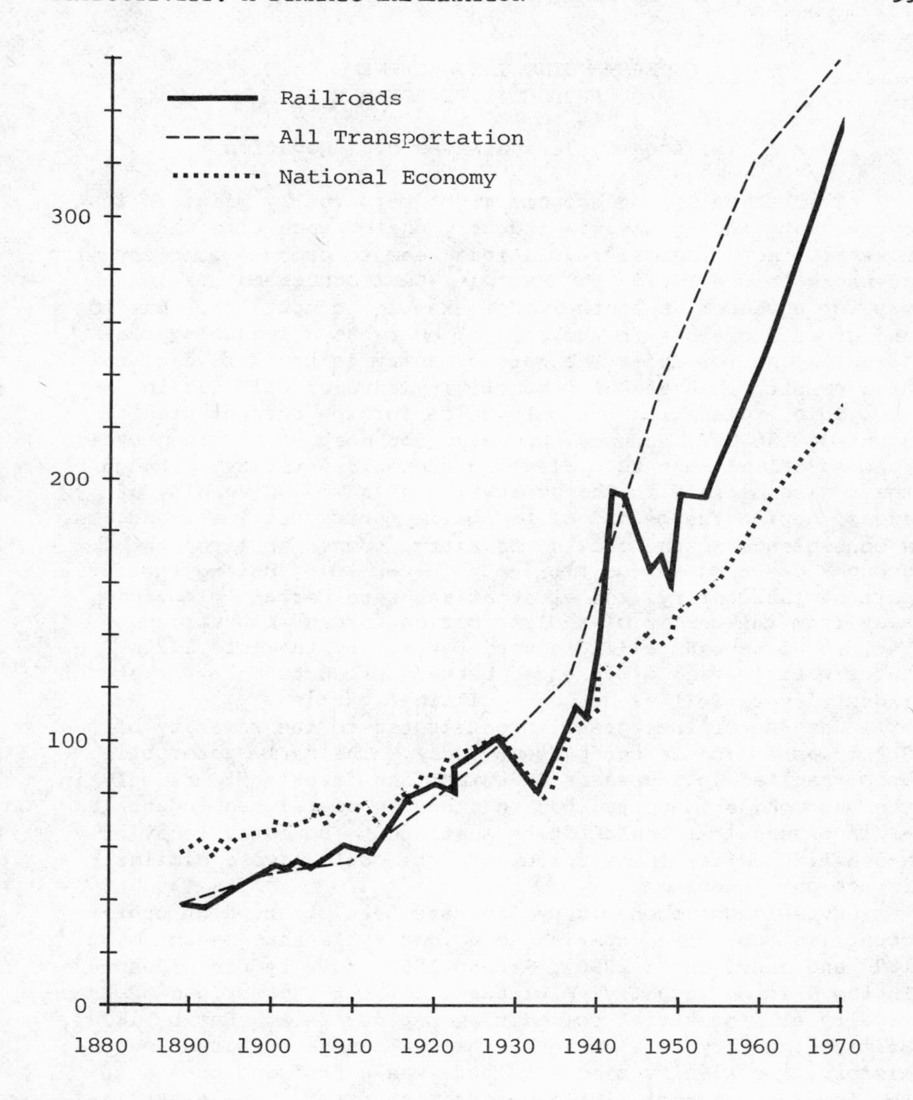

FIGURE 3.7

TOTAL FACTOR PRODUCTIVITY IN VARIOUS INDUSTRIES, 1889-1970

Source: John Kendrick, Productivity Trends in the United States, National Bureau of Economic Research (Cambridge, Mass., 1961); and Kendrick and Grossman, "Recent Trends and Cycles in Productivity of the United States Economy," unpublished.

UNDERSTANDING THE DECLINING RATE OF PRODUCTIVITY ADVANCE

The Long Cycle and Waves of Innovation

Theoretically, an economy might perpetually maintain its dynamism by adding one new industry during each time period. However, technological revolutions seem to come in bunches, with advances in one field (for example, semiconductors) paving the way for advances in another (for example, computers). Due to the great increases in the cost of entry as a technology matures, sooner or later the rate of entry is bound to decline. As a result, the rate of productivity advance will decline.

This explanation not only holds for the current predicament of the U.S. economy, but also for most of the long cycles (the so-called Kuznets cycles) in economic activity. Though the cycles begin with the generation of a wide diversity of ideas, during the period of prosperity entry declines, and, as a consequence of the decline in entry, sooner or later the economy descends to equilibrium. For example, during the first part of the century, the alternatives were certainly moving away from the center of a distribution in many industries. Yet, if we take Hotelling's word for it, by the late 1920s there were trivial differences between products in a variety of industries as well as between political parties.

One factor that greatly contributed to the severity of the Great Depression is not present today. The early automobile boom resulted in increases in output and investment not only in the automobile industry, but in a number of related industries -- increases that could not be sustained. However, today only a drastic decline in petroleum imports could have a similar impact on the economy.

Even longer economic cycles have been observed in other countries. Britain experienced a long cycle that began about 1800 and ended about 1880. Around 1850, rivalry was widespread in the British industry -- witness the fact that prices of practically all industrial commodities rapidly fell. But by 1890, British industry as a whole had become famous for its slow history. As rivalry died out there was a profound change in the internal characteristics of British firms. While earlier British entrepreneurs came from a wide variety of backgrounds, by the end of the century, British captains of industry were almost entirely lawyers and accountants. For this reason, the British economy never fully recovered from the great depression of the 1800s. Since 1900, Britain has had very low productivity growth, a chronic balance of payments problem, and a very significant decline in living standards compared with other major industrial countries.

Recovery from a long cycle is by no means automatic. Indeed, as an economy acquires a larger proportion of older industries in which firms do their best to isolate themselves from feedback (whether by product differentiation, advertising, or protective tariffs), the average degree of openness in the economy declines. When this happens, recovery from the long cycles becomes progressively more difficult. Wars can alter this result. Consider, for example, the remarkable recoveries of the Japanese and German economies after World War II. However, no one would seriously propose that war be provoked to accomplish what politicians are unwilling to do.

Mergers

A large increase in the number of mergers is an indicator of a downward phase of a long cycle. For example, during the late nineteenth century, there was an enormous increase in mergers in British industry. During the 1920s, the United States experienced its second largest merger movement; during the 1960s and 1970s, a merger movement of similar proportions occurred, with the new factor being the emergence of international conglomerates.

From the point of view of the firm, mergers are easy to explain. Obviously, growth by acquisition is easier than growth by entrepreneurship and luck. Moreover, to the extent that mergers result in diversification, they provide a way to ensure firms against risks. However, what is good for an individual firm is not necessarily good for an entire economy. Almost inevitably, mergers result in a decline in risk taking, and, by so doing, contribute to a poor showing in productivity gains.

Vertical integration almost inevitably leads to a decline in risk taking, because it leads to a decline in both Type I and Type II uncertainties. Consider, for example, the British aircraft industry during the period 1950 to 1960. All aircraft companies, except Rolls Royce, were vertically integrated firms that developed and produced both aircraft and aircraft engines. Yet, the number of times a vertically integrated firm was able to mate one of its own engines to one of its own aircraft was less than the number that would be determined by chance, [20] because in about three out of four cases an airplane used a Rolls Royce engine. Rolls Royce had an advantage because it enjoyed a relatively high degree of Type I uncertainty; that is, unless it developed superior engines, it was in a relatively poor position to sell engines at all. To deal with this uncertainty, it had to put more emphasis on developing basic engine techology than did other companies. On the other hand, it enjoyed a greater degree of Type II uncertainty because, as far as the customers were concerned, it had more options. When one of its engines was successful, a vertically integrated aircraft company would rather switch to it than sell no airplanes. Consequently, with Rolls Royce in the picture, they too had a greater incentive to take risks. Chrysler is a good example of an automobile firm that had a significant advantage in the 1930s because it was less vertically integrated than its competitors. According to William Abernathy, this provided the flexibility needed for Chrysler's pioneering of high-compression engines, streamlined cars, disk brakes, and power steering. [21]

When all firms in an industry are vertically integrated, the degree of risk taking will almost inevitably be smaller (because both buyers and sellers have to deal with a smaller degree of uncertainty). Indeed, the main reason for vertical integration is not the much-advertised reason that it will result in a gain in static efficiency. Rather, vertical integration is a protective mechanism that minimizes the probability of creative destruction of physical capital and a highly interrelated administrative structure. In short, vertical integration protects the lazy monopolist from feedback and unpredictability.

Horizontal integration reduces risk taking because it reduces Type I uncertainty. Consider a hypothetical example of a large company that purchases a highly successful small company in the same industry that is still owned and managed by its founders. By becoming part of a larger firm, the smaller one is obviously better insured against uncertainty. The former head of the small company may retire or may develop a taste for fishing. But, if he remains, his incentives will not be the same. Moreover, in the name of sound management practices, the parent company will almost always impose a variety of controls on the acquired company: controls that will maximize short-term profits at the expense of longer-run profits. Mergers between smaller companies do not always lead to this result. A small company is likely to be more aware than a large one of the cost involved in imposing many onerous constraints upon entrepreneurship.

It is often argued that without the possibility of merger, the incentive to establish a new business would not be as large as it is today. However, selling to a large firm is not the only way of selling a small one. A company can be purchased by a group of individuals. They may have difficulty in raising the capital, but it can hardly be argued that existence of imperfections in the capital markets is a valid reason for mergers.

The principal economic argument against mergers is not that they lead to a greater concentration of economic power; but, rather, that mergers have a negative effect on risk taking. Because they seriously jeopardize the preservation of an economy with a high degree of dynamic stability, mergers between large firms should have been outlawed many years ago.

Foreign Competition

There is another sense in which today's highly predictable "new Victorian" American state is like the older British Victorian state. Both "Victorian" periods reveal a marked decline in the ability of a nation to compete successfully with other countries. Until 1965, the United States traded with the rest of the world by exporting newly developed technologies and importing products made by technologies developed here earlier, as well as by exporting and importing raw materials (the so-called Vernon trade cycle discussed by Graham in Chapter 4). However, the U.S. comparative advantage shifted among products over time -- it was a dynamic advantage. During the 1880s, when rivalry was still thriving in the steel industry, the United States exported steel to Britain and Scotland. Just as Americans are now accusing the Japanese of dumping, the British were then accusing Americans. Later, our comparative advantage shifted to automobiles and automobile machinery; and still later, to commercial aircraft, synthetic fibers, computers and semiconductors. The U.S. advantage in the newer technologies resulted from the relative ease of starting new firms. The primary reason the United States was disadvantaged in the export of older technologies was that, generally speaking, the foreign firms were newer and more energetic. However, since about 1965, the competitiveness of the U.S. economy with respect to both the newer and older technologies has steadily eroded. (See Chapter 1 by Hill and Chapter 4 by Graham for further discussion of this decline.)

The Decline in Rivalry

Although there can be no question that rivalry in the U.S. economy has declined, it is unfortunately not possible to measure it. Data on changes in market shares are publicly available only for a relatively few industries. The U.S. Census Bureau has the needed information, but since it cannot release information on individual establishments to the general public, it will have to be persuaded to make the required calculations itself.

However, much information on the internal characteristics of firms, which suggests indirectly that the degree of rivalry is not what it once was, is available. For example, a 1979 Battelle report, "Probable Levels of R&D Expenditures in 1979," listed among the internal barriers to innovation the following:

1. A growing insistence on certainty of profits in the short term.

2. The "not invented here" syndrome.

3. Growth of a professional management class which has no entrepreneurial stake in the business.

4. Formalization of short-term executive tours of service -- e.g., up, down, sideways, or out every three to five years -- discouraging longer-term, innovative projects.

5. Use of executive incentive programs which emphasize accounting concepts of achievement.

6. Organizing for steady profitability rather than innovation and risk taking.

7. Tendency to try to buy corporate growth through acquisitions rather than through innovation and/or internal expansion. [22]

Relating High Inflation to the Decline in Rivalry

Dynamic theory contributes not only to understanding the productivity slowdown, but also to understanding the relationship between demand-pull (buyers) inflation and cost-push (sellers) inflation. Typically, these are discussed as if they were entirely different subjects. For example, the proponents of demand-pull inflation argue that the large public deficit associated with the war in Vietnam was the basic cause of the beginning of inflation in the United States. On the other hand, the proponents of cost-push inflation point to the fact that inflation began in the steel industry, when wages in that industry started to go up more rapidly than productivity gains -- wage increases that were closely followed by price increases. However, these factors are intimately related, since the less rivalry there is in an economy, the more likely it is to be prone to inflationary shocks.

Rivalry not only acts as a stimulus to reducing costs, and to improving productivity, but also acts as a deterrent on wage and price increases. The reason rivalry performs the second function is that when labor unions push for rapid increases in wages in an industry that possesses a significant degree of rivalry, they will cause a loss of jobs in the less profitable firms. Since the labor unions cannot afford to ignore the impact of their actions on the less profitable firms, they must moderate their wage demands.

In industries with little rivalry, a relatively small degree of constraint on wage increases can be expected. For example, because firms can pass on higher wage costs as higher prices, in the railroad industry the union benefited most from a system of economic regulation that minimized feedback in that industry. Prior to the recent deregulation of the airlines, when there was little rivalry in that industry, its very large gains in productivity were not reflected in lower prices. Even though total factor productivity increased by about 80 percent between 1955 and 1965, average passenger rates did not decline, they increased. During the same period, there was no significant increase in airlines' profits. [23] This industry featured cost-push inflation of prices, even when there was relatively little inflation in the economy as whole.

Demand-pull and cost-push inflation are related in the following way. Even when there is little excess demand in the economy as a whole, there is some degree of cost-push inflation in industries with little rivalry, but the inflationary effect is highly localized. As excess demand increases, industries with the least amount of rivalry will be the first to take advantage of the situation by charging what the market will bear. However, when industries such as the steel industry act in this manner, the shocks from excess demand become more general.

If my hypothesis is correct, in response to excess demand, inflation will gain momentum most rapidly in those industries with the least amount of rivalry, and cost-push inflation will be absent longest in those industries with the most rivalry. As indicated above, measures of rivalry for a wide number of industries are not available. Thus, there is no direct way to test this hypothesis. However, Bureau of Labor Statistics data are available on productivity wage changes, as are price changes for some five hundred industries for the period 1958-1976. These provided an opportunity to make a rough test of my hypothesis. The details of this work are in preparation for publication. [24] To provide a rough test of the hypothesis, it was assumed that the rate of productivity increase in various industries corresponds with their degree of rivalry.

Two relationships were examined. First, during the period 1965-1972, there was a negative correlation between the rate of productivity advance and the rate of price increase, confirming that rivalry holds down prices. Second, in the high productivity industries, wages increased only about half as fast as productivity advances, as compared with the low productivity industries. This confirms that rivalry does a good job of constraining wage increases as well. In other words, a low degree of rivalry and a high degree of unstatesman-like behavior on the part of labor union leaders go hand in hand.

As might be expected, there is a good deal of scatter around the average relationships just discussed. However, the relationships are nonetheless quite impressive.

On the other hand, in the period 1972-1976, the differences in price and wage behavior among three groups of industries with low, medium and high productivity advances become insignificant. Wages and prices in the three major groups increased at exponential rates. There apparently is nothing like inflation to overcome the desire of business and labor union leaders to engage in anticompetitive behavior. However, in a few highly competitive industries like semiconductors, radio and TV sets, and knitting mills, price increases remained insignificant through 1976.

Needless to say, a great deal of work remains to be done before there is a satisfactory explanation of modern inflation. However, one thing is crystal clear: while a policy of fiscal restraint will certainly help to mitigate inflation, it cannot restore economic stability. Just as important is the necessity to promote competition.

Relating Unemployment to the Decline in Rivalry

A decline in rivalry results in an economy in which a particular employment rate is associated with a higher inflation rate. (Technically speaking, a decline in rivalry results in an upward shifting of the Phillips curve.) Clearly, an economy with a high degree of rivalry will feature a good deal of short-term frictional unemployment, due to the fact that when some firms experience "prosperity," others experience "recession." Almost invariably, however, industries with a high degree of frictional unemployment exhibit a rapid increase in total employment. When there is so little rivalry in an industry that it cannot compete with foreign firms or with other industries producing substitute products, employment opportunities are increased in other countries or industries at the expense of an increasing degree of long-term unemployment in the country or industry concerned. Consider, for example, the chronic unemployment problems in the Scottish steel and textile industries; or consider the fact that with 20 percent of U.S. steel supplies now being imported from Japan, Youngstown and Pittsburgh are facing chronic unemployment problems.

Officials of the steel industry put the blame for low capacity utilization on their inability to raise the capital to modernize their facilities. But, while their ability to compete in foreign markets was steadily eroding, U.S. steel firms were expanding into such fields as cement, railroads, ocean transport, seabed minerals, pipelines, oil-drilling equipment, titanium, and real estate. This example provides a vivid illustration of the price the United States is paying for not having outlawed mergers years ago. Many of our political leaders do not seem to understand that competition not only stimulates productivity gains, it is an important instrument for preserving an economy that does not require a huge amount of unemployment in order to restrain inflation.

However, the relationship between unemployment and price changes during the 1974-75 downturn is the most impressive evidence of what is in store for the United States if better ways to promote competition cannot be found. At that time, industrial prices continued to rise, although the unemployment rate rose to almost 8 percent of the total labor force. Moreover, long-term unemployment (unemployment of fifteen weeks and over) rose from 16 percent of total unemployment in 1970 to 32 percent in 1975, and declined only to 29 percent in 1977. [25]

Actually, this experience is reminiscent of the Great Depression, when the prices of many industrial commodities proved to be unresponsive to increases in the degree of Type I uncertainty. Economists tried but could not explain this stickiness as a function of concentration ratios or monopoly power. This is not surprising in the light of dynamic theory, because high concentration ratios are not necessarily correlated with a low degree of feedback (as measured by changes in market shares). Thus, even though the aircraft and steel industries were both highly concentrated, progress in reducing costs was more rapid in aircraft than in steel.

A General Observation on Dynamic Capitalism

Although this chapter opened as an attempt to explain the slowdown in productivity advances, it should be apparent now that this subject cannot be studied in isolation from what is happening in the economy as a whole. Though the hidden foot is the major factor in explaining the rate of productivity gains, it performs other important functions as well. Indeed, if a dynamic economy is defined as an economy which, by virtue of having an effective hidden foot, can rapidly move alternatives away from the center of a distribution, then it should be apparent that the major advantage of dynamic capitalism over socialism is that dynamic capitalism has a superior ability to make economic progress. On the other hand, a static capitalist economy that can generate only trivial differences between alternatives has no important advantage over socialism, and, to his great credit, Harold Hotelling was the first economist to recognize this fact of life. Nevertheless, it is ironic that today those people who go furthest in extolling the virtues of capitalism are, generally speaking, the people least willing to prove the effectiveness of competition.

RESTORING A DYNAMIC ECONOMY: POLICY IMPLICATIONS

There is simply no way to return to the golden age of 1955-1965 -- when President Kennedy complained that this country was not moving fast enough. But the price of waiting to take effective measures to restore competition can be very high. Not only is the productivity performance of the U.S. economy likely to worsen further, but it may witness sharper oscillations between inflation and recession.

My first two recommendations for restoring a dynamic economy are both concerned with lowering the barriers to entry in activities in which new firms are sorely needed. The third recommendation is concerned with changing the antitrust regulations to make life in well-established firms somewhat more challenging.

New Automobile Engine Firms

There is no doubt that environmental regulations have greatly complicated the task of developing more efficient automobile engines. Developing automobile engines under a double set of constraints is not an easy task, nor is there any doubt that a real need exists to move forward on both fronts simultaneously. Reducing environmental degradation is as important a way to improve productivity as is reducing gasoline consumption. Indeed, if the GNP were correctly measured, it would include the benefits from reducing environmental degradation.

The present form of environmental regulation buys progress in the short-run by limiting the possibilities for more rapid progress in the longer-run. Regulation has this effect because it supplies perverse incentives for the generation of a diversity of alternatives. Firms are motivated to concentrate on the same technical approach, because this is the best strategy for getting the regulations relaxed. (This point is discussed in considerable detail by Ashford, Heaton, and Priest in Chapter 5.) Honda manufactures one of the best internal combustion engines available today. Curiously, U.S. regulations supply better incentives to foreign than to domestic car manufacturers.

What might be done to remove the disadvantage of the American automobile manufacturers? Government R&D contracts should be used to establish new firms to work on promising technological possibilities. There is a precedent for such action -- during the 1920s, military R&D contracts were used to establish firms to develop new types of aircraft engines. If it was politically possible to take such action when Calvin Coolidge and Herbert Hoover were running this country, is it not possible today?

However, automobile firms should not be discouraged from working on new types of engines. Obviously, the more competition, the greater the probability of making more rapid progress. If, as officials from the automobile companies argue, the industry is already operating at its peak dynamic capacity, no additional challenge is required. If this is the case, the automobile industry should be as thankful for technology developed on the outside as are computer companies. On the other hand, if the people in charge of engine development in the automobile industry are correct in asserting that they could make more rapid progress were it not for the constraints imposed on them by top management, then new entrants would make it easier for them to argue with their bosses. In either event, government action to help establish new firms will make the automobile industry better off than it is today.

Decoupling the Generation of Electricity from Other Activities of Public Utility Companies

This proposal would involve separation of generating electricity from transmitting and distributing electricity, while encouraging independent firms to compete in the generation of electricity on the basis of cost. While there are some outlying areas in which such competition should be precluded, encouraging the need to preserve scale economies cannot be given as a reason for not having such competition. In fact, public utilities commonly buy electricity from each other today.

The importance of such decoupling stems from two factors. The first is that the incentives of public utility companies to search for low-cost alternatives are probably weaker than in any other industry in the entire economy. Not only is there little or no price rivalry between utility companies, but in normal times, public utility rate-making procedures provide about the same incentives as cost-plus contracts. In times of inflation, the fact that regulatory commissions are loath to pass on cost increases holds prices down. However, public utility commissions do not provide the kind of feedback that would be provided by having independent companies compete on the basis of cost.

The second reason for the proposed decoupling is that, when it comes to developing exotic new technologies, the public utilities are no more capable than the railroads would have been to develop airplanes.

As a consequence of poor incentives and an almost zero ability on the part of public utilities to engage in dynamic behavior, the country has had to pay an exhorbitant price for the development of nuclear energy, in terms of both cost and safety. Public utility companies cannot be blamed for all that has happened. They did not develop either the design concepts or the detailed designs for power plants, nor did they establish the safety regulations. However, the diffusion of responsibility is in part a reflection of the fact that the utility companies do not possess the required competence to play a significant role in the generation of new alternatives.

Consider the very different situation in the field of commercial aircraft development. Just as safety requirements are imposed on nuclear power plants by a public agency, so are safety requirements imposed on commercial aircraft by a public agency. Moreover, the aircraft companies have no less a stake in the safety of airliners than utility companies have in the safety of power plants. Yet, the aircraft companies themselves have played an important role in the development of safe aircraft. For example, when the British Comet airplane developed serious problems, Boeing sent a team to Britain to help overcome these problems. Is it conceivable that nuclear power companies will ever develop the same competence to deal with their power plant safety problems?

To provide another example, when chemical companies build new plants, they, like the public utility companies, have them built by architectural-engineering companies. However, chemical companies take complete responsibility for providing the design concepts. Moreover, when built, such plants are operated by hand picked crews who, in turn, provide valuable feedback for the design of other new plants. Is it conceivable that utility companies will ever acquire the competence to engage in such behavior?

I do not mean to suggest that the United States ought to build more nuclear power plants. In the next decade or two, when it becomes possible to deliver solar energy into a central grid, or when windmills provide an important source of energy, their economical development will also be seriously jeopardized if it is entrusted to the utility companies.

My point is a general one: all new plants for generating electricity should be operated by independent companies. Utility companies' incentives for discovering clever ways to cut costs always have been, and still are, very poor. The principal difference between the past and the present is that it is now more important to provide better incentives. Given this country's predicament in generating an adequate rate of productivity gain, the luxury of making power cost even 15 to 20 percent more than it need be can no longer be afforded.

Putting a Hidden Foot in the Antitrust Laws

If this country is ever to escape from stagflation, entrepreneurs represent an indispensable asset. But, while the United States possesses quite as many entrepreneurs as it ever did, this scarce resource is being squandered because the economy as a whole is becoming ensconced in a static equilibrium. There are, of course, a few industries in which the hidden foot continues to play an effective role. However the hidden foot function is being performed today mainly by foreign firms.

As has been seen, when business executives get to know each other very well, and because the actions in which they engage are highly predictable, there is no necessity for them to collude in order to preserve the equilibrium. To be sure, in Adam Smith's world, entrepreneurs in a particular trade were constantly conspiring against the public at large. However, that was long before the game of product differentiation had been discovered. This game is played so many times over that entrepreneurs have the opportunity to learn from their experiences. Consequently, the behavior envisioned in the prisoner's dilemma response -- when a firm rushes to cut its prices to prevent its rival from striking first -- is ruled out, because if firms in an industry ever did engage in such behavior, they would have learned their lesson.

The problem, therefore, is how to change the antitrust laws to discourage behavior resulting in trivial differences in prices and products and to encourage the generation of a wider diversity of ideas. In other words, the problem is how to create better incentives for dynamic behavior.

As matters stand today, the incentives provided by the antitrust laws are better calculated to minimize risks in the legal profession than they are to encourage risk taking in the business world. Lawyers feel more at home with legal conspiracy theory than they do with economic theory, and they tend to specialize in cases for which they are best prepared. Consequently, when business firms ask their legal counsels in which kind of behavior they can engage without risking an antitrust suit, the normal answer is: "Be just a little different, but do not collude." Providing better incentives for competitive behavior requires changing the antitrust laws so that firms seeking advice on antitrust matters will be told the following, "We know that if you collude, you are in real danger. And we also know that the more your products and prices resemble those of your competitors, the greater is the danger of an antitrust suit. But we cannot tell you how far you can go to be just little bit different. Just as Presidents have to operate within a broad band of uncertainty in predicting how far they can go before being accused of an impeachable offense, so will you have to operate within a broad band of uncertainty."

How might the Antitrust Division of the Justice Department (or its successor) provide the required incentives? First, if a concentrated industry as a whole fails to meet the test, its major firms would be broken up. Furthermore, the reorganized firms would be required to select their new top managements from outside the industry in question. In other words, the reorganization would not only result in less concentration of economic power, but it also would result in bringing fresh thinking into the industry concerned. (To be sure, the government has no right to tell firms which particular people should be selected for top management positions. But, since a very narrow interpretation of self-interest can jeopardize the public interest, the government does have a right to insist that when an industry's dynamic performance is bad, the firms look outside the industry for their new managements.)

The principal criterion to be employed in bringing dissolution suits would be a dynamic one -- that is, whether the alternatives continue to remain in the middle of a distribution or whether they are being pushed out at a significant rate. However, firms would not know just how slowly alternatives need move from the center before an industry is in the danger zone. Just as automobile drivers do not know how far they can go in exceeding the speed limit, firms would not know precisely at what point they would enter the danger zone. Economists, engineers, and lawyers involved in bringing the cases would have to be given a certain amount of discretion on just where to draw the line.

When industries are in the danger zone, changes in market shares tend to become smaller and smaller, and the industry in question experiences greater and greater difficulties in meeting foreign competition. However, these are not infallible indicators. For example, significant changes in market shares can occur because of advertising, competition in style, and some newly discovered means of competing in terms of fakery. The ability of American firms to meet foreign competition is also not an infallible indicator, because many goods and services do not enter international competition. For example, housing does not, and, for that reason, housing costs have risen enormously in nearly all countries.

It is because statistical tests of competition are not infallible that engineers will have to play an indispensable role in the new or renamed "Department to Promote Competition" envisioned under this change. Neither economists nor laywers possess the competence to distinguish between a minor and a significant change in an alternative, particularly when the change involves an improvement in quality. Competition in terms of price is, of course, easier to measure.

From the point of view of fairness, the enabling legislation should contain two provisions. First, to provide firms with an opportunity to adjust to the new law (the new law would replace the Sherman Antitrust Act), a three year period would be allowed to elapse before any case is tried under the new Act. Second, constraints on technology-based productivity gains imposed by labor unions, when they cannot be justified for safety reasons, shall be regarded as reason for dissolution of the union and the selection of new labor union leaders. However, the goverment should not try to control wages, whether by wage-price guidelines or direct control. Wages should be determined by the workings of the market. Under the new law to promote competition, labor union leaders, too, will face real risks: if prices increase because of wage escalation, they will risk causing a loss in jobs; either that or major firms in the industry will risk dissolution.

The major purpose of this proposal is twofold: (1) to promote a higher rate of productivity growth by providing better incentives for risk taking, and (2) to discourage wage increases from becoming more rapid than productivity gains by providing firms with better incentives than they have today for resisting wage increases. These are almost inevitable requirements if the United States is to enjoy more competition and a greater degree of price stability.

I recognize, of course, that at first glance this may appear to be a somewhat radical proposal. However, it really is not. First, in the antitrust action against the tobacco industry during the 1930s, when it engaged in parallel actions with respect to products and prices, the major firms in the industry were found to be engaging in anticompetitive behavior even though there was no direct evidence of collusion. Apparently, lawyers were more willing to take risks in the aftermath of the Great Depression. However, the same type of industry behavior is no more excusable today than it was during the Great Depression.

Second, I do not go nearly so far as Kaysen and Turner did in their pathbreaking book on antitrust in suggesting a wholesale breaking up of firms throughout American industry. [26] I agree with their views on the importance of limiting the concentration of economic and political power. I am not concerned about possible losses in static efficiency, because they would be by far offset by gains in dynamic efficiency. However, I feel that under current conditions, as contrasted with the conditions in the 1950s when the book was written, a very strong case can be made for undertaking dissolutions in a manner that will improve incentives for competitive behavior.

An Observation on Competition and Democracy

Although generating a more rapid rate of productivity gain and restraining inflation are important objectives in themselves, there is an even more basic reason for promoting a dynamic competitive society: it is an important adjunct for promoting Thomas Jefferson's concept of democracy.

Although there is no generally agreed upon definition of Jeffersonian democracy, those familiar with his letters to Madison will agree that the following does not widely miss the mark: a democratic society is one that by virtue of generating a wide diversity of ideas can adapt itself to new circumstances. To be sure, I have been unable to find precisely such a statement in Jefferson's writing, but both his actions and his statements showed him to favor a society that could generate a wide diversity of ideas. Moreover, Jefferson yearned for a society that could adapt to new circumstances -- witness his plea for a small revolution every twenty years -- and what is the purpose of such a revolution other than to put a hidden foot into politics? In their actions before the Revolutionary War, the British did put a hidden foot into politics, and the ideological revolution that ensued did generate a wide diversity of ideas. If it did, why not, then, have a small revolution every twenty years?

A dynamic, competitive society complements Jefferson's concept of democracy, because highly interactive firms, in which people assume that they do not know all the answers, provide excellent training grounds for preserving a democratic society. Conversely, the highly authoritarian nature of cartelized firms in prewar Germany, primarily staffed by "yes-men," hardly provided an ideal training ground for the blossoming of democracy, and the same was no doubt true in Japan before World War II. Moreover, today both the Japanese and the Germans seem to be far more aware than the people of the United States of the relationship between political democracy and democracy at the level of the firm.

Bruce Cain has been studying electoral volatility in Britain. [27] He has found that the British political system has an amazing inability to deal with new circumstances. Entirely new issues such as devolution and immigration do arise, but the political candidates and the party bureaucracies remain almost as predictable as the planets. It is, of course, impossible to know which is the cause and which the effect: the economic system or the political system. In fact, the arrows of causation run both ways. Thus, highly bureaucratic labor organizations and businesses, or highly bureacratic party organizations modeled in their image, provide the training ground for members of Parliament. By thinking that one of these highly predictable bureaucratic bodies is the only way of life, members of Parliament mold the economy in the same image.

However, according to Morris Fiorina, American politics in its static orientation seems to be a full step ahead of British politics. By dispensing favors to both individuals and interest groups, Tammany Hall has in effect moved to Washington. Moreover, Fiorina has found that congressmen, in order to increase the effectiveness of their operations, have been building up the staffs of their home offices at a very impressive rate. Moreover, by making entry more difficult, these practices have greatly contributed to a decline in political competition as measured by the longer and longer periods congressmen remain in office. [28]

In short, a principal difference between the behavior of the congressional and business establishments at the present time and that of the 1920s, the period in which Hotelling wrote his famous article, is that now both establishments go much further to immunize themselves from feedback by undertaking a high degree of "product differentiation."

The more basic reason for the similarity in behavior is that both are responding to short-run incentives. How might congressional incentives be changed? This can be accomplished by limiting congressional tenure, as well as presidential tenure, to maximum periods of six years. This would change congressional incentives, because, when in office, politicians would think less about, "What can I do to please this or that voter?" and more about, "How would I like to be remembered for what I did for my country?". Moreover, just as the business establishment could generate a wider diversity of ideas when the cost of entry is relatively low, so could the congressional establishment. With better incentives, a wider diversity of people would run for office. A wider diversity of ideas would provide the country with a higher degree of congressional productivity. A higher degree of congressional productivity would result in a higher degree of Jeffersonian democracy. Why should a country wait for twenty years to have a political revolution? Why not, instead, have a country that is continuously able to adapt itself to new circumstances?

NOTES

1. To argue that productivity gains cause growth in an industry, it is necessary to assume tht price competition leads to a search for ways to reduce costs, and that demand for the product of the industry is elastic. It is no accident that nearly all industries show the most growth when these conditions are fulfilled.

2. Burton H. Klein, Dynamic Economics (Cambridge, Mass.: Harvard University Press, 1977).

3. Zvi Griliches, "Returns to Research and Development Expenditures in the Private Sector," (Conference on Research in Income and Wealth, 1975); and Nestor Terleckyj, "Effects of R&D on the Productivity Growth of Industries," (National Planning Association, 1974).

4. Except for semiconductors, the following examples were taken from Klein, Dynamic Economics, Chapter V. The semiconductor example is based on research now in progress.

5. Klein, Dynamic Economics, p. 118.

6. Harold Hotelling, "Stability in Competition," The Economic Journal, March 1929, p. 41.

7. Roger W. Klein, "Decisions with Estimation Uncertainty," Econometrica 46 (November 1978): 1363-87.

8. Klein, Dynamic Economics, pp. 48-50.

9. Harvey Leibenstein, Beyond Economic Man (Cambridge: Harvard University Press, 1976), Chapter 1.

10. Richard M. Cyert and James G. March, A Behavioral Theory of the Firm (Englewood Cliffs, N.J.: Prentice-Hall, 1963).

11. John Kenneth Galbraith, The New Industrial State, 2nd rev. ed. (Boston: Houghton Mifflin Co., 1971).

12. Paul MacAvoy, The Economic Effects of Regulation: The Trunkline Railroad Cartels and the ICC Before 1900 (Cambridge, Mass.: MIT Press, 1965).

13. Frederick M. Scherer, Industrial Market Structure and Economic Performance (Chicago: Rand McNally & Co., 1971), pp. 376-78.

14. For further discussion see Klein, Dynamic Economics, pp. 17-18.

15. For an excellent discussion of the post-World War II challenges to AT&T, see Bruce Owen and Ronald Braeutigam, The Regulation Game (Cambridge, Mass.: Ballinger, 1978).

16. Richard Schmalenzee, "Entry Deterrence in the Ready-to-Eat Breakfast Cereal Industry," Bell Journal of Economics, Autumn 1968, pp. 437-57.

17. While not all spatial models exhibit equilibria, in the real world, the absence of feedback makes an equilibrium stable.

18. Edwin Mansfield, "Innovation and Technological Change in the Railroad Industry," Transportation Economics, publication of the National Bureau of Economic Research (Princeton, N.J.: Princeton University Press, 1965).

19. A rough estimate based on fragmentary information.

20. Klein, Dynamic Economics, pp. 60-61.

21. William Abernathy, *The Productivity Dilemma: Roadblock to Innovation in the Automobile Industry* (Baltimore: Johns Hopkins University Press, 1978), pp. 36-7.

22. "Probable Levels of R&D Expenditures in 1979: Forecast and Analysis," Battelle Memorial Institute (Columbus, Ohio, December 1978), pp. 5-6.

23. An estimate based on John W. Kendrick, *Postwar Productivity in the United States, 1948-1969*, National Bureau of Economic Research (Washington, D.C., 1973), pp. 94-5, and information supplied by the Department of Commerce on the implicit deflation.

24. Burton H. Klein, unpublished research.

25. *Economic Report of the President* (Washington, D.C.: U.S. Government Printing Office, January 1978) Table B-1, p. 293.

26. Carl Kaysen and Donald Turner, *Antitrust Policy: An Economic and Legal Analysis* (Cambridge, Mass.: Harvard University Press, 1959).

27. Bruce Cain, "Electoral Volatility in Britain," unpublished manuscript.

28. Morris P. Fiorina, *Congress: Keystone of the Washington Establishment* (New Haven, Conn.: Yale University Press, 1957).

4 Technological Innovation and the Dynamics of the U.S. Comparative Advantage in International Trade

Edward M. Graham

INTRODUCTION

This chapter has three tasks. The first task is to review what is generally understood to be true about the role of technology in determining patterns of international commerce between the United States and other nations. "International commerce" here includes the trading of goods and services between the United States and the rest of the world, and the overseas activities of multinational corporations (including both those based in the United States and those based outside of the United States). The second task is to assess whether any changes are occurring in global patterns of generation, transfer, and the use of technology; and if so, to determine whether these changes are likely to work to the net advantage or disadvantage of the United States. Underlying such an assessment is the assumption that any change which serves to increase the wealth of the United States over that which would have prevailed had the change not occurred is to the nation's advantage. [1] The third task is to consider policy options for the U.S. government that would serve to improve the position of the United States in the world economy, and those that would not.

INTERNATIONAL TRADE, COMPARATIVE ADVANTAGE, AND TECHNOLOGY-INTENSIVE PRODUCTS

The role of technology in international trade and investment has been the focus of a certain amount of controversy, much of it turning on the question of what determines comparative advantage among trading nations.

Central to the controversy is the question of exactly how changes in technology make themselves felt in the economy. [2] In classical and neoclassical thought, technological advances are viewed as primarily affecting the processes by which goods are produced, allowing reductions in the amounts of inputs needed to produce given quantities of standardized, homogeneous outputs. Stated somewhat more succinctly, this view holds that technological advances reduce the cost of manufacture of standardized goods. An alternative view is that the technological advance is embodied in the output itself, resulting in new or improved end products which fulfill needs that otherwise would be unmet. The two views can be partially reconciled by noting that what is an end product for one firm might be utilized as an input by some other firm. Thus, while advances in technology might result in the creation of new or improved end products, these products themselves result in cost reductions to firms which utilize them as inputs. The principle holds even if the new or improved product is destined for household consumption, for in that context, any product that lessens the labor requirement for a necessary task results in an implicit cost savings. Thus, the partial reconciliation of the two views is that the economic consequence of advances in technology is primarily a reduction in the cost of production of final goods and services. [3]

While this reconciliation might simplify the treatment of the economic consequences of technological change in the aggregate, it unfortunately does not help resolve the issue of the role of technology in the determination of a nation's comparative advantage in international trade. If, for example, technological advances result primarily in a reduction of the costs of traded goods, the implications for trade are quite different than if technological advances are embodied in the traded goods themselves. In either case, the net result of the advances ultimately might be to reduce the global cost of goods and services produced, but the particulars of the trading relationships among the nations of the planet would be quite different for each of the cases.

In order to explore this terrain further, it is useful to review quickly what have been the principal streams of thought regarding the determinants of comparative advantage and the consequent international specialization in the production of goods and services.

The Principle of Comparative Advantage: The Ricardian View

In the early nineteenth century, David Ricardo first detailed the elementary principles of comparative advantage. [4] Virtually ever since, there has been controversy over the reasons why one nation should possess a comparative advantage over another in the production of goods and services. In the Ricardian view, comparative advantage results from international differences in the relative costs of the production of goods. To illustrate, suppose that two nations (Nations 1 and 2) produce two goods (Goods A and B), but that in Nation 1 the relative cost of A to B is lower than in Nation 2. Nation 1 has a comparative advantage in Good A; conversely, Nation 2 has a comparative advantage in Good B. Ricardo's contribution was to demonstrate that, in such a case, it would be in each nation's interest to specialize in the production of the good in which it held a comparative advantage, and to trade that good for the other one to meet its consumption requirements. This would be true even if one nation held an absolute cost advantage in the production of both goods. If the other nation could produce one of the products at a lower relative cost than it could produce the other, it would still be in the interest of the first nation to specialize in the production of the good in which it held a comparative advantage and to trade this good for the other. [5]

More concerned with the effect to which nations exercise comparative advantage, Ricardo paid relatively scant attention to its origins. In his famous example of wine and cloth trade between Portugal and England, he assumed that labor was the only variable input into the production of both commodities and that Portugal utilized labor more efficiently than did England for the production of both. [6] Portugal, however, could utilize labor even more efficiently for the production of wine than for cloth and, therefore, enjoyed a comparative advantage in wine. Ricardo attributed Portugal's greater efficiency to its more favorable climate, which was even better for growing grapes than for the growing of cotton. Later analysts, however, noted that the differences in efficiency could be as easily attributed to differences in technology as to differences in climate. So, "Ricardian" models of international trade were developed showing that comparative advantage between two trading nations results from their technological differences, as manifested in the relative efficiencies of the productive processes by which traded goods are manufactured in each. [7]

The Principle of Comparative Advantage: The Factor Proportions Theory

During the twentieth century, emphasis in economic thought regarding the causes of comparative advantage turned away from the Ricardian explanation and towards factor proportions as the primary causal element. [8] Under the factor proportions theory, production of a traded good requires the input of at least two factors -- as capital and labor. [9] Manufacture of different products requires inputs of these factors in different proportions -- some products will require relatively more capital than labor, and others relatively more labor than capital. The former can be labeled as "capital intensive" products, the latter as "labor intensive" ones. Utilization of productive factors entails some cost to the producer. If the same choices of productive processes are available to producers in all nations, and if patterns of consumption are similar in all nations, it can be shown that, in the absence of international trade, relative factor costs are inversely related to relative factor endowments. Therefore, in a nation where the population is large but where capital is scarce, the cost of using labor will be relatively lower, and the cost of using capital relatively higher than in a nation well endowed with capital but underpopulated. It follows that the well populated nation will possess a comparative advantage in the production of labor intensive goods, and the well capitalized nation a comparative advantage in the production of capital intensive goods.

As in the case of the Ricardian model, it is possible to demonstrate under the factor proportions theory that if each nation exports goods in which it possesses a comparative advantage and imports goods in which it is disadvantaged, there will be gains for both nations in the form of increased national incomes. (In the special case in which one nation is very much smaller than its trading partner, the small nation will register gains from trade but the larger one will not. However, the larger nation will not suffer an income loss from trade.) World income is maximized if no restrictions are placed on trade. Either nation may increase its own share of the gains, at the expense of reducing the total gains, by unilaterally imposing tariffs or other restrictions on its imports. This works to the nation's advantage, however, only if the trading partner nation does not retaliate by imposing trading restrictions of its own. If both nations impose restrictions, both will lose some of the potential gains from trade. That substantial losses can occur was demonstrated early in this century. Trade "wars" resulting from successive rounds of tariff increases among the world's trading nations contributed in substantial measure to the world economic depression of the 1930s. [10] Much of U.S. trade policy during the post-World War II era has been aimed at preventing nations from once again pursuing policies similar to those of the 1930s policies which might destroy the collective gains that expanded world trade has created.

It is not difficult to find fault with the factor endowments approach as an explanation of comparative advantage. The explanation depends upon the assumption that all nations utilize identical technologies and that consumption patterns are similar in all nations -- assumptions that no world traveler would accept. Yet, it is important to note than even if these assumptions are proven wrong, the general argument that trade brings gains to trading nations is not invalidated. That comparative advantage can be exploited by trading nations to their mutual benefit holds true, whatever the source of the comparative advantage. All that is invalidated by disproving the assumptions is the argument that relative factor endowments are the primary determinants of comparative advantage among trading nations.

The factor proportions model can be adapted to deal with the case of differing technologies among trading nations, provided that the technological differences are manifested in the costs of production of traded goods and not in the design of the goods themselves. [11] Such a model can be used to explore possible consequences of international technology transfer. In this model, technology is transferred in disembodied form from a technologically advanced nation to a less advanced one. Within the context of the model, world income invariably increases by virtue of the transfer, but income in the advanced nation may increase or decrease depending upon specific circumstances. Within the advanced nation, income is transferred between labor and capital, but the direction of the transfer also depends upon specific circumstances.

A Challenge to the Factor Proportions Theory: The "Leontief Paradox"

In the early 1950s, Wassily Leontief published findings that have come to be known as the "Leontief paradox." [12] Leontief set out to test whether U.S. comparative advantage in the international trade of manufactured goods was determined by the nation's relative abundance of capital over labor. [13] He found that U.S. imports of manufactured goods typically were less labor intensive than its exports, a finding which at first blush would appear contrary to the factor proportions theory. Two possible explanations, however, were advanced to reconcile Leontief's empirical findings with the theory. The first, and the one Leontief embraced as being a likely explanation of his "paradox," was that U.S. labor was more productive than non-U.S. labor. If this increased productivity were sufficiently great, it would enable U.S. industry to produce labor intensive goods more cheaply than overseas industries, giving the United States a comparative advantage in these goods. The reason why U.S. labor might be more productive than labor abroad, it has been argued, is that U.S. labor is better educated and more highly trained. Because an investment has been made by U.S. society in this education and training, some economists assert that U.S. labor embodies "human capital." [14] If each U.S. worker is equivalent to one unskilled worker plus some amount of "human capital," and if "human capital" can be aggregated with physical capital, then it is claimed by certain economists that the Leontief findings are not at all inconsistent with the factor proportions theory. This latter reasoning (but not the possibility that U.S. workers do embody high levels of skill) has been dismissed as "Ptolemaic" by other economists.

The second reason centered on the possibility of factor intensity reversals. A factor intensity reversal between two products occurs if, as the scale of output of the product increases, the product which is relatively labor intensive at low volumes of output becomes relatively capital intensive at higher volumes. Most economists have dismissed this second reason as being theoretically possible but in fact highly implausible. [15]

A third reason for the paradox, not wholly reconcilable with the factor proportions theory, is that U.S. producers possess technologies different from those of its trading partners' producers. If the United States were to possess more efficient technologies for the production of certain labor intensive goods (but not for capital intensive goods) than did other nations, the United States might hold a comparative advantage in these goods despite the nation's relative scarcity of labor. Such a possibility could be incorporated into the analytic framework of the factor proportions theory, and the theory could be revised to meet the changed assumptions. [16]

The possibility also exists that certain goods exported from the United States embody technologies which are not known outside of the United States. That products play a special role in international trade is not really a recent idea. Bertil Ohlin mentioned the possibility in his 1933 treatise best known for its definitive statement of the factor proportions theory. [17] In a 1956 article, Irving Kravis put forth the idea that a nation generating new products possesses trade advantages over non-innovating nations, for at least as long as the innovations remain unimitated. [18] Writing in 1958, Erik Hoffmeyer suggested that the constant ability of the United States to innovate might have been contributing to the "dollar shortage" of the period. [19] When the dollar shortage ended in the early 1960s, Albert Hirschmann posited shortened lags between U.S. product innovation and overseas imitation as the cause. [20] Charles Kindleberger, in a 1960 article, suggested that U.S exports to Europe during the 1950s expanded much faster in new and dynamic industrial categories (most of which could be characterized as "technology-intensive") than in mature industries. [21] He attributed the U.S. balance of payments deficit to protectionist policies in Europe, policies which he felt served to force U.S. firms operating in dynamic industries to manufacture in Europe rather than to export from the United States. Kravis' ideas were expanded by Michael Posner, who suggested that foreign imitation of a product innovation would lag foreign demand, which in turn would lag demand in the innovating nation. [22] An opportunity for export by the innovating nation would occur during the time between the realization of foreign demand and foreign imitation. These concepts were tested by Christopher Freeman and Gary Hufbauer. [23] Freeman showed that the locus of production and export of new plastics in advanced nations was more a function of technical progress than of relative factor costs. Hufbauer demonstrated that national share of world exports of synthetic materials could be better explained by national market size and imitation lag than by factor costs.

Technology-Intensive Products in World Trade

Several other industry studies performed during the 1960s demonstrated the importance of technological innovation in U.S. export patterns of the industries in question. In particular, G. K. Douglas showed that innovation in the U.S. motion picture industry was frequently followed first by a surge of exports, and later by foreign imitation. [24] Seev Hirsch demonstrated a similar phenomenon in the electronics industry. [25]

Results of a number of empirical studies published during the 1960s and 1970s indicated that the United States indeed possessed a general comparative advantage in the introduction of new technology-intensive products into world trade. In a 1967 article, William Gruber, Dileep Mehta, and Raymond Vernon demonstrated that U.S. exports tended to be concentrated in industries in which the ratio of research and development expenditures to value added was high. [26] Noting that "the industries with the strongest research effort are also those with the strongest new product orientation," they tentatively concluded that a sizeable percentage of U.S. exports, at the time of the writing, consisted of products recently introduced into the market and available only from U.S. firms. These authors also showed the export industries to be characterized by high seller concentrations, but not by high measures of capital intensity, a fact consistent with the Leontief paradox. Additional tests by Gruber and Vernon subsequently served to characterize U.S. industries with a high propensity to export as technology-intensive, but not intensive in capital or in the use of low skilled labor. [27] The authors suggested that the basis for U.S. comparative advantage in the research and development industries is a continual ability to develop new products.

Empirical studies by Donald Keesing corroborated the Gruber, Mehta, and Vernon findings. [28] Keesing showed that industries in which U.S. exports were concentrated, relative to those in which U.S. imports were concentrated, employed high percentages of scientists and engineers. The findings of Keesing and of Gruber, Mehta, and Vernon serve partially to reconcile the "human capital" and the new product technology explanations of the Leontief paradox. A vital input into the development of new products is, of course, research and development, which is performed primarily by highly trained scientists, engineers, and technicians -- persons who embody high levels of so-called human capital.

Building upon techniques he had published earlier, Bela Balassa in 1978 examined the "revealed" comparative advantage of the United States for the years 1963-1971. [29] Balassa developed an index which ranked the export competitiveness of manufacturing industries by nation. The index for a particular product and nation was calculated by dividing that nation's share of exports of a particular category of manufactured goods by its share of total exports of manufactured products. One finding was that the United States had actually increased its trade advantage in most categories of research intensive goods. Exceptions were observed in a number of areas, notably in the medical and pharmaceutical industries and in certain categories of electrical goods. In the medical and pharmaceutical fields, the challenger nations were Denmark, West Germany, the Netherlands, and the United Kingdom. The loss of U.S. comparative advantage in the electrical goods field was attributable mostly to the radio and television subsectors, subsectors which are no longer really technology-intensive.

None of the results of these studies should be construed as demonstrating that U.S. exports consist only of new technology-intensive products. The evidence points to the fact that some substantial proportion of U.S. exports in recent decades has consisted of such products. Nor should the studies be interpreted as suggesting that U.S. comparative advantage in international trade is entirely based on the ability to generate new products. Although this ability is clearly shown to be one major determinant of U.S. comparative advantage. [30] The importance of technology-intensive goods in U.S. trade is indicated in Table 4.1.

Two questions emerge from the empirical data. The first question is "Why should the United States possess a comparative advantage in international trade based on new product development?" This question becomes more intriguing when one considers that the principal trading partners of the United States in manufactured goods are other developed nations, many of which themselves possess substantial internal research and development capabilities. The second question is, "Can the United States maintain the advantage in trade, or is there evidence to suggest that the nation is already losing it?" In evaluating the second question, one must exercise caution. It has been emphasized throughout this section that the advantage accruing to this nation from the innovation of any particular product is a fleeting one. Inevitably, the new technology will be imitated by overseas competitors, who may in time gain a comparative advantage. U.S. comparative advantage in new technology-intensive products lies not in the ability to forever dominate world trade in these products but rather in the capacity to continually upgrade them or to create entirely new ones.

TABLE 4.1

U.S. TRADE IN MANUFACTURED GOODS, CATEGORIZED AS "TECHNOLOGY-INTENSIVE" AND "NON-TECHNOLOGY-INTENSIVE" 1967-1977 $ Billions

Year	Technology-Intensive			Non-Technology-Intensive			Total		
	Exports	Imports	Balance	Exports	Imports	Balance	Exports	Imports	Balance
1967	8.0	3.1	4.9	12.8	12.7	0.2	20.8	15.8	5.1
1968	9.6	3.9	5.7	14.2	16.7	-2.5	23.8	20.6	3.2
1969	10.7	4.7	6.0	16.1	18.3	-2.2	26.8	23.0	3.8
1970	12.3	5.7	6.6	17.0	20.2	-3.1	29.3	25.9	3.4
1971	13.2	6.6	6.6	17.2	23.8	-6.6	30.4	30.4	0.0
1972	14.1	8.5	5.6	19.6	29.3	-9.7	33.7	37.8	-4.1
1973	19.0	10.6	8.4	25.7	34.4	-8.7	44.7	45.0	-0.3
1974	26.6	12.9	13.7	37.0	42.4	-5.4	63.5	55.2	8.3
1975	28.0	12.3	15.7	43.1	38.8	4.3	71.0	51.1	19.9
1976	31.2	17.0	14.2	46.1	47.8	-1.7	77.2	64.8	12.4
1977	33.4	19.6	13.8	47.1	57.7	-10.6	80.5	77.2	3.3

Note: Negative Balance Figures Indicate that Imports Exceeded Exports

Source: Regina Kelly, "The Impact of Technological Innovation on International Trade Patterns," (Monograph ER-24, Bureau of International Economic Policy and Research, U.S. Department of Commerce, Washington, D.C., December 1977); revised data and data for 1977 supplied to the author by Regina Kelly; "Technology-Intensive" data include goods in SITC categories, 51, 531, 532.3, 533.1, 541, 551.2, 561, 581, 599.2, 711 (except 711.1, 711.2, 711.7, and 711.9), 714, 719.63, 722, 724.1, 724.9, 729.3, 729.5, 729.7, 86, 891.1, 891.2.

EXPLANATIONS FOR U.S. COMPARATIVE ADVANTAGE IN WORLD TRADE OF NEW TECHNOLOGY-INTENSIVE PRODUCTS

Three explanations for U.S. comparative advantage in new technology-intensive products will be presented. One is based on a set of ideas, developed primarily by Raymond Vernon and colleagues of his, that has come to be known as the "product life cycle" model. The second explanation, developed by Burton Klein, is a modification of the "product life cycle" model based upon theoretical and empirical work regarding the relationships between rivalry among firms and their propensity to innovate. The third explanation is derived from the "demand similarity" model of Staffen Linder. The "product life cycle" models seek to explain both why the United States has come to be the world's leading nation in product innovation and what part this leadership plays in determining patterns of U.S. trade and international development. By comparison, the "demand similarity" model focuses not so much on the role of the United States in world trade as on why the developed nations of the world engage in trading similar, but differentiated, manufactured products of advanced design. Although the explanations for patterns of trade offered by the three models are not mutually exclusive, they do offer alternative explanations for the results of the empirical studies discussed earlier.

The Product Life Cycle Model

While the product life cycle model was first outlined in comprehensive detail by Raymond Vernon during the 1960s, a number of individuals -- notably Irving Kravis -- had earlier published key ideas which were to be incorporated into its formulation. [31] The model was tested and refined by a number of persons, including Louis Wells, Robert Stobaugh, and John E. Tilton. [32] M.P. Claudon contributed to restating the model in a more mathematical form than that in which it was originally put forth. [33]

In the Vernon formulation of the model, the extraordinary internal market characteristics of the United States are deemed to be the causal element behind U.S. advantage in product innovation. Three characteristics can be identified.

First, the United States is characterized by very high per capita income and by a large number of high income consumers. This characteristic, it was argued, induces U.S. entrepreneurs to develop new types of consumer products for which demand is highly income elastic. These products, generally classifiable as "luxury" goods, would include such high cost personal goods embodying new technologies as the transistor radio of the 1950s, the electronic hand-held calculator of the late 1960s, and the electronic watch of the middle 1970s. For each of these examples, it might be noted, the good ceased to be a luxury within a few years of its introduction, a point which will be taken up shortly.

Second, the relative cost of labor to capital in the United States has been high compared with that of other nations of the world. Vernon reasoned that this condition would induce U.S. firms to develop labor saving capital goods. The possibility that factor cost relationships would induce factor biased innovation has long been discussed by economists, who have theorized that in nations where the relative cost of specific factors of production is high, technological innovation will be biased toward use of the cheaper factor. [34] Significantly, an empirical study demonstrating that technological innovation in the United States is indeed biased toward the use of capital (i.e., is labor saving) has yet to be published, although one done by William Davidson is clearly suggestive of this possibility. [35] Vernon assumed that high labor costs in the United States would induce labor saving innovation and, importantly, that much of this innovation would be reflected in the development of labor saving goods.

Third, the United States is a very large market -- large in terms of both aggregate consumption and geographic extent. This, Vernon argued, would induce U.S. firms to develop new products which would be characterized by, enable the achievement of, or embody economies of scale. Two categories of products can be identified. The first category emcompasses products which can be economically manufactured only on a large scale and are characterized by high price elasticities of demand. Many petrochemicals would fall into this category, for example. The second category subsumes products that enable the achievement of scale economies. These would include new capital goods that enlarge the minimum economic scale of production for some other good, and, in addition, might include goods that enable producers to link two geographically separate markets and service them from single manufacturing or distribution centers. Among the latter goods would be transport and telecommunications equipment.

These three general characteristics of the U.S. market are unique in the sense that the nation has, throughout most of this century, been richer and better endowed with capital on a per capita basis and has possessed a larger aggregate internal market than any other nation on the earth. Under the product life cycle model, then, these characteristics underly the high rate of product innovation in the United States. Overseas demand for these new products exists, and in the period of time following their initial appearance in the United States, this demand would be filled by means of U.S. export. This reasoning is in accordance with the hypothesis of Irving Kravis discussed previously.

The export role of the United States would not be everlasting, however, due both to dynamic changes that would take place in the economic characteristics of the product and to changes in the nature of overseas supply and demand.

Changes in the economic characteristics of the product are manifested in changes in the design of the product itself, changes in the cost of producing it, and changes in the structure of both the supplying industry and the demand for the product.

Changes in the product design involve both product improvement and product standardization. Each results from information feedback from the marketplace, enabling designers to determine exactly the needs of the consumer. As these needs become better understood, the product can be modified so as to better satisfy them. When the needs become fully understood, the number of design changes per unit time is reduced, and the product is standardized. Standardization of the product allows changes to be made in its manufacture. Manufacture of a nonstandardized product requires a process which utilizes flexible, general-purpose production equipment and skilled labor, which can adapt or be adapted to frequent changes in product design. As the design becomes standardized, specialized production equipment can be substituted for the flexible equipment -- a substitution which permits reductions in the marginal cost of manufacture. In addition, less skilled labor may be substituted, particularly if the specialized equipment can be automated. Highly specialized, automated production processes typically require less labor input per unit of output than do less specialized processes. Hence, the substitution of the former process for the latter also entails a substitution of capital for labor for a given level of output. The net result of these changes is to reduce the marginal cost of producing the product.

Changes in the structure of the supplying industry come about as competing firms enter the market for the newly innovated good. The innovator of a new product that has been successfully launched in the marketplace will typically earn a monopoly rent on sales. Indeed, it is argued that the prospect of the rent is what induces the innovation in the first place. [36] Seeking to capture some of the rent, competing firms will imitate the innovation and enter the market. Because the innovator firm will have had to overcome many of the uncertainties associated with the introduction of a new product, imitator firms will not have to bear the costs associated with these uncertainties. Thus, the "demonstration effect" of successful innovation by one firm serves as an incentive for other firms to imitate the innovation. One counteracting disincentive might be patents held by the innovator. In most cases, however, the patents can be "invented around" by the determined imitator. The competition engendered by new firm entry will drive down the price of the product. The rent to the innovator may or may not be reduced, however, depending upon whether the innovator can lower the cost of the product faster than his competitors. Thus, innovation and subsequent new entry stimulate cost reduction of the new product.

Demand for the product will change as information about it becomes diffused. Initial demand for a new product may be constrained by uncertainty (or just plain ignorance) about the uses to which the product can be put or its performance. As the product becomes more widely used, however, demonstration effects by early consumers may induce additional potential users of the product to consider its purchase. In the longer run, the structure of demand might be affected by rising per capita incomes. Louis Wells has suggested, for example, that demand for U.S. innovations in Europe has lagged U.S. demand because incomes in Europe have lagged those in the United States. [37]

These changes in the economic characteristics of the product are interactive. For example, if demand for a product is price elastic, falling prices resulting from manufacturing cost reductions and new entry will cause the quantity of the product sold per unit of time to increase. Increases in sales, in turn, might induce further cost reductions and additional new entry. Cost reductions would be induced by sales increases if the fixed to variable cost ratio of new processes was higher than that of old processes. This is often the case if the new processes involve the substitution of specialized equipment for general purpose equipment. Additionally, rising sales could induce suppliers to substitute larger scale processes for smaller scale ones, allowing the achievement of scale economies. [38]

Changes will come about in overseas markets as well as domestic ones. As information about the product becomes more widespread, demand for the product will grow overseas. If overseas demand is price elastic, a decline in the price of the product will result in further growth in volume of overseas sales. As the overseas sales volume grows, local firms in foreign nation markets will be induced to imitate the product and commence local production. Vernon has postulated that in order to avoid loss of the overseas market share to local producers, U.S. exporters might themselves become local producers by establishing manufacturing subsidiaries within the foreign markets. Therefore, he argued, U.S. manufacturing firms become multinational for defensive reasons. [39]

Based on the above reasoning, Vernon and Wells developed a temporal sequence of stages of the product life cycle in international trade. Early on, the product would be manufactured and consumed solely in the United States. With the passage of time, demand would appear in other advanced nations (Europe and Japan). Overseas demand would at first be filled by U.S. exports, but eventually these exports would be displaced by local production in other advanced nations. Some local production might be commenced by subsidiaries of U.S. firms. Demand would begin to appear in developing nations and would be met by either European or U.S. export. With the passage of still more time, exports from advanced nations other than the United States would capture a share of the U.S. market. Ultimately, however, the locus of production might shift to developing nations, and these nations might export to both the United States and to other advanced nations.

It should be noted that under the product life cycle model, when a product reaches full maturity and its associated technology becomes widely diffused, the locus of production of the product will be determined by relative factor costs. It is consistent with the product life cycle model, then, that in the long run the factor proportions theory is the correct predictor of patterns of international specialization and trade. The model suggests, however, that the "long run" is, indeed, long and that in the meantime, predictable transitional patterns of specialization and trade will occur. The product life cycle model, therefore, is not so much an alternative to the factor proportions theory as it is a supplement.

Klein's Rivalry Hypothesis

While Vernon and his colleagues look primarily to the characteristics of demand in the United States to explain the country's historically high rate of innovation, Burton Klein focuses on the characteristics of supply. [40] Klein hypothesizes that the firm, when confronted with the decision of whether to invest resources in innovation, faces two types of uncertainty. "Type I uncertainty" is the uncertainty intrinsic to innovation. A would-be innovator is uncertain of levels and elasticities of demand for a new product. The innovator is also uncertain of the costs involved in perfecting the new product. If the product represents a major departure from existing products, the innovator might even be uncertain of the design and performance characteristics of the product. These uncertainties pose risks to the firm since there is some possibility that the investment in innovation will yield unsatisfactory or negative returns. Thus, the uncertainties act as disincentives to innovation. Offsetting these "Type I uncertainties" is the potential of reward should the innovation prove successful.

"Type II uncertainty" is uncertainty with respect to the actions of competitors. If a competitor firm is first to develop a new technology, that firm will possess an advantage in the marketplace over its rivals. The possibility that a competitor firm will develop a new technology acts as an incentive for other firms to engage in innovation.

Klein argues that both the potential of economic reward for overcoming "Type I uncertainty" and a high level of "Type II uncertainty" are prerequisites for innovation to occur at a rapid pace in a market economy. "Type II uncertainty" is high if a great degree of rivalry exists within an industry. Rivalry is itself a behavioral concept. It exists when competing firms are unable effectively to collude with one another on either an overt or a tacit basis. Klein believes that the most important prerequisite for a high degree of intraindustrial rivalry is easy entry into the industry. Generally, entry into an industry is facilitated if (1) the industry is rapidly growing; (2) scale economies are not of crucial importance; (3) key technologies are not closely held by one or a few firms; and (4) other important intangible assets (such as brand names) are not held by one or a few firms.

Klein notes that highly innovative firms are characterized by internal organizations that are not rigidly hierarchical. Innovation is most likely to be forthcoming from an organization in which managerial and technical employees are able to interact freely with one another on a personal basis. Firms that operate in highly rivalistic industries typically possess interactive internal organizations, and thus, it is difficult to ascertain the direction of causality: do interactive internal organizations lead to high degrees of intraindustrial rivalry, or is it that intraindustrial rivalry forces firms to develop interactive internal organizations?

It should be noted that Klein sees his hypotheses not so much as an alternative but as a supplement to the product life cycle. The Vernon version of the product life cycle, in Klein's view, places too much emphasis on the demand side of the U.S. economy in its explanation of U.S. innovation. Klein argues that the characteristics of the supply are an equal, if not more important, determinant of innovation. In particular, Klein points out that Vernon's arguments do not explain why some sectors of the U.S. economy are technologically dynamic while other sectors are virtually moribund. Klein's reasoning fills this gap in the product life cycle model.

Linder's Demand Similarity Model

Staffan Linder's demand similarity model, by contrast, is definitely an alternative to the product life cycle as an explanation of international trade. [41] The essence of the model is that trade among nations in manufactured goods is occasioned by product differentiation. The concept is best illustrated by an example. Suppose that two nations -- the United States and West Germany -- are characterized by similar levels of per capita income and possess advanced industrial capabilities. Within each nation there exists a large demand for automobiles, and, in response to this demand, each nation has developed a large automotive industry. The tastes of consumers in the United States are such that the majority prefers a big, heavy automobile. This will be called, somewhat whimsically, the "Behemoth." In West Germany, by contrast, most consumers prefer a lighter, tighter vehicle with a small, high-performance engine -- the "Neuroticmobile." In the United States, however, a sizeable minority of the consumers prefer Neuroticmobile to the Behemoth, and in West Germany a minority prefers the Behemoth. The Linder theory asserts that the United States will export Behemoths to West Germany and import Neuroticmobiles from that nation.

If this reasoning is correct, several consequences follow. First, international trade in manufactured goods will be most intensely conducted between nations having similar tastes and levels of per capita income. The gains from trade, following Linder's reasoning, would be much less than under the factor proportions theory because the determinant of a nation's comparative advantage is not a relative cost advantage, but rather a fine difference in consumers' preferences.

Under Linder's reasoning, trade in new technology-intensive goods would take place among all nations with advanced technological capabilities. The major difference between this reasoning and that of the product life cycle lies in its treatment of the role of the United States. Linder sees the United States as but one of the advanced nations, exporting goods embodying its new technologies to other advanced nations and, simultaneously importing different new technology-intensive goods from them. [42] The United States would not necessarily, as under product life cycle reasoning, lose comparative advantage in the export of its differentiated products with the passage of time.

The demand similarity model has stood up well in empirical tests. For example, empirical testing of Linder's hypothesis with respect to growth of trade within the European Economic Community has been performed by H. G. Grubel, who found the data to be corroborative of the hypothesis. [43] Gary Hufbauer notes that empirical work by Gruber and Vernon, cited earlier, also supports the Linder hypothesis, as well as the product life cycle model. [44] Econometric studies of world trade by H. Linneman also yielded results which could be interpreted as supportive of the hypothesis. [45]

Which of the two models -- the product life cycle or the demand similarity model -- best explains trade? Referring again to Table 4.1 suggests that the United States does simultaneously export and import high technology goods, a fact which at first hand would appear to support the Linder hypothesis. The data also indicate that the United States exports more of these goods than it imports. The positive balance of trade supports the product life cycle theory if one assumes that this trade surplus is generated by the constant exportation of innovative products. If the export composition of these goods does not change, however, the surplus might simply reflect the Linder notion that comparative advantage in these goods does not shift with the passage of time. The data do not conclusively support one model over the other. In a detailed econometric analysis of the composition of world trade in manufactured goods, Gary Hufbauer reached a similar conclusion: the data support both the product life cycle model and the demand similarity model. [46] Hufbauer also noted, however, that observed patterns of the trade of manufactured goods demonstrate at least some elements of factor specialization among nations of the world and, therefore, that this trade can not be entirely explained by fine-line product differentiation.

It is important to note that one implication of Linder's model is that the gains from trade are not nearly so great as those which are implied by the product life cycle model (or, for that matter, by the factor proportions model, of which the product life cycle model is a special case). This is because in Linder's model trade does not bring about as much specialization of production as is predicted by the other models. Thus, in Linder's model, resources are not reallocated to any great degree from less efficient uses. If Linder is correct, the importance to world welfare of an open international trading system for manufactured goods is not as great as most economists believed, and protection policies could be justified in more cases than under the other models of trade. As has been noted, however, empirical evidence does not fully support the Linder model.

CHANGES IN GLOBAL PATTERNS OF TECHNOLOGICAL INNOVATION AND IN THE U.S. POSITION IN THE INTERNATIONAL ECONOMY

At least four trends are significantly affecting U.S. comparative advantage in world trade of goods embodying new technologies.

First, the technological capabilities of a number of industrialized nations other than the United States have advanced very rapidly during the past decade or so. In certain sectors these nations, and not the United States, are the leading sources of new technologies. Second, the industrial sectors of a limited number of the so-called developing nations have grown very rapidly in recent years. Several of these nations are beginning to emerge as important exporters of certain types of manufactured products. Third, the ability of multinational firms to transfer new technologies across international lines has been growing. Additionally, the ability of local firms in a number of foreign nations to quickly imitate new technologies has increased markedly. These conditions reduce the relevance of the product life cycle theory as an explanation of international trade and investment in new technology-intensive goods. Fourth, in the last ten years or so, there appears to have been a worldwide slowdown in the rate of development of new industrial technology. In the following discussion, the causes and consequences of each of these trends are explored in this section, with implications and recommendations for U.S. official policy spelled out in the next section.

It is important to note that neither the causes nor the consequences of the identified trends are independent of one another. Similarly, the effects of nontechnological developments in the world economy (such as the oil price increase of 1973, to choose the most apparent example) are difficult to separate from the effects of technology-related developments. For instance, it is argued later in this section that the real terms of trade for manufactured goods have, on the whole, moved against the United States (i.e., compared to U.S. imports of manufactured goods, U.S. exports of manufactured goods now generally command lower prices on world markets than they did during the 1960s). This deterioration in the terms of trade can be attributed to a number of causes, among them the oil price increase, the rapid advancement of the economies of West Germany and Japan during the 1960s, the recent economic slowdown in Europe, and the emergence of Third World exporters of manufactured goods. It is beyond the scope of this discussion to attempt to identify all such causes and determine the relative importance of each. Rather, the more modest goal is to identify a limited number of causes -- those which are related to technology -- and to suggest how they affect the position of the United States in world trade.

Changing Technological Capabilities of Industrialized Nations

That the technological capabilities of a certain number of industrialized nations other than the United States have advanced rapidly in recent years should not, at base, be considered a surprising development. [47] Nations such as Germany and France have historically been major sources of technological innovation. Many modern technologies were pioneered in France, for example, production of hard goods using standardized parts was begun in France in the 18th century, more than one hundred years before such techniques were employed in North America. Early in the 20th century German firms technologically dominated such important sectors of the world economy as chemicals, pharmaceuticals, and branches of the metallurgical industries. Japan emerged as a major new industrial power early in the century. However, the economies of all of these nations were dealt crippling blows by World War II, and the United States emerged as the dominant source of new technological innovaton even in the sectors in which these nations had been important innovators prior to the war. But, by the middle 1950s, it became evident that the economies of these nations would recover and that their previous eminence would be in all likelihood reattained.

In order to stimulate an economic resurgence in the war-devastated nations, the United States initiated a number of key policies during the late 1940s and early 1950s. Among them was the Marshall Plan, which provided for direct U.S. assistance for the rebuilding of European industry. The Bretton Woods Agreement provided a system of fixed exchange rates among the world's major currencies, with the currency of the United States serving as the major international reserve medium. Importantly, major European currencies were deliberately undervalued against the dollar. The dollar overvaluation was designed to stimulate U.S. investment in Europe and exports of European manufactured goods to the United States. During the 1950s and early 1960s, the overvaluation caused no problem for U.S. trade. The U.S. economy enjoyed the benefit of low cost imports, while its exports consisted, to a large degree, of goods embodying technologies that could not be duplicated elsewhere and, therefore, could command high prices on international markets.

The recovery process in the war devastated nations was fairly complete by the late 1960s. Table 4.2 indicates that by this time, the postwar dominance of the United States as the locus of innovative activity had ended, and a large percentage of the world's research and development activity was being performed outside the United States. To be sure, research and development is but one input into the process of technological innovation, and the tables do not indicate innovative output. Nonetheless, to the extent that input can be taken to be a proxy for output, the table shows that the relative importance of the United States as an innovator has declined since the 1950s.

Of the war-devastated economies, the two that clearly emerged as the strongest were those of the principal foes of the United States during World War II -- West Germany and Japan.

The redeveloped economy of West Germany is in many regards similar to that of the old Germany. Sectors in which Germany had been historically strong reemerged as those in which modern Germany has become a leading innovator. These include the chemical and pharmaceutical industries, precision and heavy machinery, heavy electrical goods, metallurgy, and surface transport equipment. West German firms have not emerged as leading innovators in the very high technology industries like aircraft and aerospace, advanced electronics, and high speed electronic computation, although some German firms have achieved excellence in specialized pockets of these industries. (Exceptions may lie in the nuclear power and telecommunications areas, areas in which German firms seem to be performing quite well.)

Japan has been one of the major success stories of the latter half of the 20th century. In the early postwar period,

TABLE 4.2

TOTAL EXPENDITURES ON RESEARCH AND DEVELOPMENT OF SIX MAJOR NATIONS, 1963-1975

$ MILLIONS

Nation	1963	1967	1969	1971	1975
France	1,428	2,505	2,697	2,884	5,982
W. Germany	1,443	2,084	2,668	4,472	8,857
Japan	897	1,694	2,592	4,532	8,762
U.K.	2,159	2,694	2,597	2,900 (est.)	4,706
U.S.	17,371	23,613	26,169	27,336	35,200
TOTAL	23,298	32,590	36,723	42,124	63,507
U.S. as a % of total	74.6	72.5	71.3	64.9	55.4
U.S. as a % of Japan + W. Germany + U.S.	88.1	86.2	83.2	75.2	65.5

Source: Data for 1963, 1967, 1969, and 1971: Organization for Economic Cooperation and Development (OECD), Patterns of Resources Devoted to Research and Experimental Development (Washington, D.C., 1975). Data for 1975 for West Germany and Japan from scientific attaches in these nations' U.S. embassies. U.S. data for 1975 from U.S. Department of Commerce, Statistical Abstract of the United States, 1977. All data converted to dollars at average yearly spot rates as per International Monetary Fund. 1975 industrial data from OECD, Science and Technology Policy Outlook, 1978.

Japan rebuilt her traditional industries, becoming a major exporter of cotton textiles and simple consumer products during the 1950s. Rebuilding of the steel, shipbuilding, and heavy machinery industries followed, and in these sectors Japan became a very efficient producer during the early to middle 1960s. The automotive and consumer electronics industries were built up rapidly during the 1960s, and by the early 1970s, Japanese firms were making advances into such high technology industries as semiconductor and computer manufacture. The Japanese now expect that a major source of future growth in domestic output will come from the most technologically advanced sectors of the economy -- an expectation exemplified, in one instance, by a major effort to surpass the United States in the development of ultra-sophisticated microelectronic technologies.

Whether Japan will succeed in this latter effort is a major question. The technological strength most evident in Japan throughout the past two decades has been a superb ability to imitate and improve upon existing technologies. At first, the Japanese focused on improvement of mature technologies. In recent years, however, the Japanese have demonstrated a growing capability to imitate and improve upon very young and sophisticated technologies. While Japanese firms during the postwar period have generally allowed other nations to perform the basic research and development leading to major innovation, in some technologies they are leading innovators. For instance, they probably lead the world in the innovation of highly automated assembly processes. Nevertheless, the cases of innovation are relatively few, and whether the Japanese can parlay their exceptional ability to imitate into an ability to truly innovate is a question that will be answered only with the passage of time. It should be noted, however, that prior to World War II, the United States itself was often viewed as being primarily an imitator nation rather than an innovator nation.

There has been considerable discussion about the source of vitality of the Japanese. Norman Macrae, writing on January 4, 1975, in The Economist, notes that he visited Japan three times at approximately seven-year intervals beginning in 1960. On each visit he was told that Japan's economic growth rate was about to decline permanently, and each time the rate of growth accelerated instead. Macrae attributes the vitality of the Japanese economy to the organizational structure of the typical Japanese business firm (which, by Macrae's description, is a highly interactive one, having the characteristics claimed by Burton Klein to distinguish a dynamic firm from a nondynamic one) and to the attitude of Japanese society towards the business community. Other observers attribute Japanese strength variously to the ability of the government to centrally allocate resources to specific sectors of the economy; to intense rivalry among the major business groups of the nation; or to the lifelong employment policies pursued by large, diversified

Japanese firms, which facilitate the transfer of workers by those firms from declining sectors to growing ones. High educational attainment by Japanese workers has doubtlessly also contributed to that nation's economic performance; a higher percentage of Japanese over the age of seventeen are enrolled in schools or training institutes than are citizens of any other nation.

The implications for the United States of advances in other industrial nations' technological capabilities have been straightforward. The percentage of the world's commercial technological innovation that originated in the United States began declining during the early 1960s, and this decline has continued into the 1970s. In a number of sectors in which the United States held a monopoly over the introduction of new technologies in the early post-World War II years, the monopoly has since been broken. In a few sectors, including steelmaking, shipbuilding, and the manufacture of shoe machinery, the United States has ceased to play much of an innovative role at all. The monopoly rents that U.S. firms could achieve from export sales in these industries have been reduced substantially or in some cases have disappeared entirely.

An inevitable consequence has been that the terms of trade in manufactured goods have turned against the United States. Dollar exchange rates established under the Bretton Woods Agreement enabled the United States to import many manufactured goods cheaply. At the same time, the nation was able to charge high prices for exports of its technologically unique products. As the uniqueness of U.S. exports diminished, a shift in the terms of trade put pressure on the parity of the dollar. This pressure was one reason for the collapse of the Bretton Woods system during the early 1970s.

So as not to paint too bleak a picture, it must be noted that U.S. manufacturing firms in many sectors remain quite innovative and internationally competitive. This is particularly true in industries characterized by very new and rapidly changing technologies. [48] For U.S. firms operating in these industries, the economies of Western Europe and Japan offer significant opportunities since they continue to require the products of advanced U.S. technologies. In these industries, however, some problems do exist. An immediate one is that U.S. exports of goods embodying high technology are income elastic, and current slow rates of economic growth in Europe have reduced growth in demand for such products. Furthermore, even in the most technologically advanced sectors, U.S. firms are facing some new competition from Japan and other advanced nations. This new competition need not necessarily result in major loss of U.S. export market share, but U.S. high technology firms may have to try a little harder to succeed in export markets in the future.

The devaluation of the dollar has provided opportunities for export for a number of U.S. industries. While it has been frequently noted in the financial press that U.S. exports have not surged following periods of decline of the dollar against other major trading currencies, there are signs that the U.S. is regaining exports in some industries in which competitiveness was clearly lost during the 1950s and 1960s. Exports of U.S.-made automobiles, for example, are now growing after more than two decades of steady decline. This has become possible because of the revaluation of the dollar and the popularity of new product designs recently introduced by the automotive manufacturers. There is a time lag between changes in relative prices and changes in world trade patterns in many categories of manufactured goods, and U.S. export performance in a number of other industries is beginning to improve. Again, U.S. exports in a number of sectors -- especially exports of intermediate products manufactured by multinational firms -- are suffering because of sluggish economic growth in Europe.

It should be pointed out that although the economic resurgence of Europe and Japan has resulted in a number of headaches for the United States, the economic vitality of these areas is ultimately in the nation's best interest.

It must also be pointed out that the international dispersion of innovative capabilities does not unequivocally work to the disadvantage of the United States. Innovation is achieved at some real resource cost to an economy, and throughout the post-World War II era, the costs of innovation have largely fallen upon the U.S. economy. [49] The benefits of innovation, by contrast, can be captured by an economy irrespective of where the innovation first takes place. One clear implication is that U.S. firms should strive quickly to imitate foreign technologies. This would, of course, be particularly advisable for firms operating in industries in which the innovative impetus has moved abroad. Unfortunately, it would seem that many such U.S. firms are slow to adopt foreign technologies even after these have proven to be workable.

The rise in the technological capabilities of such large, industrialized nations as Japan, West Germany, and France serves to reduce the relevance of the product life cycle model as an explanation of U.S. trade, at least as the model applies to Europe and Japan. The economies of West Europe and Japan no longer have to wait for the fruits of U.S. innovation. Innovation can and is accomplished by firms in these nations in response to the needs of their consumers. It might happen, of course, that an innovation which first appears in the United States later finds use in Europe, in accordance with the precepts of the product life cycle model. But there is also a high and growing probability that a major innovation will first appear outside of the United States and later find use in the United States. If, as is likely, the types of innovation that occur separately in the United States, West Germany, Japan, France, the United Kingdom, and other advanced nations result in similar but differentiated end products, patterns of international trade in newly innovated goods may come increasingly to resemble those postulated by Linder's demand similarity model.

A further reduction in the relevance of the product life cycle model results from the increasing probability that multinational corporations, when they first begin to produce a newly innovated good outside of their home market, will locate the production directly in the nation of least cost manufacture, irrespective of where the product is to be marketed. This possibility is discussed later.

Rapid Industrialization of Developing Nations

Rapid industrialization in developing nations poses a different set of implications for the United States than does the technological progress of other advanced nations. Rapidly industrializing nations do possess labor forces for which wages are much lower than in the industrialized nations, even after adjustment is made for differences in skill and productivity. A small number of developing nations, however, are becoming significant exporters of manufactured products, generally characterized as embodying mature technologies. [50] Although there are well over one hundred sovereign states in the world which can be termed "developing," only a handful now possess rapidly industrializing economies, including Brazil, Mexico, Taiwan, South Korea, Singapore, India, and perhaps several other nations. Furthermore, only a few of these nations -- most importantly Brazil, Mexico, and India -- are, at this time, developing broadly diversified industrial bases; in the others, industrial activities tend to be quite specialized. The major manufactured products of the rapidly industrializing group currently include textiles, apparel, shoes, and standardized consumer electronics. Small but growing product categories include steel and steel products, standardized machinery, ships, and automotive components.

Industrialization of developing nations presents the United States with a number of dilemmas and important choices. Because of low labor costs, developing nation industries which are likely to become highly competitive in world trade of goods are those based on mature, standardized technologies. It is clear that the more labor input required per unit of output for a good, the more competitive the developing nations can become as producers of that good. In the absence of governmentally imposed barriers to trade, goods from developing nations will inevitably capture a growing share of world markets, including the U.S. domestic market. Equally inevitable will be calls for high import tariffs or restrictive import quotas to protect threatened industries. Such calls should not be heeded. To protect these industries would be to force U.S. consumers to pay high domestic prices for domestically produced goods which could be more cheaply purchased from abroad. This would be tantamount to taxing U.S. consumers in order to subsidize producers. Furthermore, protection of a domestic industry shields producers in that industry from the stimulus of foreign rivalry. The shield acts as a disincentive to innovation within the industry, eventually causing the industry to lag behind world standards in the introduction and use of new technology. Thus, protection of a domestic industry ultimately brings in tow technological obsolescence.

Politically, the effects of large-scale curtailment of imports of manufactured goods into the United States would be highly negative for the nation. Throughout the post-World War II era, the United States in both its rhetoric and its actions, has been an advocate of economic development in the so-called Third World. Most of the rapidly industrializing Third World nations are highly dependent upon export markets in the developed nations, and particularly in the United States, in order to gain foreign exchange needed for future growth. For the United States to adopt protectionist policies toward these nations would undermine their prosperity and create levels of resentment much greater than those which now exist. The United States, more than it is already, would be seen as a powerful nation willing to "beggar its neighbors" in order to serve its own narrow interests. Ironically, by doing so, the United States would not in fact be serving its own interests.

Multinational Firms and International Transfer of Technology

Empirical data recently collected by Raymond Vernon and William Davidson indicate that the time lag between the introduction of a new technology in the United States by a U.S. multinational and its first use abroad by a foreign affiliate has steadily decreased during the past three decades. [51] This demonstrates that U.S. multinational firms have increased their capabilities to transfer technology rapidly to foreign subsidiaries. Similarly, non-U.S. firms, especially those of Japan, but also those of Western Europe, and certain of the developing nations, have exhibited increasing capabilities to rapidly imitate new technologies first introduced in the United States. Both trends suggest that for any given new technology, the length of time that the nation will hold a monopoly over the use of a new technology will decrease. Thus, the length of time that the United States can expect to export products embodying new technologies prior to the commencement of production abroad will also decrease.

Two implications follow. The first is that the relevance of the product life cycle hypothesis as an explanation of U.S. comparative advantage is reduced yet further. If new technologies become internationally diffused within a short time of their introduction, the locus of the innovation will cease to become an important determinant of the locus of production of goods embodying those technologies or of patterns of world trade of these goods. The second implication is that patterns of international trade will come to resemble more closely those predicted by the factor proportions theory. Production and exportation of goods will be concentrated in those nations which are, on a world basis, the least cost producers of the goods. [52] International differences in levels of technological capabilities will not be as important determinants of trade patterns as they have been in the past.

Negative effects on the U.S. economy of rapid diffusion of technology can be identified. [53] The U.S. economy bears most of the costs of the domestic creation of new technologies but may yield much of the benefit to those nations to which the technologies are transferred without receiving compensation for the transfer. If the total benefit that can be appropriated by a U.S. firm is further reduced by the rapid diffusion, there will be less incentive for innovation to occur. This is likely to be the case if the diffusion is accompanied by rapid imitation of the technology by the worldwide competitors of the innovating firm. If, however, the innovating firm is able to utilize its own network of subsidiaries to diffuse the technology and is able to exclude the technology from competitors for a certain length of time, the diffusion might act as a stimulus to innovation. This might be the case, for example, if the innovating firm were able to achieve very rapid cost and price decreases by quickly transferring the technology to subsidiaries located in low-wage areas. [54]

There have been numerous proposals to regulate the transfer of technology out of the U.S. economy, but such proposals have invariably run into problems of practicality. Most efforts at controlling technology outflow, viewed with historical perspective, have proven to be exercises in closing the barn door after the horses have fled. [55] Past efforts apart, whether diffusion of technology can actually be retarded is a valid question. In essence, technology is human knowledge, and to prevent one determined human being from acquiring the knowledge of another is not easy. A national effort to block the transmission of knowledge from U.S. citizens to foreigners, to be at all efficacious, would require a state policing apparatus so restrictive as to be intolerable to the nation's citizenry and so elaborate as to almost certainly cost more than it could save.

The Worldwide Deceleration of Technological Innovation

It is widely felt that there has been a slowdown in the rate of technological innovation during the past ten or so years, in the sense that this rate has been below that which prevailed during most of the post-World War II era. Tangible evidence that a slowdown has occurred is actually quite slim. One fact in this direction is that the rate of measured increases in labor productivity has diminished during the past decade. (See chapters by Hill and Bourdon.) Productivity increases are one output of technological innovation, and a diminution of the rate of productivity increase is consistent with the hypothesized slowdown. Furthermore, the total number of patents issued annually in the United States has declined steadily since 1971. While this fact is suggestive of a slowdown in technological innovation, it may also be that U.S. corporations are choosing to patent a lower percentage of new technologies than had been the case in the past. No evidence presently exists to prove or disprove this possibility. Most additional evidence used to demonstrate a slowdown is anecdotal in nature. [56]

If a slowdown has indeed occurred, it has not made itself felt in all sectors of the economy as Bourdon shows so vividly in Chapter 6. In the highly sophisticated semiconductor industry, for example, advances in technology have resulted since 1970 in almost a one hundredfold decrease in the cost of producing hand-held calculators, quartz timepieces, and other products utilizing microelectronic components. In the automotive industry, the rate of innovation in end products probably accelerated in the 1970s over rates that prevailed earlier in the postwar period.

Nonetheless, it is doubtlessly true that in many sectors either the rate of technological innovation has fallen off in recent years or the rate of embodiment of new technologies in the productive process has not proceeded at satisfactory levels. It is probably true also that the unsatisfactory performance can be linked to low rates of capital formation. Jacob Schmookler and others have demonstrated that high rates of capital formation within an industry tend to stimulate technological innovation. [57] That the U.S. economy has been marked by low rates of new capital formation in recent years is virtually undeniable. Symptomatic of the problem are low rates of return on invested capital (according to U.S. Treasury estimates, the average rate has fallen from about 8 percent in the middle 1960s to about 4 percent at present), low rates of creation of new business ventures, and a decline in the public issuance of new equity securities.

The problem is particularly acute among new business ventures in the high technology industries. Small companies have contributed a disproportionately high share of innovation in these industries, and the rivalry engendered by these companies serves to stimulate innovative activities of larger companies. The rate of formation of new, small, high technology ventures has seen a substantial reduction in recent years, a condition that can probably be attributed in large part to the virtual "drying up" of the venture capital market.

It is important that the United States seeks means by which to stimulate a higher rate of technological innovation. While it is beyond the scope of this discussion to present detailed recommendations as to how this might be done, it is important to note that only by maintaining a high rate of domestic innovation can the stature of the United States in world commerce ultimately be preserved. Once an exporter of a wide range of natural resources, the United States has become a net importer of most raw materials. Because of domestic labor costs which are high relative to world standards, even after adjusting for productivity differentials, the nation cannot in most instances be a competitive exporter of standardized manufactured goods. The nation does export its substantial surplus of agricultural commodities -- a surplus made possible by application of advanced technologies to the agricultural sector -- and manufactured goods embodying advanced technologies. The trends identified here all point to the fact that simply to hold its place in the world economy, the United States will have to become even more innovative. In light of this fact, it is especially disturbing that the rate of domestic innovation may be on the decline. Thus, much of the focus of the next section is on reversing the decline.

POLICY OPTIONS FOR THE UNITED STATES GOVERNMENT

This section presents policy options for the U.S. Government. Policy options that have already been proposed but which the author feels would not serve to further the position of the United States in the world economy are discussed first. These are followed by a discussion of options which he feels would serve that interest.

Policy Options Not in the Nation's Best Interests

Policy options that aren't in the best interests of the nation include most proposals either to restrict imports of foreign-made goods into the United States or to restrict the transfer of technology out of this nation.

There are several reasons why import restrictions would not enhance the U.S. position in the world economy. First, to recapitulate, import restrictions, either in the form of high tariffs or import quotas, act as a tax on the American consumer to subsidize inefficient sectors of the economy. When protected by import restrictions, these sectors are shielded from rivalry with dynamic overseas competitors, and thus the stimulus to technological innovation that this rivalry creates is removed. Protectionist policies encourage technological stagnation, not technological innovation, in the protected sectors.

Another reason the U.S. Government should not try to restrict technology transfer out of the United States is that restrictive efforts are by and large unlikely to succeed. Even if technology transfer from the United States were deemed to be generally undesirable, it is not at all clear that an effective means to stop it could be devised. Most efforts in the past have been unsuccessful because they were either tardy or too easily circumvented.

A third reason for avoiding restriction is that U.S. firms, especially multinational firms, earn a positive return on foreign use of sales of technology. It would not be in the interests of the United States to restrict the abilities of U.S. firms to achieve a commercial reward on the overseas utilization of technology. Such restriction would reduce the incentive to private firms to create new technology.

It must be recognized, however, that the U.S. Government in certain cases may legitimately restrict the dissemination of particular technologies for national defense reasons. Such technologies are typically ones that have specific and specialized military applications and generally lack potentially widespread commercial use. Control of the dissemination of such technologies, therefore, is facilitated by the limited uses to which they can be put. It is important that national security arguments not be invoked in attempts to restrict widely used commercial technologies or technologies readily available from non-U.S. sources. Attempts to restrict technologies that fall into these categories will not succeed in preventing their use by hostile powers. Furthermore, such efforts will reduce the opportunities of U.S. suppliers to gain commercial rewards from selling the technologies to non-U.S. buyers.

There is, however, one general case in which government intervention in the sale of commercial technology by private firms would be justified. This occurs when U.S. firms face a monopsonistic state buyer of technology like the Soviet Union or the Eastern bloc nations. The monopsonistic buyer can in some instances force sellers to bid against one another and by so doing force all sellers to reduce their prices. To counteract the monopsonistic power of the state-buying agency, the U.S. Government could establish a minimum price on a case-by-case basis below which no firm would be allowed to sell. The government might also seek to persuade other Western industrialized nations to pursue similar "command price" policies when dealing with buyers in the Eastern bloc. It should be noted that government command price action is not necessary if the private seller of technology holds a monopoly or a near monopoly over that technology. Pragmatic considerations suggest that government intervention should be limited to reasonably large commercial transactions and to sales of commercially important technologies.

Policy Options in the Nation's Best Interests

In the most general terms, the options that would improve the international position of the United States fall into three categories: (1) those which would serve to increase rates of domestic technological innovation; (2) those which would lower for U.S. exporters or sellers of technology barriers imposed by foreign governments; and (3) those which would help the U.S. economy to adjust to changing world conditions.

The first category does not really pertain directly to the international economic position of the nation, but rather to the health of the nation's domestic technology. As has been stressed, however, U.S. competitiveness in the international markets is to a large extent determined by the state of innovation in the domestic economy. The vitality of the economy is the primary determinant of the state of domestic innovation and of the nation's long-run competitiveness in international trade of manufactured goods. Any measure designed to bolster the nation's competitiveness in the world economy that does not also serve to improve the climate for domestic innovation is, at best, a measure of short-run or secondary importance. Of primary importance is the need to reverse the current slump in U.S. innovative output. The other chapters in this book all address this vital issue.

It was noted earlier that rivalry among business firms is a major stimulus to technological innovation. Some analysts believe that a diminution in rivalry among major business enterprises in the United States might be a cause of reduced rates of technological innovation. It might be argued that a solution to this problem lies in stricter enforcement of the antitrust law, accompanied by an occasional breakup of large and senescent organizations. This argument, however, is rejected by many U.S. industrial leaders, who note that scale is an important factor in much industrial research and development. It should be noted that two firms now facing prosecution on antitrust grounds -- American Telephone and Telegraph and International Business Machines -- have consistently been highly innovative organizations. Whether more or less innovation would have resulted in the telecommunications and computer industries had they not been characterized by a single, dominant firm is an unsettled issue. In this general area, more understanding is required before meaningful policy recommendations can be advanced. One recommendation that can be stated, however, is a reiteration of a previous point: firms facing competition from imports should not be sheltered or protected. To do so would be to remove them from the innovative stimulus of rivalry from foreign sources. The question of whether firms should be granted temporary relief in order to adjust to rapidly changing circumstances in the world economy complicates this issue, however, and is discussed below.

Technological innovation, it has been noted, is associated with periods of rapid rates of new capital formation. New capital investment in the nation has lagged in recent years, and this has doubtlessly had a deleterious effect on innovation. As is the case with the slowdown in the development of new technology, the slowdown in the rate of capital formation is not fully understood. Measures that might stimulate a higher rate of investment include lower rates of taxation on capital gains, faster allowable depreciation of new investment for tax purposes, higher investment tax credits for new investments, and other tax incentives for investment. To the extent that such measures work to stimulate capital formation, they are likely also to stimulate technological innovation.

One proposal designed to foster innovation is a specifically international one. The U.S. Government should do everything that it can to encourage foreign firms to make direct investments in the United States. The presence of U.S. subsidiaries of foreign companies in the domestic economy creates additional rivalry within U.S. industry. These firms bring the technologies of their parent organizations to the U.S., technologies which can be superior to those utilized by domestic firms.

The complex subject of government efforts to persuade foreign governments to lower barriers to U.S. exports has received a great deal of attention. U.S. efforts in this direction have largely been successful throughout the postwar period, beginning with the General Agreement on Trade and Tariffs (GATT) in the 1940s and up to the most recent round of multilateral tariff negotiations. Yet, a number of thorny issues in the trade area remain.

It has become the position of the U.S. labor movement, along with a number of other constituencies, that the nation makes too many concessions in international trade negotiations without extracting sufficient quid in the quid pro quo. Japan and certain of the faster growing developing nations are often singled out as nations that have unilaterally restricted U.S. imports while having relatively unimpeded access to the American market. It must be pointed out that this is not an entirely fair claim. The Japanese in particular have been subjected to a number of trade restrictions, including the trigger-pricing scheme in steel, the Long Term Agreement in cotton textiles, and so-called voluntary agreements in television receivers, noncotton textiles, automobiles, and other goods. In spite of these restrictions, exporters in Japan and other nations undoubtedly have readier access to U.S. markets than U.S. exporters have to Japanese markets. This is true because of substantial non-tariff barriers to trade which exist in Japan. If this situation continues to prevail, domestic political pressure for retaliations is likely to grow.

One area where much needs to be done is that of differences in national philosophies and approaches to health, safety, and environmental regulations. It is the point of view of the U.S. labor movement that regulation standards in the United States tend to be much stricter than in most other nations, and that the effect is to subject U.S. producers to a "hidden tax" which, among other things, gives multinational firms based in the United States an incentive to relocate production abroad. Rather than having U.S. health, safety, and environmental standards revoked, the labor movement favors legislation to prevent U.S. firms from shifting the locus of production to escape regulation and legislation to restrict imports of goods from countries where standards are lax.

The author finds himself in sympathy with much of the labor position. While the merits and nonmerits of the regulations themselves are beyond the scope of this chapter (see Chapter 5), the author does believe that nonexistent or laxly enforced standards in foreign nations constitute subtle government subsidies to producer firms. The costs of the subsidy are borne by the worker who is subjected to health and safety hazards and by the citizen who must live in a befouled environment. It is not practical, however, for the United States to insist unilaterally that domestic U.S. standards be accepted by all nations. What is in order is a set of internationally agreed upon minimum standards for the health and safety of workers and a system of sanctions that could be applied against noncomplying nations. The sanctions most easily applied would be countervailing duties which would be applied under the new subsidies/countervailing measures code negotiated under the recent multilateral trade regulations. It should be a matter of priority for multilateral talks to be initiated on the subject of instituting a set of minimum health, safety, and environmental standards, the violation of which could subject the offending nation to sanctions under the new code.

Another area in which international cooperation is needed is that of performance requirments imposed on multinational firms by host governments. Performance requirements are often stipulated as preconditions for a firm being allowed to do business in a particular nation -- the local government requiring the firm to perform minimum local value-added and/or minimum levels of exportation. While performance requirements may be an effective means by which some developing nations can foster growth in their local economies, extensive use of these requirements can significantly distort patterns of international trade and investment. Given the growing use of performance requirements by numerous nations, it should be a matter of priority for governments to agree on a set of ground rules to govern their use.

A number of policies (or nonpolicies) of the U.S. Government serve as barriers to U.S. exports. For example, the government does little to disseminate information about foreign nations' product standards. This deficiency, dealt with in some detail in the chapter by Hollomon, may act as a deterrent to U.S. exports to Japan and other nations. All governments of other major industrialized nations have programs to help local firms meet the product standards of foreign nations. The policies and programs of Japan and West Germany are outstanding in this regard, and the U.S. Commerce Department would do well to significantly extend its activities in this area. The Commerce Department also would do well to strengthen the programs by which it identifies export opportunities for small companies.

A vital component of the nation's international economic policy is a workable adjustment assistance program (see Chapter 6). The necessity for effective adjustment assistance derives from the fact that changing patterns of international trade will inevitably cause the relative importance of some sectors of the economy to decline. If the decline is rapid, there will be loss of employment opportunities in these sectors. Concurrently, other sectors will gain in importance, and new job opportunities will exist. If the rising sectors are marked by a high rate of technological innovation, as has generally been the case in the United States during the post-World War II era, the new jobs would typically be better in terms of compensation and working conditions than would be jobs in the displaced sectors. It is, however, likely to be of little comfort to the worker who faces a job loss in a Massachusetts shoe factory to know that there are employment opportunities at a Texas semiconductor plant. The Massachusetts worker might not want to move to Texas, and even if he did, he might not have the skills necessary to move into a desirable new position there.

Whether some sort of adjustment assistance or relief should be given to firms or industries threatened by imports, is a matter of controversy. Such assistance, as practiced by many nations, typically takes the form either of temporary tariffs or quotas on imports or of temporary government operating subsidies to the affected firms. In principle, the assistance should last only as long as is reasonably required for the affected firms to adjust to changing circumstances. In practice, however, firms more often than not will be unwilling or unable to adjust. If this is the case, political pressure for the assistance to be extended will mount. Thus, a program of temporary assistance can metamorphose into a program of wholesale protection or subsidization of a noncompetitive industry. An example of this is the British steel industry, which has been the beneficiary of a series of "temporary" relief programs since the 1950s and has, as a consequence, become a major burden on the British taxpayer.

It has been argued that if an import-threatened firm could be granted relief for a short length of time, it would be able to develop and adopt the new technologies that would enable it to become competitive once again in world markets. The argument may sound plausible, but it is in fact unfounded. There is no greater impetus to technological innovation than a dose of competition from a rival firm. Any program of relief, temporary or otherwise, serves to reduce rivalry and, therefore, is likely to act as a disincentive to innovation or to the adoption of new technology.

There may be good arguments for creating adjustment assistance programs for firms or industries for which the pace of change in world markets exceeds their ability to adjust, but only so long as it is understood by all that the program is temporary and will end on schedule. The author, however, feels that once such a program is initiated, self-interested parties will probably find ways to perpetuate it. Furthermore, once the program becomes self-perpetuating, it brings with it the chief evil of protectionism: de facto taxation of the public to subsidize an inefficient, technologically moribund sector of the economy. For this reason the author believes that the correct approach to adjustment assistance lies in direct assistance to workers on a scale much greater than that to which the U.S. Government is already committed.

Government adjustment assistance is provided for under the Trade Act of 1974. The present program is of a modest scope, designed to give workers whose jobs are displaced by imports both financial assistance of limited duration and job retraining. [58] To be eligible for adjustment assistance, unemployed workers must meet a number of restrictive criteria, criteria that eliminate most workers whose jobs are lost because of indirect effects of imports. The U.S. labor movement rightfully scorns the present program as constituting little more than "burial insurance." More imaginative and far-reaching programs are needed. In addition to enlarging the existing program and liberalizing the eligibility criteria, at least three additional steps should be taken: (1) the granting of lifelong pensions to import-displaced workers who, because of age and skill, cannot reasonably expect to find new work; (2) encouragement of business firms operating in expanding industries to locate plants in regions where import competition has created pockets of severe unemployment; and (3) the sponsoring of extensive programs to retrain and relocate displaced workers who can reasonably expect to move into new careers. It must be recognized that a change of career is both costly and traumatic for the individual whose job is lost because of imports. While it is doubtful that the government can do much to reduce the trauma, it can and should bear the cost. The alternative would be for the nation to subsidize and protect declining industries, an alternative which would be far costlier and would ultimately stifle the vitality of the economy.

The policies governments should pursue, as outlined in this chapter, have come to be known collectively as "positive adjustment policies." "Positive adjustment" implies that governments should not protect or subsidize industries that are threatened by imports or otherwise noncompetitive internationally, but should take concrete steps to encourage the transfer of resources from less competitive into more competitive industries. It has been noted by numerous analysts that while it is easy to advocate "positive adjustment policies," under the conditions for slow economic growth and high levels of inflation characteristic of the late 1970s, it is difficult for governments to pursue such policies actively. Yet, unless these policies are pursued, the slow growth, high inflation problem is likely to remain.

NOTES

1. The assumption that increased wealth is to the advantage of the United States can be challenged on several grounds. For example, it could happen that the effect of a change would be an increase in the wealth of both the United States and its trading partner nations, but at a more rapid rate in the case of the trading partners than in that of the United States. Thus, even though the United States became wealthier in absolute terms, its relative wealth would have decreased. Then it might be argued that diminished relative wealth would reduce the global political power of the nation, and this would not be in the nation's best interest. As an alternative example, it might be felt by some individuals that the United States is reaching a limit on its ability to accumulate wealth, and that the nation should be pursuing a policy of limited (or zero) economic growth. While there may be merit to these arguments against the assumption that increased wealth is always in the nation's best interest, the author will leave it to other analysts to advance and defend them.

2. For a review of thought regarding the relationships between technological progress and economic growth, see Charles Kennedy and A.P. Thirwall, "Surveys in Applied Economics: Technical Progress," The Economic Journal 82 (March 1972): 11-72.

3. The reconciliation is particularly well developed by Nathan Rosenberg in his "Factors Affecting the Payoff to Technological Innovation," mimeographed, National Science Foundation (Washington, D.C., 1977).

4. The original work is David Ricardo, On the Principles of Political Economy and Taxation (1817). The ideas of Ricardo are developed in most elementary international economics textbooks. See, for example, Richard Caves and Ronald Jones, World Trade and Payments, 2nd ed. (Boston: Little, Brown and Co., 1977).

5. By specializing in the one good in which it held a comparative advantage and trading away some of this for the other good, each nation would be able to consume at least as much of either one of the goods as it could under autarky and consume more of the other good. The gains to both nations result from the fact that under trade the total productive resources of the world are more efficiently utilized than would be the case were both nations to be autarkic.

6. Thus, Ricardo (and not Karl Marx) is generally credited with the earliest development of the "labor theory of value."

7. For a modern example of a "Ricardian" approach to international trade, see R. Dornbusch, S. Fischer, and P.A. Samuelson, "Comparative Advantage, Trade, and Payments in a Ricardian Model with a Continuum of Goods," American Economic Review 67 (December 1977): 823-39.

8. The factor proportions theory of international trade is frequently identified as the "Heckscher-Ohlin" or the "Heckscher-Ohlin-Samuelson" theory, after the three persons credited with the most significant contributions to the development of the theory. (The persons are Eli Heckscher, Bertil Ohlin, and Paul Samuelson.) Somewhat ironically, the work considered to be the keystone of the theory, Bertil Ohlin's Interregional and International Trade, first published in English by the Harvard University Press, Cambridge, Mass., in 1933, and reissued in a revised form in 1967, takes a much more eclectic view of the origins of

comparative advantage than does modern interpretation of the theory.

9. Under formal treatment of the theory, it does not matter what the two factors are called. Each simply is treated abstractly as an input variable to a production function. Paul Samuelson commonly identifies the two factors as "land" and "labor."

10. For an account of the consequences of the Depression era trade "wars," see Charles P. Kindleberger, The World in Depression 1929-1939 (Berkeley and Los Angeles: University of California Press, 1970). Kindleberger believes that trade "wars" were a principal causation of the length and depth of the 1930s depression.

11. See, for example, Harry G. Johnson, "Economic Expansion and International Trade," The Manchester School of Economic and Social Studies 23 (May 1955): 95-112; Ronald Findlay and Herbert Grubert, "Factor Intensities, Technological Progress, and the Terms of Trade," Oxford Economic Papers, 11 (February 1959): 111-21; and R.W. Jones, "The Role of Technology in the Theory of International Trade," The Technology Factor in International Trade, ed. R. Vernon, National Bureau of Economic Research (Cambridge, Mass., 1970).

12. Leontief published his findings in "The American Capital Position Reexamined," Economia Internationale (February 1954): 9-38. See also Leontief, "Factor Proportions and the Structure of American Trade," Review of Economics and Statistics 38 (November 1956): 386-407; R.E. Baldwin, "Determinants of the Community Structure of U.S. Trade," American Economic Review 61 (March 1971): 116-146; and R.E. Baldwin, "Determinants of Trade and Foreign Investment: Further Evidence" Review of Economics and Statistics 61 (February 1979): 40-48.

13. It should be noted that Leontief's test applied only to manufactured goods and not to natural resources or to agricultural commodities. Comparative advantage in these latter categories clearly would be dependent upon geographic distribution of the resources and of arable land.

14. That "human capital" might be a determinant of U.S. comparative advantage in international trade is a topic well explored. See, for example, Peter B. Kenen, "Nature, Capital and Trade," Journal of Political Economy 73 (October 1965): 437-60; Kenen and R. Lawrence, eds., The Open Economy: Essays in International Trade and Finance

(New York: Columbia University Press, 1968), especially the essays by Kenen, Donald Keesing, Helen Waehrer and Merle J. Yahr contained therein; Mordechai Kreinin, "Comparative Labor Effectiveness and the Leontief Scarce Factor Paradox," American Economic Review 55 (March 1965): 131-40. The "human capital" hypothesis is reviewed thoroughly by Jagdish Bhagwati in the Addendum to Tariffs, Trade and Growth (Cambridge, Mass.: MIT Press, 1969). Bhagwati also looked into human skills as a factor in the trade of India in a study in collaboration with R. Bharadwaj, "Human Capital and the Pattern of Foreign Trade: The Indian Case," Indian Economic Journal 2 (October 1967): 117-42; see also J. Harkness and S. Kyle, "Factors Affecting U. S. Compartive Advantage," Journal of International Economics 5 (May 1975): 153-65.

15. One reason for the rejection is that a factor intensity reversal would be impossible in a world characterized by linear homogenous production functions. Linear homogenous production functions are loved by economic theoreticians, who would find much of their efforts to be in vain were the validity of use of these functions ever to be shown wanting. Empirical evidence against factor intensity reversals has been published by Hal B. Lary in Imports of Manufacturers from the Less Developed Countries, National Bureau of Economic Research (Cambridge, Mass., 1969).

16. This would be easy if the technologically advanced nation were to forever possess an unchanging, superior technology. The theory would not accomodate easily the more likely possibility that the less advanced nations constantly were able, with a lag, to learn the technology of the more advanced nation, and that the latter constantly renewed its advantage via the creation of new technologies. For one effort to do this, see Ronald Findlay, "Factor Proportions and Comparative Advantage in the Long run," Journal of Political Economy 78 (January-February 1970): 27-34. See also John S. Chipman, "Induced Technical Change and Patterns of International Trade," ed. Vernon, The Technology Factor in International Trade.

17. Ohlin, Interregional and International Trade.

18. Irving B. Kravis, "Availability and Other Influences on the Commodity Composition of Trade," Journal of Political Economy 64 (April 1956): 143-55.

19. Erik Hoffmeyer, *Dollar Shortage* (Amsterdam: North Holland Publishing Co., 1958); see also Sir Donald MacDougall, *The World Dollar Problem* (New York: St. Martin's Press, 1957).

20. Albert O. Hirschman, "Invitation to Theorizing About the Dollar Glut," *Review of Economics and Statistics* 42 (February 1960): 100-2.

21. Charles P. Kindleberger, "The Cause and Cure of Disequilibrium in the Balance of Payments of the United States" (Paper presented at a Conference on Trade Policy organized by the Committee for a National Trade Policy, January 27-28, 1960); reprinted in Kindleberger, *Europe and the Dollar* (Cambridge, Mass.: MIT Press, 1966).

22. Michael Posner, "International Trade and Technical Change," *Oxford Economic Papers* 13 (October 1961): 323-41.

23. Christopher Freeman, "The Plastics Industry: A Comparative Study of Research and Innovation," *National Institute Economic Review*, no. 26 (November 1963): 22-62. Gary Hufbauer, *Synthetic Materials and the Theory of International Trade* (Cambridge, Mass.: Harvard University Press, 1966).

24. Gordon K. Douglas, "Product Variation and International Trade in Motion Pictures" (Ph.D. diss. in economics, Massachusetts Institute of Technology, 1963).

25. Seev Hirsch, "The United States Electronics Industry in World Trade," *National Institute Economic Review*, no. 34 (November 1965): 92-97.

26. William Gruber, Dileep Mehta, and Raymond Vernon, "The R&D Factor in International Investment of United States Industries," *Journal of Political Economy* 75 (February 1967): 20-37.

27. William Gruber and Raymond Vernon, "The Technology Factor in a World Trade Matrix," ed. Vernon, *The Technology Factor in International Trade*.

28. Donald B. Keesing, "Labor Skills and Comparative Advantage," *Papers and Proceedings of the American Economic Review*, 56 (May 1966): 249-58; and Keesing, "The Impact of Research and Development on United States Trade," *Journal of Political Economy* 75 (February 1967): 38-48.

29. Bela Balassa, " 'Revealed' Comparative Advantage Revisited: An Analysis of Relative Export Shares of the Industrial Countries, 1953-1971," The Manchester School of Economics and Social Studies (December 1977).

30. The United States, for example, is a major exporter of wheat, indicating that the nation holds a comparative advantage in wheat production. Wheat, however, is not really a new product, new strains of the grain not withstanding. U.S. export performance in agriculture is a function of the favorable factor endowments of the nation (specifically in this case of arable land); of favorable climates; and also of the technological input in the agricultural sector.

31. The "product life cycle" model was first developed by Raymond Vernon in the article "International Investment and International Trade in the Product Cycle," Quarterly Journal of Economics 80 (May 1966): 190-207. Further developments by Vernon appeared in Sovereignty at Bay: The Multinational Spread of U.S. Enterprises (New York: Basic Books, 1971) and in "The Location of Economic Activity," in Economic Analysis and the Multinational Enterprise, ed. John Dunning (London: George Allen and Unwin, 1974). The earlier works of Kravis include "'Availability' and Other Influences on the Commodity Composition of Trade," and "Wages and Foreign Trade," Review of Economics and Statistics 38 (February 1956): 14-30.

32. See Louis T. Wells, Jr., "Test of a Product Cycle Model of International Trade: U.S. Exports of Consumer Durables," Quarterly Journal of Economics 83 (February 1969): 152-62; Robert B. Stobaugh, "The Neotechnology Account of International Trade: The Case of Petrochemicals," Journal of International Business Studies, Fall 1971; John E. Tilton, International Diffusion of Technology: The Case of Semiconductors, The Brookings Institution (Washington, D.C., 1971).

33. Michael P. Claudon, International Trade and Technology: Models of Dynamic Comparative Advantage (University Press of America, 1977).

34. See, for example, Charles Kennedy, "Induced Bias in Innovation and the Theory of Distribution," The Economic Journal 74 (September 1964): 541-47. See also bibliography in Chipman, "Induced Technical Change and Patterns of International Trade."

35. William H. Davidson, "Patterns of Factor-Saving Innovation in the Industrialized World," *European Economic Review* 8 (October 1976): 207-18.

36. This argument is well developed by Harry G. Johnson in "The Efficiency and Welfare Implications of the International Corporation," *The International Corporation*, ed. Charles P. Kindleberger (Cambridge, Mass.: MIT Press, 1970).

37. Louis T. Wells, Jr., "Product Innovation and the Directions of International Trade" (Ph.D. diss. in business administration, Harvard University, 1966).

38. Thus, substitution of capital for labor might be accompanied by growth of scale of the individual productive unit. For exploration of the relationships between capital-labor substitution and scale of output, see J.M. Utterback and W.J. Abernathy, "A Dynamic Model of Product and Process Innovation," *Omega*, vol. 3, no. 6 (1975).

39. The postulate is developed in detail in Chapter 3 of Vernon, *Sovereignty at Bay*.

40. Burton H. Klein, *Dynamic Economics* (Cambridge, Mass.: Harvard University Press, 1977).

41. Staffan B. Linder, *An Essay on Trade and Transformation* (Stockholm: Almqvist and Wiksell, 1961). Similar ideas are presented by Jacques Dreze in "Quelques Reflexions sereines sur l'Adaptation de l'Industrie Belge au Marche Commun Europeen," *Competes Rendus des Travaus de la Societe Royale d'Economie Politique de Belgique* (December 1960). The Linder hypothesis is developed and tested in Gary C. Hufbauer, "The Impact of National Characteristics and Technology on the Commodity Composition of Trade in Manufactured Goods," ed. Vernon, *The Technology Factor in International Trade*.

42. In fact, because of the much greater size and level of *per capita* income which prevailed in the United States at the time of his writing, Linder did see the United States as playing a unique role in trade of high technology goods. Linder did, however, expect the income and technological differences between the United States and Western Europe to diminish with the passage of time.

43. H.G. Grubel, "Intra-industry Specialization and the Pattern of Trade," *Canadian Journal of Economics and Political Science* 33 (August 1967): 374-88.

44. Hufbauer, "The Impact of National Characteristics of Technology on the Commodity Composition of Trade." Hufbauer's interpretation of the Gruber and Vernon findings is, however, somewhat at variance with the authors' own interpretation, who hold that their statistical results are not consistent with the Linder hypothesis.

45. H. Linnemann, *An Econometric Study of International Trade Flows* (Amsterdam: North Holland Publishing, Co., 1966).

46. Gary C. Hufbauer, "The Impact of National Characteristics of Technology on the Commodity Composition of Trade."

47. John H. Dunning, for example, presented evidence that the "technology gap" between the United States and Europe was closing quickly in his "European and U.S. Trade Patterns, U.S. Foreign Investment, and the Technological Gap," mimeographed, Western Economic Association (1969).

48. For evidence, see Bela Balassa, "'Revealed' Comparative Advantage Revisited."

49. Edward M. Graham, "Technological Innovation, the Technology Gap, and U.S. Welfare," *Public Policy* 27 (Spring 1979): 185-202.

50. See J.B. Donges and J. Riedel, "The Expansion of Manufactured Exports in Developing Countries: An Empirical Assessment of Supply and Demand Issues," *Weltwirtschaftlichesarchiv* 113 (1977): 58-87.

51. Raymond Vernon and William H. Davidson, "The International Spread of U.S. Based Technology-Intensive Firms, 1945-1975," Harvard University Graduate School of Business Administration Working Paper No. 40 (Cambridge, Mass., 1978).

52. The implications of this are explained in R.E. Lipsey, I.B. Kravis, and R.A. Roldan, "Do Multinational Firms Adapt Factor Proportions to Relative Factor Aides," mimeographed, National Bureau of Economic Research Working Paper No. 293 (Washington D.C., 1978).

53. For a mathematical exploration of this proposition, see Rachel McCulloch and Janet Yellen, "Technology Transfer and the National Interest," Harvard Institute of Economic Research Discussion Paper #526 (December 1976).

54. See E. Mansfield, A. Romeo, and S. Wagner, "Foreign Trade and U.S. Research and Development," *Review of Economics and Statistics* 61 (February 1979): 49-57.

55. For an account of the experiences in England in trying to contain technology, see Charles P. Kindleberger, "An American Climacteric?," *Challenge* 16 (January-February 1974).

56. See, for example, National Academy of Engineering, *Technology, Trade and the U.S. Economy*, National Research Council (Washington, D.C., 1978) for a collection of eclectic views of possible reasons for a "slowdown" in innovation.

57. Jacob Schmookler, *Invention and Economic Growth*, (Cambridge, Mass.: Harvard University Press, 1966).

58. The limitations of the present programs are detailed in Chapter 4, National Academy of Engineering, *Technology, Trade, and the U.S. Economy*.

5 Environmental, Health, and Safety Regulation and Technological Innovation

Nicholas A. Ashford
George R. Heaton, Jr.
W. Curtiss Priest

A FRAMEWORK FOR VIEWING THE EFFECT OF REGULATION ON TECHNOLOGICAL CHANGE

This chapter offers an overview of the important and complex relationship between environmental, health, and safety regulation, and technological innovation. In the authors' view, this relationship is often misperceived and generally oversimplified. Our intent is to critically review the existing evidence on this topic and to provide an analytical framework within which discussions of the issue can occur and from which new policy initiatives may proceed. It should be emphasized that our conclusions concerning environmental, health, and safety regulation may not be generalizable to other regulatory areas.

Technological innovation is particularly relevant to environmental, health, and safety regulation in three different ways:

1. Technological change in the past has been a source of environmental, health, and safety problems that created the demand for the existing regulatory efforts. (For example, the rapid postwar growth of the petrochemical industry changed both the amount and the toxic nature of chemicals produced.) [1] Similarly, in the future, new technology may require additional or new regulatory responses.

2. Technological innovation is an important determinant of economic growth and welfare. Regulation may affect the future rate, direction, and nature of that innovation.

3. Technological innovation can be an important pathway to solving environmental, health, and safety problems, i.e., innovation is one way to save ourselves from ourselves. Innovative solutions can be stimulated by regulation or by other demands for solution of these problems. The limited ability of regulators to deal with problems on a case-by-case basis means that industrial innovation must be redirected to anticipate and prevent environmental, health, and safety problems.

Thus, technological innovation is of concern both to those whose primary interest is industrial production and to those principally involved in improving the environment or worker and consumer health and safety. The recently enacted Toxic Substances Control Act is the clearest legislative expression of the relationship between these two societal goals. Section 2(b)(3) states:

> It is the policy of the United States that... authority over chemical substances and mixtures should be exercised in such a manner as not to impede unduly or create unnecessary economic barriers to <u>technological innovation</u> while fulfilling the primary purpose of this Act to assure that such innovation and commerce in such chemical substances and mixtures do not present an <u>unreasonable risk</u> of injury to health or the environment. (emphasis added) [2]

Regulation is an emotional issue and discussions of its effects on technological change are often clouded by fierce debates about whether regulation is good or bad. Care must be taken to separate the assessment of the <u>effects</u> of regulation <u>on technological innovation</u> from the <u>valuation</u> to society of regulation, i.e., a comparison of its costs and benefits. Indeed, many of the costs and benefits of regulation have little or nothing to do with technological innovation. In addition, distinctions must be made between short- and long-term effects, and between adjustment effects and permanent changes.

The chapter examines research concerning the relationship between regulation and innovation, drawing heavily from three of the most studied industries -- the chemical, pharmaceutical, and automobile industries. These three industries have been the targets of much regulatory action, are important sources of technological innovation and significant economic activity, and probably contain examples of most of the effects of environmental, health, and safety regulation on innovation that might be found in other industries. Since more has been written about adverse than positive effects of regulation on innovation, this chapter devotes significant attention to elucidating arguments on the other side.

Although there is an extensive commentary on the effects of regulation on innovation, only a few studies have actually attempted to measure these effects, or even to model the relationship between regulation and innovation. This conclusion was reached by one research group in 1975, and it is still true today. [3] In addition to the literature on the industries mentioned above, there are a great many articles in the popular press on regulation's effect on innovation, as well as much work concerning the costs and benefits of regulation in general. Most of this literature is limited by its lack of empirical support and excessive generality. It provides evidence which is, for the most part, suggestive rather than definitive.

The chapter is organized into four sections. The remainder of this introductory section establishes a conceptual framework for viewing the effects of regulation on technological change and distinguishes this issue from attempts to evaluate regulation in an overall sense. Using the framework developed, the second section considers the effects of regulation on both "main business innovation" and on development of compliance technology. The third section considers various policy instruments that affect technological innovation, and the final section provides a brief summary of the chapter and discusses policy implications for specific actors.

The Characteristics of Regulation

The kinds of environmental, health, and safety regulations most relevant to the industrial process include controls on: air quality, water quality, solid and hazardous waste, pesticides, food additives, pharmaceuticals, toxic substances, workplace health and safety, and consumer product safety. Regulation is a complex stimulus which redirects the normal (or unregulated) course of industrial production. Regulation may control different aspects of development or production, rely heavily on informal government action, change over time, and be "technology-forcing" to different degrees. Thus, the effects on technological innovation may be different for regulations which:

- require product safety to be demonstrated <u>prior</u> to marketing (pesticides, food additives, pharmaceuticals, and new chemicals);

- require the efficacy of products to be demonstrated <u>prior</u> to marketing (pharmaceuticals);

- require safety to be proved or require the control of the use of products after products have been marketed (existing chemicals under the Toxic Substances Control Act, worker protection, and consumer products);
- require the control of production technology to reduce workplace safety and health risks;
- require effluent, emission, or waste control (air, water, and hazardous waste regulation); or
- require the safe transportation of hazardous material.

The perception of the need to change course in industrial product development and production may precede the actual promulgation of a regulation. Most regulations are promulgated after a rather extended scrutiny of a potential environmental, health, or safety problem by government, citizens and workers. A chemical industry innovation study done at the Massachusetts Institute of Technology's Center for Policy Alternatives (CPA) found these informal stimuli are often more important than formal rulemaking, since the anticipation of regulation is usually the beginning of the stimulus for change. [4] For example, the formal regulation of polychlorinated biphenyls (PCBs) came years after the government's initial concern about these substances. Aware of this concern, both the original producer and other chemical companies had begun to search for substitutes prior to actual regulation. This informal preregulation period may allow an industry sufficient time to develop compliance technologies or product substitutes, while giving leeway for adjustments to be made to ensure continued innovation for commercial purposes.

The initial show of concern by the government is often an uncertain stimulus to technological innovation. Uncertainty as to the ultimate regulatory requirement may be caused by both technical uncertainties and by the extent to which the regulated industry, labor, or consumers can provide further relevant information to the government and apply pressure for accommodation through both formal and informal means. Regulatory uncertainty is often necessary and beneficial. It is a necessary consequence of administrative flexibility, which allows regulations to be improved. Although too great a regulatory uncertainty may result in inaction on the part of industry until the outcome is definite, too much certainty about the final standard is not likely to result in the development of technology which exceeds the minimum requirements. George Eads makes the point that although regulatory outcomes may be uncertain, business is often the instigator, via litigation, of this uncertainty. [5] He sees major opportunities for the regulatory process to be influenced to the advantage of the firm.

Regulations differ greatly in the extent to which they "force" the development of new technology for compliance. For example, regulation may be based on environmental, health, and safety considerations alone; existing technological capability for compliance; or technology within reach of a vigorous research and development effort. Throughout the early history of environmental, health, and safety regulations, existing technological capabilities dominated the level at which standards were set. [6] This fact may reflect the limits of legislative authority at that time and the substantial industry input which influenced the drafting of standards. More recent initiatives, however, may require responses which go beyond current technological capabilities. [7]

The Characteristics of Technology

The characteristics of both the management and the technology of a firm are determinants of a firm's response to regulation. Management characteristics, though certainly important, tend to vary greatly from firm to firm and, therefore, are difficult to study in a systematic way. By contrast, the characteristics of the technology associated with a particular hazard are easier to generalize. Focusing on the technology of the regulated industry can aid in designing regulations and explaining the response to regulation.

In any given industrial segment, innovation tends to proceed along rather predictable lines. [8] These patterns are closely tied to the characteristics of the technology and its state of development. Similarly, the reaction to regulation will be strongly influenced by the technology and its evolution. The CPA chemical industry innovation study has shown that technological responses to regulation within a given industrial segment are highly uniform and that they often resemble the kinds of innovations occurring before regulation. [9] William Abernathy's work in the automobile industry, however, indicates a dramatic change in the nature of innovation after regulation. [10]

The Framework

It is useful for analytical purposes to separate the impacts of regulation on innovation into those which affect innovation for ordinary or "main business" purposes and those which affect compliance. In the first case, regulation affects a traditional, ongoing activity of the firm; in the second, regulation demands technological changes not previously within the ordinary scope of firm activity.

Regulation may be viewed as a perturbing force on "business as usual." [11] Fig. 5.1 presents a schematic of the effects of this perturbation on technological change. Fig. 5.1.A presents a simplified version of normal business operation in the absence of regulation. Resources (including technical and managerial manpower, equipment, and candidate technologies in early stages of development) are used in the process of innovation (including activities such as basic research, applied research, development, and pilot plant operation) to achieve the first commercial application of technology. Subsequently, the new technology is used to produce consumer or industrial products.

Fig. 5.1.B shows that innovation ordinarily is motivated by a combination of the recognition of market needs -- "market pull", and the technological capabilities of the firm -- "technology push." Figs. 5.1.C and 5.1.D show that regulation acts in a complex way to alter the motivations and opportunities for innovation and for production. In Fig. 5.1.C, arrows are drawn from the regulatory stimulus to the stage of the development or production process where the stimulus is perceived. The dashed lines in Fig. 5.1.D illustrate how, once the stimulus is perceived, it may lead to changes in production, in the use of resources, in the motivation for innovation, and in the innovation process itself.

Evaluating Regulatory Consequences

This chapter does not investigate the kinds of effects (other than those on technological innovation) regulation can have, such as changes in the levels of production, employment, prices, and wages, or effects on the environment, health, and safety. Nonetheless, it is important to understand the relationship between these other consequences and the effects on innovation.

Fig. 5.2 shows how the effects of regulation on innovation, on the economy, and on environmental, health, and safety problems are related. Changes in main business innovation lead to changes in economic growth, in the standard of living, and in some aspects of social utility (symbolized by ΔX). Changes in compliance-related technology lead to changes in environmental quality, health, and safety, in the "quality of life", and in other aspects of social utility (represented by ΔY). Other changes that do not necessarily affect technological innovation are represented by ΔZ. Purchasing low sulfur fuels or giving workers in noisy factories rest periods are examples. This chapter is concerned only with the effects shown in the first boxes in the first and second lines of the figure. Further, it only identifies and describes these effects in terms such as a change in the numbers of products brought to market or in the nature of compliance technology.

Deciding whether regulation is "good" or "bad" requires more than a description of effects; it requires a valuation of those effects, i.e., an assignment of values to the changes represented by ΔX and ΔY in Fig. 5.2. It is in this area that regulation precipitates the most emotional discussion, as illustrated in Fig. 5.3.

A complete assessment would include all the changes represented by ΔX, ΔY, and ΔZ. A common criterion for evaluating a regulation is to determine whether its benefits exceed its costs (or the marginal benefits exceed the marginal costs). However, traditional cost-benefit analysis has serious limitations in assessing environmental regulation. [12]

Comparing the elements of costs and benefits in their natural units, rather than all in dollars, avoids some of the pitfalls of inappropriate valuation. This demands performing trade-off analyses -- for which no unique decision rule exists -- to determine the "correctness" or appropriateness of a regulation. [13]

This chapter does examine the possible trade-off between compliance innovation and main business innovation. However, it is not always necessary to make such a trade-off. Indeed, when business profits from compliance innovation, the valuation of the trade-off becomes particularly difficult, since new compliance technology may become a main business activity. This might occur, for example, when the development of a substitute for a banned product allows the innovating firms to capture a bigger market share.

Valuation of the overall effects of regulation, or of its impacts on technology, is meaningful only against the comparative bases of alternative regulations, of changes that would have occurred in the absence of regulation, or of the intent of the legislation giving rise to the regulation. In addition, care must be taken to distinguish between short- and long-term effects, and between changes during the transition to a new regulatory regime and permanent changes at equilibrium. The long-term effects on innovation are difficult to discern and the existing evidence mostly relates to costs during initial adjustment periods.

THE EFFECTS OF REGULATION ON TECHNOLOGICAL CHANGE

Conclusions about the effects of regulation on technological change are likely to differ depending on whether the focus is on main business innovation or compliance responses. Possible effects in each of these two contexts are explored in the following discussion. The work referenced includes diverse

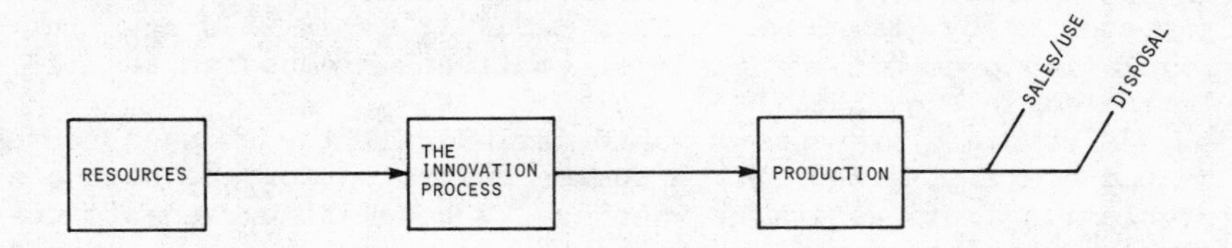

FIG. 5.1.a

A flow diagram for industrial production

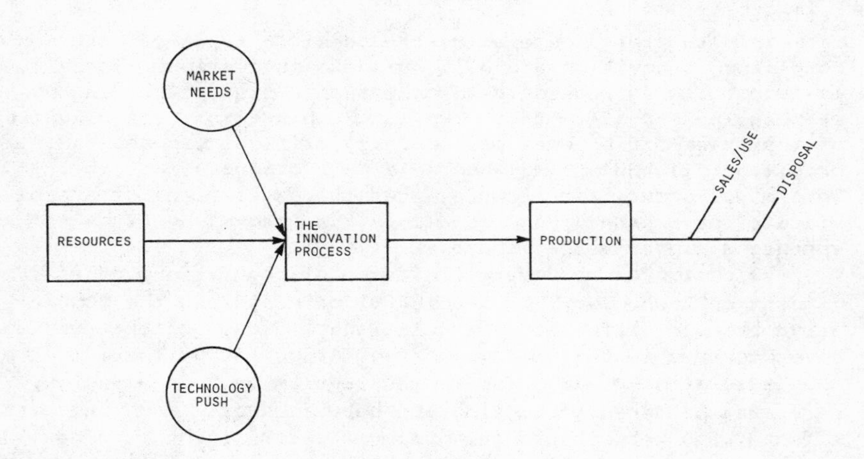

FIG. 5.1.b

The recognition of market needs and technology push as motivations for commercialization

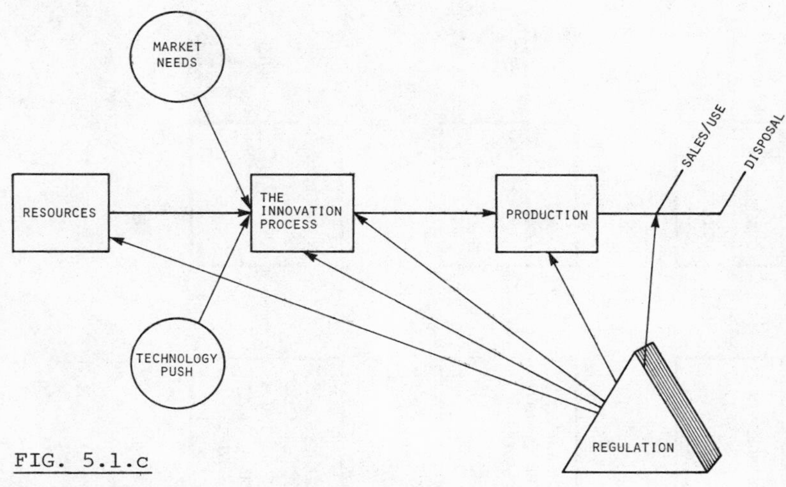

FIG. 5.1.c

Places in product development where regulation is perceived

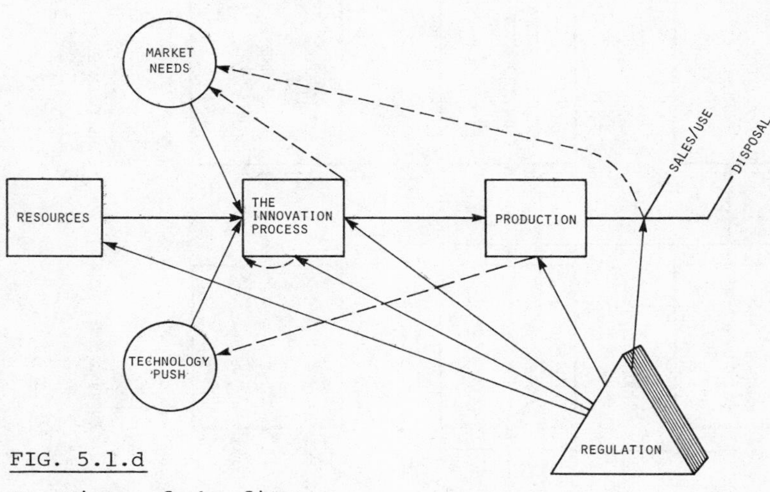

FIG. 5.1.d

Reactions of the firm to the perceived regulatory stimulus

FIGURE 5.1

EFFECTS OF REGULATION ON TECHNOLOGICAL INNOVATION

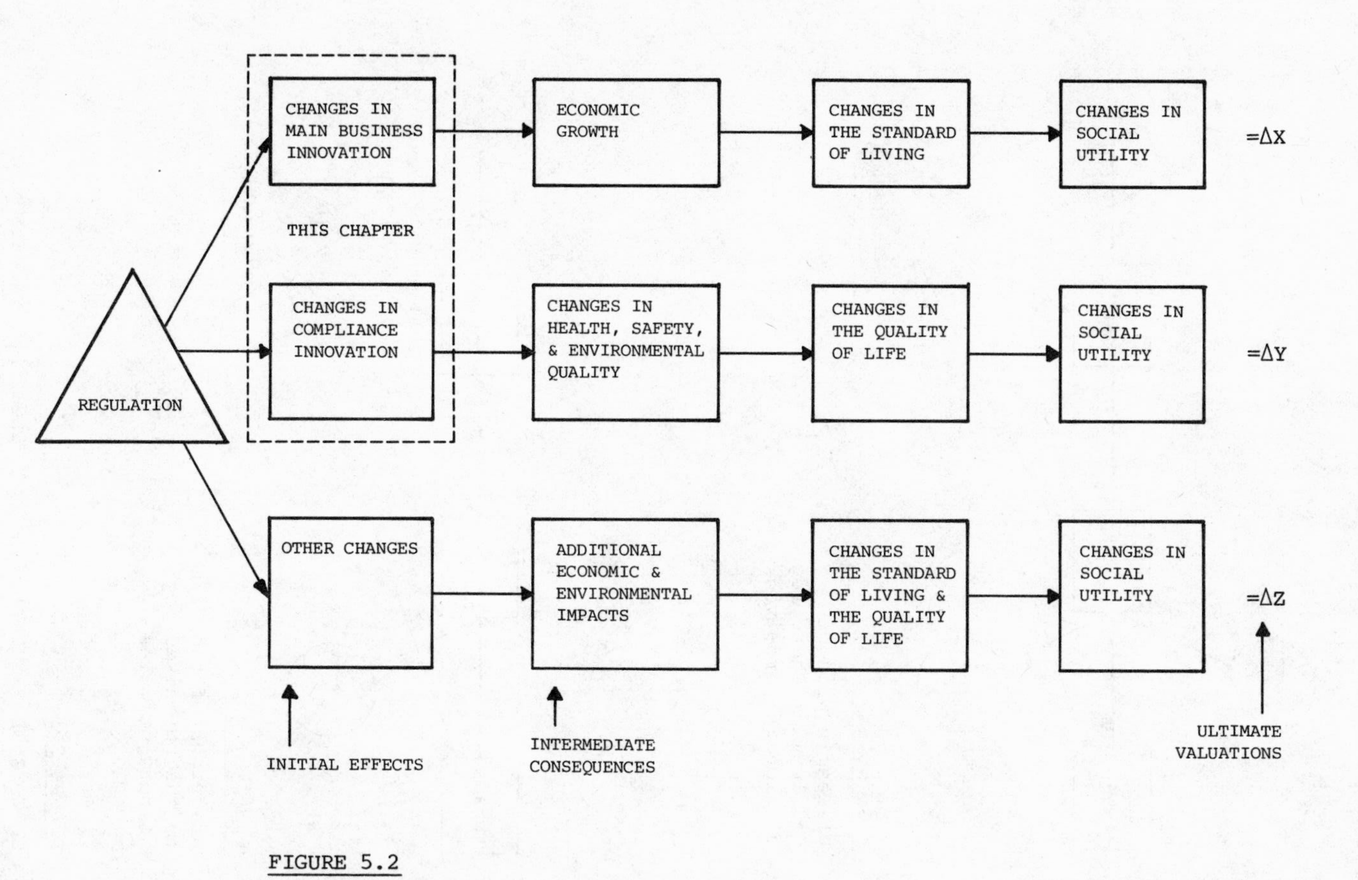

FIGURE 5.2

THE EFFECTS OF REGULATION ON INDUSTRIAL SOCIETY

approaches which document: changes in the number of new products developed during a period of changing regulation, changes in the level and pattern of investment in R&D during a period of changing regulation, the growth of the pollution control industry, and compliance technology development in specific firms.

Courtesy of Washington Star Syndicate.

FIGURE 5.3

While the effects of regulation on technology are theoretically measurable, the problem of causality remains a crucial, unresolved issue. For example, observed falls in main business innovation during a period of regulation do not prove a causal relationship. Indeed, both regulation and research plateaus have been thought to contribute to the observed decline in the number of new drugs produced by the pharmaceutical industry. [14] Similarly, the observation that R&D expenditures are directed toward solving environmental, health, or safety problems is not conclusive evidence that R&D is less than it should be for business purposes, especially after a history of high R&D budgets. [15] The time lag in the innovation process between stimulus and response may make regulatory effects difficult to measure. For example, Edward Greenberg, Christopher Hill, and David Newburger found a six year lag in the ammonia industry and were not able to use an econometric model to isolate regulatory effects. [16] Another difficulty arises from the fact that there is often no appropriate measure of innovation -- for example, the number of new chemical entities, without adjusting for therapeutic and functional importance, may be a poor indicator of innovation in the pharmaceutical industry. [17]

Much of the work is overly general and fails to take account of sector differences. Often the research is limited to effects that are immediately traceable to compliance (i.e., transient effects) and ignores longer-term institutional changes. [18] Whatever the effects, their proper assessment demands a comparison against a realistic alternative basis. Robert Leone suggests, for example, that industry probably would have devoted resources to improving worker health and safety even if there had been no Occupational Safety and Health Act (OSHAct); this is because other social and legal pressures existed, such as large tort or products liability damage awards. [19] Most of the research surveyed has not attempted this kind of comparison.

Effects on Main Business Innovation

Regulation can affect innovation for main business purposes in several ways, as illustrated in Fig. 5.4. Each of the major pathways of impact is discussed in the following sections.

Changes in Expected Profitability

Regulation may change the profitability expected from a portfolio of R&D investments by affecting either the expected rate of return or the perceived risk. As a result, a firm may modify its level of investment in R&D. A firm's response to the market pull stimulus for innovation will be modified by actual changes in R&D costs (see Fig. 5.4). In addition, to the extent that R&D is perceived to be less profitable or to be less certain to pay off, investment may be cut back and fewer main business innovations may be produced.

Profitability may decrease as a result of regulation-induced costs, delays, or uncertainty. Perhaps the greatest costs imposed by regulation on new products have occurred in the pharmaceutical industry, where testing for both safety and efficacy are required. David Schwartzman has cited several studies which indicate an increase of 100-1000 percent in R&D costs per new chemical entity. [20] These costs are attributed to toxicological testing, premarket testing, and increased paperwork required by the Food and Drug Administration (FDA) for registration and approval. Ronald Hansen found approximately equal economic burdens in the costs of increased expenditures and the costs of tying up capital. [21] Others

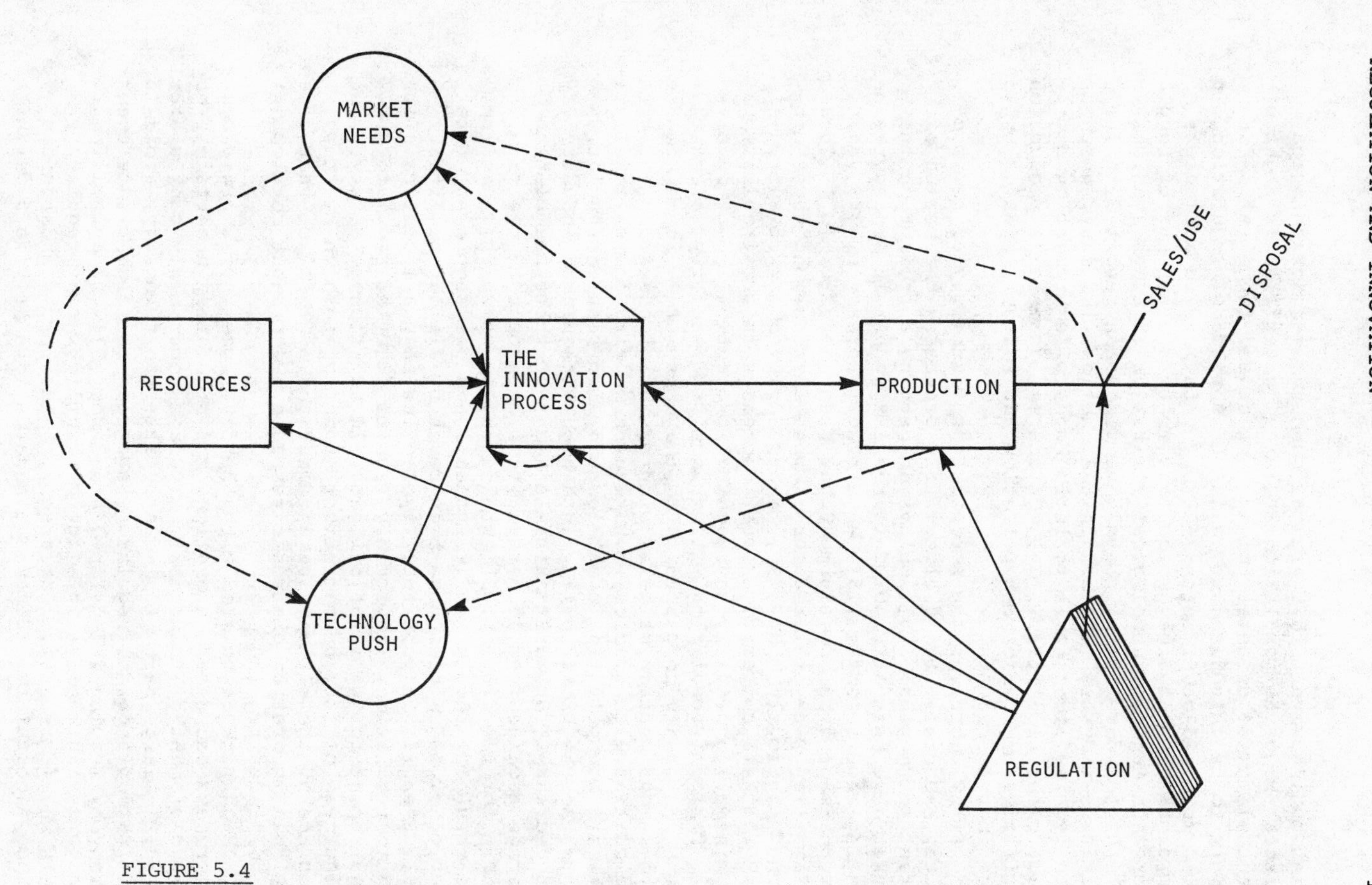

FIGURE 5.4

EFFECTS OF REGULATION ON MAIN BUSINESS INNOVATION

point out, however, that greater concern over toxicological effects among the population at large, especially in light of the thalidomide disaster, might have caused these tests for safety to be introduced, even in the absence of regulation, in order to avoid court suits. [22] The failure to articulate a proper baseline against which to compare the effects of regulation is a major methodological limitation of these studies.

Higher development costs have also been reported in the pesticide industry [23] and in the chemical industry, which has been required to install pollution control equipment on some pilot plant facilities. [24] Of course, commercial production facilities must also meet pollution and safety standards, which also increases costs.

Delay decreases the potential profitability of new products, especially if it effectively decreases the period of patent protection. [25] Delays in getting a product on the market have been attributed to regulatory permit procedures and testing requirements. [26] An average increase of 3.9 years for pharmaceutical development has been reported. [27] These delays tie up capital and thus impose costs beyond those mentioned previously. However, because delayed market entry is a risk that all competitors face, decreases in profitability may be minimal, especially where patent protection is not involved.

Costs of regulation will reduce profitability unless these costs can easily be passed on to the consumer -- as may be done with unique pharmaceuticals, with essential chemicals, or with commodities with no close substitutes or with low price elasticities. Monopolistic industries can also pass on regulatory costs, so the profitability in a highly regulated industry may not decrease. Schwartzman has noted that there has been an average decline from 15 percent to 6.3 percent in the rate of return for new drugs. [28] The extent to which this decline is the result of regulation is unclear, and one should not infer that other industries have been beset by similar declines. [29] Even if the profitability decline in pharmaceuticals is due to regulation, it probably results from the regulatory requirement for proof of efficacy, in addition to safety. The uniqueness of the pharmaceutical industry is shown by the presence of the efficacy requirement and by the fact that serendipitous innovation characterizes much new product development in the industry. [30]

High risk associated with the development of a new product or process can deter innovation, and regulatory uncertainty may be one element of this risk. Although uncertainty can be attributed, in large part, to industry attempts to modify, avoid, or eliminate standards, regulatory agencies can also create uncertainty by issuing ill-defined standards [31], or by failing to recognize and resolve interagency conflicts over regulatory goals and procedures. Conversely, regulatory standards can reduce uncertainty. Because standards both provide a definite statement of legal requirements and encourage the development of safer products, they may limit highly unpredictable products liability suits.

Agency actions, other than promulgating individual standards can change the profitability of investment in innovation in subtle ways. For example, requiring the submission to agencies of confidential, technical data may disturb trade secret protection. The reduction in trade secret protection is seen as penalizing technological innovation because it decreases the legal protection (and hence the rewards) available to new technologies. Whether such fears are real or imagined, they may chill the desire to develop new products or processes, especially if they are not patentable. On the other hand, it must be recognized that some technologies present enough risk to the public that their components must be disclosed. In such cases, whatever chilling effect occurs toward innovation may be justified by the public benefit of disclosure. Moreover, such disclosure is likely to provide an incentive to redirect innovation along competing, but safer, technical lines.

Changes in the Number of Innovations that Fail for Environmental, Health, and Safety Reasons

Otherwise successful innovations may ultimately fail if they are found to pose unacceptable environmental, health, or safety problems. Regulation can increase the number of such failures by imposing new requirements on products. On the other hand, regulation may actually reduce the number of innovations which fail in the marketplace for environmental, health, or safety reasons. Indeed, the stated purpose of some regulation is to increase the rate of failure for products which are unsound. On the other hand, regulation that requires premarket testing can eliminate such failures of fully developed products by catching problems early. Practolol, a cardiovascular drug, which has not been approved for use in the United States, was found to cause long-term toxic effects, including blindness, during its sale in Britain. [32] It was subsequently removed from the British market. Pesticides in use for some time have often been removed from the market for environmental and health reasons. [33]

Care must be taken to distinguish observations of decreased innovation during the period of transition to new regulatory demands (when existing, but never-before-scrutinized products are taken off the market), from an equilibrium or final state (when the developer scrutinizes products more thoroughly for possible problems during the development process). Overall, the change in failure rate is likely to reduce the output of harmful new products. It is not clear to what extent regulation leads to a compensating effect by bringing safer products to market. This effect may vary with the innovativeness and characteristics of the particular responding industry.

Changes in Investment Opportunities Due to Increased Environmental, Health, or Safety Risks

The chance that a new product or process might be unable to enter or remain on the market due to environmental, health, or safety problems may discourage investment in innovation. This may be particularly true for products with limited market potential, such as specialty chemicals. Schwartzman suggests that a shift is occurring in the nature of pharmaceutical innovation, with new applications for demonstrably safe technologies being preferred to open-ended searches for new concepts. [34]

While the existence of regulation may increase commercial risk and discourage investment in some cases, in other cases the lack of regulation itself can deter new investment. For example, development of new, high-risk, large-scale processes, such as shale oil production, may be hindered by the fact that environmental or workplace regulations are undefined. Here, regulations that specify acceptable emission targets are needed to reduce uncertainty.

Regulation undoubtedly changes investment opportunities; however, the ultimate effect is not generalizable across all industries. Industries which historically have been highly innovative may merely shift the type of products developed. On the other hand, noninnovative industries may find themselves competing with more innovative new entrants. Thus, government is both a creator and destroyer of business opportunities. [35]

Regulation may also change the rate at which technologies are diffused. Because regulatory standards are often based on the best practice in an industry, regulation can force the adoption of those technologies throughout the industry. This clearly results in widespread technological change, although it may be largely noninnovative adoption of existing technology.

Diversion of Managerial Personnel

Managerial resources are one of the factors in the innovation process that are affected by regulation (see Fig. 5.4). To the degree that important advances in marketing, financing, strategic planning, and corporate organization depend on actions by management, the diversion of management can have serious implications for the innovative performance of firms.

It is important to distinguish between two types of diversion: transition diversion and administrative diversion. Transition diversion occurs as the emergent regulations create new and taxing problems with which management must deal. Once management has decided on a strategy to address these problems, the transition diversion will disappear. What remains is a need to monitor the compliance efforts and regulatory developments that are likely to follow. This administrative diversion will accompany management as long as the problems to which regulation is addressed remain. Much of the literature about the diversion of management is addressed to the transition problem, rather than to the long-run effects of regulation on management.

Diversion of R&D Resources

Regulation causes some firms to redirect resources away from conventional innovative activities into compliance-related activities, as noted in Fig. 5.4. This resource reallocation will tend to reduce main business innovation.

The impact on the overall budget has been the major focus of attention in the press. [36] A number of studies cite the decline of effective R&D budgets in the chemical industry [37] and attribute this effect to regulation. Although regulatory actions may be responsible for some of this decline, it is important to recognize that changes in corporate strategy, [38] changing technological opportunities, [39] and other factors may also be responsible. Some have suggested that there was overinvestment in R&D during the 1960s and that now there is a return to more realistic levels. [40] As a result, the regulatory impact is unclear, although its presence is widely noted.

When compliance diverts resources which might be used for innovation, the diversion represents an opportunity cost. If, as some say, the long-term marginal rates of return on R&D investment are as high as 30-50 percent, [41] this highly productive use of resources is not likely to be significantly reduced by the firm. Instead, other ways to reduce spending will be found, and the opportunity cost of regulation will be more likely reflected in a cutback in outlays for expansion and acquisition. Moreover, early scrutiny of the environmental effects of new technologies can offset much of the diversionary impact on R&D. For example, Edwin Mansfield points out that in the chemical industry, 83 percent of the costs of new product development occur after the applied research stage and 57 percent after the pilot plant stage. [42] This finding would imply that rejection of new products early in their development would not be especially costly.

One study has indicated that compliance with air pollution regulations halted a $3 million research program undertaken by one company. [43] Reported instances like this are rare and do not appear to reflect a general industry-wide effect. Rather, they are limited to firms which suffer extraordinarily high compliance costs, lack the ability to pass these on to consumers, and are either undercapitalized or choose to forego the higher long-term R&D returns for short-term profits. Furthermore, they often represent retrofit responses for old plants in industries which, to begin with, were noninnovative [44].

When R&D resources are diverted as a result of regulation, it does not follow that there is a corresponding proportional decrease in total innovative output. In small firms, incremental reductions in R&D could have significant results, especially if the firms have limited access to capital. On the other hand, such incremental decreases may not lead to particularly dramatic results in large firms, which may already have surpassed the advantages of economies of scale in R&D. [45]

Innovation in large firms may even increase if organizational red tape and communication barriers decrease with personnel reallocation, although this effect is not likely to be significant throughout the industry. The productivity of R&D for innovation may be improved in other ways if regulation encourages the more efficient use of resources. In sum, the effect of resource diversion on innovation is firm-specific, and aggregate effects are not well established.

On the average, approximately 3 percent of industrial R&D is spent on pollution abatement, [46] with the chemical industry allocating about 13.5 percent. [47] Thus, new commercial successes may arise in compliance technologies. Data substantiating the extent of this phenomenon are not available.

In addition to affecting the allocation of R&D funds, regulation has affected the allocation of manpower engaged in conventional R&D. Research personnel have spent more time in testing products and processes for potential environmental, health, and safety problems [48]; presenting these data at agency conferences [49]; appearing before administrative agencies [50]; and keeping up with compliance-oriented technological developments (e.g., in the prediction of carcinogenicity). [51] In addition to these changes in the nature of research jobs, there has been related investment in new equipment and facilities associated with these new tasks. [52]

A shift has been noted in R&D activity from basic to applied research. This may be a reflection of corporate strategies now focused on new applications for existing products rather than on developing new products. Glenn Schweitzer cites the drug and pesticide industries as examples. The implications of this shift are that new products may be minor variations of existing ones and the novelty of product innovation may be less. [53] L.H. Sarett, however, suggests that for pharmaceuticals, the shift may be away from "me-too" drug research, since extensive testing must be done for both major new products and for minor variations of existing products. [54]

Ancillary Innovation from Redirected R&D

Some research suggests that the redirection of R&D may actually result in more innovation. A study of government effects on the innovation process in five foreign countries found that innovations for ordinary business purposes (not necessarily for compliance) were much more likely to be commercially successful when environmental, health, and safety regulations were present as an element in the planning process than when they were absent. [55] In addition, compliance-related technological changes often led to product improvements far beyond the scope of the compliance effort. In an example from the five-country study, a textile manufacturer developed a new dye in order to minimize worker exposure to toxic fumes. In so doing, he arrived at a dye which was also much more colorfast and, hence, a better, more saleable product.

Another study of six countries suggested that environmental regulation does not appear to have prevented or delayed innovation to a significant extent and may even have had a "substantial stimulating influence on improving quality and performance." [56] This study was conducted both in heavily regulated industries and those subject to fewer regulatory constraints. Thus, although some of the improvements cited must have occurred to meet regulatory requirements, others were more in the nature of "ancillary" changes.

Ancillary innovations often appear, in individual instances, to be unexpected or serendipitous results of regulatory compliance efforts. Nevertheless, there are enough of them to suggest that they are, overall, a predictable phenomenon. They may arise most often in industries not previously innovative and may occur because of the necessity (brought on by regulation) to rethink established and previously unquestioned modes of operation. [57]

Regulation may also change the overall nature of the R&D effort. Although it is commonly asserted that regulation prompts "defensive" R&D, the opposite may also be the case. Klein's work suggests that the unsettling stimulus of regulation may in fact promote a more "offensive" R&D strategy (see Chapter 3). This would be necessitated by the increased competition in the market for environmentally sound products and for abatement technology created by regulation. Moreover, it appears that for some industries, the resources devoted to compliance-related R&D do not deplete existing R&D expenditures, but rather, add to them. The funds to support this additional effort may come out of profits or increased prices rather than the preregulation R&D budgets. In the automobile industry, for example, the R&D to sales ratio has actually increased since 1970. [58]

Regulation-Induced R&D and Process Improvement

Regulation appears to create opportunities for firms to make a variety of process improvements unrelated to compliance, as shown in Fig. 5.4. Process improvements appear to occur more frequently as greater technological change is required to achieve compliance. Two examples from the CPA chemical industry innovation study provide good illustrations of the pattern. [59] In one instance, the petroleum refinery industry developed improved catalysts, and, consequently, a more efficient system, as a result of the R&D which went into the effort to comply with lead-in-gasoline regulations. Similarly, the need to limit employee exposure to vinyl chloride monomer led to the creation of a tighter production system and to some increase in output.

Similar phenomena have been uncovered in other studies. Joe Iverstine reported that 33 percent of his study's respondents cited process improvements resulting from regulatory changes; these included the development of closed systems and better process instrumentation. [60] The Denver Research Institute similarly found that regulation provides an opportunity to make process improvements in areas not related to regulation. [61]

The explanation for this phenomenon is quite simple. Because it is less expensive and disruptive to make multiple changes simultaneously, rather than individually, businessmen naturally take the opportunity of regulation to introduce other improvements. Such improvements are often complementary to the regulatory purpose (e.g., safer closed systems with greater yields), and they may often be suggested by the R&D which was necessitated by regulation.

Although these improvements might have occurred eventually in the absence of regulation, regulation can be viewed as accelerating normal business innovation. For example, the improvement in polymerization technology already in development was accelerated by the vinyl chloride regulation of the Occupational Safety and Health Administration (OSHA). This accelerating effect is an increase in innovation, just as a delay in product development is a decrease in innovation.

It should be emphasized that, in the short term, the process improvements occasioned by regulation are not likely to offset the costs of regulatory compliance. Indeed, the interviewees in the cited CPA study almost uniformly qualified the admission that their process improved with the reminder that the regulation still caused a net cost increase. However, the benefits from process improvements will continue to accrue to the firm long after the adjustment compliance costs are past. In the long run, benefits might very well exceed the costs.

Rechanneling Creativity

The innovative potential of a firm is in large part a function of the creative energies and abilities of its personnel. While one effect of regulation is to divert personnel attention from the normal business of the company to regulatory problems, there also appears to be an opposite effect: the creative potential of the firm is rechannelled, augmented, or enhanced.

Because compliance with environmental, health, and safety regulations involves a large component of technical expertise, many of the people brought into firms in order to assist in the compliance effort have been highly trained professionals. Typically, they are environmental scientists or engineers. Often they possess advanced degrees. When this new source of expertise enters the normal R&D process, innovative products and processes are likely to result. One type of effect was cited by companies during the CPA chemical industry innovation study. [62] Some of the companies interviewed mentioned the new need for sophisticated analytical chemistry expertise in order to assess the health and environmental risks of both new and existing products. They felt that the sophisticated analyses gave the companies a better knowledge of the properties of their products and suggested new uses for them. In addition, they believed that this new analytical capability would be important in the future in developing new products and processes. This same phenomenon was noted by 33 percent of the interviewees in Iverstine's study. [63]

An explanation for increased creativity under regulatory conditions was offered by Thomas Allen and colleagues, who argued that regulation, by adding new dimensions to older problems, "increases the problem space of the engineer." [64] This new need to optimize along several dimensions is likely to foster more creative solutions than those which prevailed under less complex conditions. This effect is especially likely to occur in older, more rigid, industries where few external stimuli have demanded creative responses.

Change in Industry Structure

Regulation may have different impacts upon firms within an industry and may change the composition of that industry. The mix of size of firms or the competitive environment may change. The structural changes brought about by regulation will in turn affect innovation. Research suggests that regulation influences the quality and quantity of innovation insofar as it changes barriers to entry of new firms into an industry, the balance of firm size, and the extent of monopoly power. Distinguishing between effects of regulation and of other influences (e.g., changing technology, or inflation) is a difficult methodological problem, and no consensus has emerged as to the relative importance of these factors.

The relationship between industry structure and innovation has been explored with mixed results by a number of researchers. [65] Some researchers have argued that large firms are more innovative because: the costs of innovation are so great that they can only be borne by large firms [66];

innovation is so risky that only large firms can maintain a balanced research portfolio to assure financial stability [67]; economies of scale in terms of instrumentation and knowledge may exist [68]; large firms are able to utilize product and process innovations faster and more extensively [69]; and large firms have a lower cost of capital and can more easily finance innovation. Others have argued that small firms are more innovative because managers in large firms are more risk-averse due to job security [70]; the large-firm bias against risky innovation may drive out entrepreneurial talent [71]; large firms have extensive overhead costs; creative researchers in large firms may find themselves in managerial positions and no longer engaged in innovation; and implementation of innovations in a large firm is resisted because of potential displacement of employees.

Some empirical studies indicate that medium-sized firms are the optimal size for innovation. [72] However, these analyses indicate that there are large variations among industries and that other qualitative factors influencing innovation, and not taken into account, may limit this conclusion. Of notable exception to the optimal firm size result are the pharmaceutical [73] and chemical industries, in which larger firms have been more innovative, and the steel industry, in which smaller firms have been more innovative. [74] These observations were made using data from 1940 to 1960 and may not hold under present-day inflation, technological and market opportunities, and increases in multinational corporate activity.

Some studies show that, irrespective of the rate of innovation, small firms tend to introduce more radical developments. [75] Pharmaceutical firms are an exception to this conclusion. [76]

Data on industry concentration and monopoly power indicate that modest barriers to entry and some monopoly power in the form of structural concentration is conducive to innovation. [77] This balance exposes established firms to the threat of new entrants, while allowing enough profit potential for the established firms to develop and exploit innovation.

Barriers to the entry of new firms into an industry are introduced when compliance measures are expensive and subject to economies of scale. [78] A recent controversial study of regulation in the automobile industry supports this conclusion. [79] Conversely, new entrants may be able to meet regulatory requirements more effectively since they do not have to engage in costly retrofit activities. Regulation can create market opportunities that attract new entrants, especially those with a new technology. For example, there are now several competing substitutes for PCBs; whereas, before the regulatory ban, there was only one manufacturer. [80] In general, regulation is likely to make it somewhat harder for firms and industries to compete and survive in a more dynamic market. The inference to be drawn from Klein's argument in Chapter 3 is that this new competitive pressure should cause innovation to increase.

Commentators in popular literature have also been concerned that the additional costs of regulation may hurt the U.S. position in international trade. This may occur when the regulatory standards of other countries are not as strict as those in the United States; however, no systematic evidence shows this.

There is general agreement that small firms are harder hit by regulation, [81] although the evidence is primarily qualitative. On the other hand, regulatory agencies tend to concentrate their enforcement efforts on the larger firms. Small firms are limited in their ability to influence policy, and regulations are not typically designed with their special problems in mind. [82] However, most of the regulatory agencies now have programs directed at the needs of small firms. The effects of resource diversion mentioned earlier may fall more heavily on the small firm due to its limited resources.

In sum, innovation may decrease as a result of increased entry barriers and decreased competition caused by regulation. The effect will be reduced if regulations contain provisions for variances and financial assistance for small firms. There may be a compensating effect, however, since regulation provides new market opportunities for new entrants, especially those with new technologies. In addition, existing firms may try harder to retain their market share through innovative competition. These long-term effects may have the most important influence on innovation. [83] Quantification of the net effects is speculative due to the presence of additional influences, the diversity of industry structures, and the lack of suitable aggregate measures of innovation that capture both quantity and quality differences.

Effects on Technological Change for Compliance

Regulation clearly encourages technological change for compliance purposes -- i.e., all changes which occur in order to comply with specific regulatory requirements, as well as other technological changes developed for the purpose of abating environmental, health, and safety hazards, even though there is no government requirement to which they are specifically directed. Indeed, this statement is almost a tautology since regulation is intended to ameliorate the adverse consequences of technology by changing the technology itself. However, such changes will not necessarily be innovative. Regulatory mandates often elicit the adoption of technologies that are fully developed but on the shelf. Moreover, in cases where the regulatory standard is set at the level of the best practice in a few leading firms, the major effect which occurs is diffusion of an existing technology to the lagging firms -- a noninnovative, though important, response. The discussion in this section is concerned with both innovative and noninnovative changes, as long as they are for compliance purposes.

Many of the technological changes necessitated by regulation might not be expected to have important direct benefits to the firm undertaking them, unless that firm is in the pollution control industry. Although many businessmen appear to recognize that technological changes for compliance can have some indirect economic benefits to the firm, these are rarely perceived as outweighing the compliance costs for the firm.

There are, of course, better and worse ways of complying, and it is often in the best interests of both firms and society to encourage the development of innovative compliance technology. To the firm, innovative compliance technology is likely to reduce the cost of meeting regulatory goals. To society, the adoption of better, safer technologies resulting from innovation is an important addition to the benefits of regulation. Moreover, to the extent that the overall impact of regulation is to demand a long-term and widespread alteration in the nature of industrial technology, innovation is crucial.

Fig. 5.4 illustrates schematically the relationship between regulation and technological change for compliance purposes.

Redirection of Technological Capabilities for Environmental, Health, and Safety Purposes Only

The discussion here is concerned with the characteristics of compliance technology that results primarily in environmental, health, and safety benefits, with little or no benefit accruing to the firm. Its focus is on the technological changes occurring in the products or processes of regulated industries, as opposed to the pollution control industry. It describes some of the characteristics of compliance technology and addresses the issue of how innovative efforts to develop this technology can result in substantial compliance cost savings to the regulated firm.

Unfortunately, few studies attempt to portray in any systematic way the characteristics of technological changes related to compliance, and much of the evidence, to date, is anecdotal. Several studies indicate that there is indeed widespread compliance [84] and others estimate costs [85]; however, few studies attempt to assess the degree of technological change resulting from compliance efforts.

Several studies of the automobile industry have pointed out the different kinds of effects produced by different regulations. Fuel economy regulations, for example, appear to have encouraged very creative responses, which, in many cases, have surpassed the regulatory requirements. This may have occurred because the regulations complemented normal competitive pressures. [86] On the other hand, the emissions standards are sometimes seen as having the effect of even more firmly entrenching a long-established technology, the internal combustion engine. The impetus which regulation creates for transfers of technology from one industry to another is illustrated by the use of microprocessors in automobiles as a means to control fuel economy, emissions, and other performance dimensions. [87] The crucial role of firms outside the industry, especially suppliers, in developing innovative compliance responses is shown in the automobile industry, particularly for safety devices such as the "air bag" restraint system. [88]

The CPA chemical industry study looked at the compliance technologies of 50 firms (approximately 120 separate responses). [89] Most compliance technologies were found to be in a late stage of development and required only moderate additional development to meet regulatory demands when the regulatory signal was acted upon by the firm. Similarly, the great majority of responses were based on well-established technologies. Responses involving modifications of an industrial process as opposed to a product tended to be more comprehensive in scope; there were, however, significant exceptions.

There is anecdotal evidence that innovative and cost-saving compliance technologies are arising. For example, in 1977, the 3M Company reported on 19 separate innovations which saved about $11 million in averted pollution control expenditures. [90]

Saleable Compliance Technologies

Apparently the great majority of compliance technologies are developed in the firms subject to regulatory requirements. [91] However, there is also a large market for the sale of goods and services to meet compliance requirements. Some of these sales are made by regulated firms seeking to market the technologies developed to solve their own in-house control problems. Although it is not clear from the existing research whether recognition of sales potential most often predates the development of compliance technology, or is actually an after-the-fact appreciation of market potential, the attempt to sell compliance technologies appears to be fairly common. Most often, the sales are to other firms in the same industry. One study showed that a developer of a less-polluting process for the production of chlorine tried to market it to other firms in the chloralkali industry. Iverstine's study documented that a large percentage of firms were able to sell the pollution control technology they had developed. [92] On the other hand, the study also indicated that the uniqueness of each firm's environmental, health, and safety problems often makes such sales difficult.

The pollution control industry illustrates dramatically how new technologies are encouraged by a new market demand (see Fig. 5.4). In 1974, air and water pollution control was estimated to be a $1.4 billion industry. Between 1974 and 1983, the expenditures will total about $44 billion. [93] Often, these compliance technologies appear to be addressed toward meeting more than one regulatory requirement; for example, air pollution control and energy conservation are sometimes seen as complementary. [94] The 1977 report of the Environmental Industries Council contains many examples of innovative new technologies. [95] Some critics of the pollution control industry portray it as a noninnovative expansion of the old municipal waste treatment industry. [96] For the wastewater treatment industry, one study reports that the combination of regulatory requirements and federal subsidy has actually discouraged innovation. [97]

The need to create new compliance technologies, and the dynamic relationships between the regulated and pollution control industries, have restructured the innovative effort in many industries. For example, the lead-in-gas regulations appear to have encouraged diversification among lead additive manufacturers, including the development of some highly innovative new automotive technologies. [98] Similarly, the suppliers of automobile parts, perhaps more than the automobile manufacturers, have been a major responder to regulatory demands, thereby changing the balance of innovative activity within the industry. [99]

Compliance Technologies with Ancillary Benefits

Technological change to comply with regulation can provide ancillary benefits to the complying firm. These benefits are more likely when compliance responses are innovative and/or comprehensive in scope. [100]

Typically, ancillary benefits result from the ability of the firm to transfer the technologies developed for compliance purposes to other uses. For example, the use of microprocessors in automobiles to regulate fuel consumption and emissions has opened the door to other applications of this technology for improving performance.

From the point of view of the firm, the ancillary benefits of regulation may result from serendipitous, unpredictable events. However, the documentation of such effects in several studies leads to the conclusion that this phenomenon is, in the aggregate, to be expected. [101] Moreover, it appears that the ancillary benefits deriving from compliance efforts can be consciously sought and achieved by firms. For instance, Bell Laboratories announced in May 1978 a new computer-controlled electroplating system that both reduces pollution and conserves raw materials. [102] The system uses an enclosed processing method which has reduced gaseous exhaust 97 percent, chemical waste 90 percent, and gold consumption by 50 percent. Because it increases production dramatically, the pay-back period is about six months.

Similarly, the Denver Research Institute study reported that there is a strong link between management's commitment to compliance and the nature of the impact from regulation. [103] Such commitments tended to result in both lower compliance costs and favorable (or at least not negative) impacts on innovation.

Joint R&D Efforts for Compliance

In many instances, compliance with regulations yields no benefits directly appropriable by the complying firm. In other situations, the need for better compliance technologies is clear, but, because of the high cost or the absence of a perceived reward to the developer, the new technologies are not being produced. Some argue that in such situations, the most appropriate response of the firm is to undertake joint R&D with other interested firms.

Joint R&D is usually undertaken in an industry or trade association, or by private arrangement among individual companies. Industry associations are doing more R&D on environmental issues. The Chemical Institute of Toxicology was recently established to perform toxicological tests on chemical products. Similarly, the lead and petroleum industries have recently sponsored a joint research effort to assess the effects of the gasoline additive MMT on automobile functioning. One study indicates that this is the major kind of cooperative endeavor occurring in response to regulation. [104] The study shows that because compliance changes are often unique to individual firms, most of the shared research occurs in preparing environmental impact analyses and in toxicity testing. These efforts may affect the introduction of new technologies because they reduce the costs associated with regulation. However, the history of trade association R&D in other countries, where it is more common than in the United States, indicates that it is not a major source for the development of new products or processes. [105]

When joint R&D is performed among a small number of companies, it has the greater potential to yield new products of important commercial value; it also has drawbacks. These are illustrated by the context of the automobile industry, whose members were charged with conspiring to retard the development of technology to meet pollution control requirements when they undertook joint R&D efforts on this issue. As a result of this case, the threat of antitrust enforcement may loom in the minds of firms contemplating combined R&D efforts for pollution control. Although combinations of this kind in most other industries are likely to be viewed quite differently by antitrust authorities, especially if they result in important new compliance technologies, there do not appear to be many such combinations.

Firms within an industry often share the results of their compliance research -- especially with respect to difficult regulatory problems -- even when they do not jointly undertake it. One study found such sharing in 53 percent of its sample. [106] Although this phenomenon is not likely to have a major impact on the development of new compliance technologies, it may have an important impact on the diffusion of appropriate solutions.

Reorganization of Firms to Meet Compliance Requirements

It has been widely reported that regulation has fostered organizational change in companies. The CPA chemical industry innovation study found that about 65 percent of the chemical firms interviewed had formal environmental affirs groups [107], and the Conference Board recently reported that 78 percent of a sample of chemical and pharmaceutical firms have government relations units. [108]. These groups often serve primarily as a liaison between the regulators and their company. They participate regularly in the regulatory process, often indicating to the regulatory agencies the technical limits of existing compliance capability. This interaction is seen by some as a way of tempering potentially strict technology-forcing regulatory standards by considerations of "feasibility," and also as a way by which the holders of certain compliance technologies can "capture" the regulatory standard for their particular compliance method. [109] The environmental affairs unit often functions within the firm in a manner very similar to a regulatory agency. Environmental review procedures are often established, with the environmental affairs unit able to "pass" on the acceptability of various products or processes, particularly in their early stages of development. Thus, these groups will encourage the development of safer technologies.

Environmental affairs units are more common in large corporations. They are typically located in the central corporate headquarters, rather than in production facilities. They may be staffed with young environmental scientists rather than engineers. As such, it appears they often do not play a major role in the development of new compliance technology or in the engineering aspects of compliance. These functions are more typically within the realm of the engineers at the plant level, or R&D personnel. [110]

Information Sources for Compliance Technology

One barrier to effective regulatory compliance is a lack of knowledge about the best technical solutions, especially in smaller firms. Government agencies such as OSHA have programs to assist firms in developing appropriate compliance responses. The OSHAct mandates a program of assistance for small business.

The efficacy of information programs sponsored by the government is open to some question. One study reported that less than one percent of all solutions to environmental problems originated with the government. [111] Another study reported a widespread perception that EPA is deterred from establishing closer working relationships with industry for fear of legal or political reaction by environmental groups. [112]

In addition to government action, a firm's personnel assigned to the regulatory compliance function plays an important informational role. [113] In the regulatory area, these individuals provide liaison between the firm and outside technical knowledge, which can be a force for innovation in both compliance and noncompliance areas.

Multiple Effects

The discussion in the previous sections has perhaps been overly simple in the sense that it has treated the various effects of regulation on technological change as separable categories. In reality, many of the effects reviewed are likely to occur in combinarion and/or simultaneously.

Regulation -- or any market perturbation -- will result in both economic "winners" and "losers"; therefore, regulation should not be viewed as a single uniform stimulus with a uniform impact. Perhaps the most important distinction to make is between the impacts on the "regulated" industry and those on the "pollution control" industry. Nevertheless, in some industries, the distinction is not preserved -- for example, chemical companies often market the techniques they have developed for their own abatement purposes.

Regulation may alter both the "market pull" and "technology push" stimuli to innovation. In some instances, regulation increases technology push by encouraging new development based on existing, environmentally proved, products and processes rather than on radically different technologies that face uncertain prospects of regulatory approval. In other cases, regulation creates a new market demand for environmentally superior goods and services, a demand which has obviously not been filled by existing technologies.

Regulation may also simultaneously encourage and discourage new entrants to an industry. For example, in the case of PCB manufacture, the sole producer of the product chose to leave the business rather than market a substitute. Several substitutes developed and were marketed by new entrants, including one Japanese company. [114] Contrastingly, in the case of the vinyl chloride industry, where regulation was directed toward the manufacturing process rather than end products, the technological response arose entirely from within the industry.

Regulation will have positive and negative impacts simultaneously. For example, some research suggests that the negative impact on main business innovation is greatest where compliance costs are highest. However, the cost of compliance may, in such cases, be an important motivator for innovation in compliance technology. Where this is true, a trade-off may exist between main business innovation and compliance innovation. However, the fact that ancillary innovations often result from compliance efforts (especially in mature, noninnovative industries) indicates that compliance can actually benefit main business innovation.

In sum, the complexity of the regulatory stimulus, and its multiple and interactive impacts, must be recognized. Although regulation may result in negative impacts on innovation, there are also usually compensating positive effects elsewhere. There are almost always winners and losers as a result of regulation, and indeed contradictory effects may occur simultaneously within the same firm or industry.

DESIGNING REGULATORY POLICY TO ENCOURAGE INNOVATION

The primary goal of environmental, health, and safety regulation is to protect the public from hazards. Such regulation is not mainly intended to change the process of technological innovation. Nevertheless, technological changes are required by regulations, and changing the overall nature of products and industrial processes may be the only realistic long-term solution to avoiding future hazards. Thus, regulation should work toward two policy goals concerned with innovation:

- Regulation should be designed so as not to unduly hamper innovation for ordinary business purposes, as long as this is consistent with the environmental, health, and safety goals established by legislation.

- Regulations should be designed to encourage innovation in compliance. This innovation should both alleviate immediate hazards and contribute to longer-term systemic changes which can result in safer products and processes in the future.

It is hoped that regulations can be designed to further both of these goals, but it should be recognized that the two may be incompatible to some degree. For example, while a stringent, "technology-forcing" regulation is likely to cause innovation in compliance, it may also present a significant burden to main-business innovation. Thus, difficult trade-offs may have to be made between the two goals. Because the role of technology has largely been ignored in designing regulation, the achievement of these goals will require a new orientation of existing policies, as well as new perspectives on the part of policy makers.

This section explores how specific policy instruments may affect innovation. It offers for consideration alternative strategies that can be employed to facilitate innovation while still meeting primary regulatory goals. Both strategies within the existing authority, and supplements or alternatives to current approaches, are discussed.

Strategies Within Existing Authority

Policies currently used to improve the environment, health, or safety include: product labeling, education, tax incentives, R&D, and mandatory controls. Mandatory controls in the form of standards have been predominant. This section examines ways that existing regulation might be redesigned to achieve desirable effects on technological innovation within the existing legal authorities of the various agencies.

Regulatory Analysis Focused on Technology

Although, to date, the regulatory initiatives in the environmental, health, and safety area have been accompanied by voluminous analysis (e.g., environmental impact statements and economic impact statements), much of this analysis is not focused on the capabilities and limitations of technology in the regulated industries. Recently, this analytical orientation has begun to change. For example, the Federal Water Pollution Control Act focuses on best available technology as the determinant of the standard of protection, and recent OSHA standards, such as that for lead, have been greatly concerned with technological feasibility.

One major conclusion of the CPA study of the chemical industry [115] is that regulators can better predict regulatory outcomes and design more effective regulations by focusing their analysis on the technology of the industries subject to regulation. We do not mean to argue for a "technology impact statement"; rather, we are suggesting that a new perspective be introduced more explicitly into regulatory design. It should be noted, however, that in order to be successful, the analytical task will need input from the regulated industries, who hold most of the necessary information, as well as from the regulators.

The idea that technology-focused analysis can aid in predicting outcomes of alternative strategies rests on recognition of the fact that innovation in a given industrial segment tends to proceed along lines that are rather predictable. [116] These patterns are closely related to the characteristics of the technology and to its state of maturity. For example, regulations can be expected to elicit product substitutes from firms whose previous rate of product innovation has been high, but process modifications from firms whose innovations have been concentrated on standardization and mass production. Moreover, these compliance changes will be uniform across firms in a given subsector that employ a particular technology. [117]

Technology-Forcing Aspects of Legislative Standards

A few studies have considered how the relationship between regulation and innovation depends on the basic legislative approach to the design of standards. [118] D. Bruce LaPierre categorizes legislative mandates into three groups: "health-based standards" (e.g., the Clean Air Act protects public health with an adequate margin of safety), "technology-based standards" (e.g., the Federal Water Pollution Control Act's "best available technology"), and hybrid standards (e.g., the Occupational Safety and Health Act protects health to the greatest extent "feasible"). [119] His hypothesis is that the greatest amount of compliance-related innovation is likely to occur in response to strict, health-based standards. (He also finds that many health-based standards have actually been limited by court interpretation and agency implementation to requiring little more than the technological status quo.) On the other hand, OSHA health regulations have been recognized for their technology-forcing potential through court interpretations that allow OSHA to set standards requiring the development of compliance technologies not yet in use. [120]

A standard based on existing technology can provide a constant stimulus to compliance innovation if it is continually revised. Under Section 306 of the Water Pollution Control Act, EPA is to set new source performance standards on the basis of the "best demonstrated" technology. Thus, in theory, EPA could upgrade its standards each time a demonstrably better compliance technology arises. [121] In such cases, the possibility of a revised standard creates a market incentive to innovate, and implementation of the standard results in rapid diffusion. Competition may be enhanced between firms that might gain a competitive advantage by developing cheaper or more effective compliance technology.

In many cases, the regulatory agencies do not have the discretion to choose among technology-based, hybrid, or health-based standards, because Congress has done so by law. In other cases, however, the agency may have such discretion and can significantly influence a regulation's effect on innovation. For example, Section 307 of the Water Pollution Control Act gives EPA leeway to use either health-based criteria for standard setting or a "best available technology" criterion. [122] This is a unique legislative design. In cases where the agency discretion is limited, the agency should consider a technology focus when recommending changes in existing legislation.

Performance vs. Specification Standards

It is often asserted that performance standards spur compliance innovation more than specification standards. Indeed, this is often the case. One of the difficulties in generalizing this claim is that what appear to be performance standards may in fact be established with the knowledge that only one existing technology can comply. Such a standard is in effect a specification standard. Comparisons between the two kinds of standards are also confounded by the different stringencies of the standards. A strict specification standard may prompt more innovation or diffusion than a lax performance standard.

Innovation can result from a specification standard approach when the industry is confident that regulators will base those standards on the capability of a newly developed technology and will revise them as the state of the art improves. Diffusion occurs because the specified technology is not generally in use throughout industry. However, specification standards often make compliance innovation difficult because of the reluctance of agencies to grant variances. The CPA chemical industry innovation study showed that companies with new compliance technologies superior to that specified in the regulation experienced difficulty when they tried to persuade the regulatory agencies to allow them to use the new alternative. The popular literature is replete with similar examples.

However, the choice between performance and specification standards cannot be made only on the basis of the innovation consequences. A specification standard may be the only way to ensure protection due to the greater difficulty of enforcing performance standards, especially in the safety area. Thus, choosing between the two regulatory modes on criteria related to increasing innovation may not be realistic. Moreover, performance standards may create greater uncertainty about whether mandated compliance can be achieved. Therefore, many firms actually prefer the more definite specification standards.

Timing, Time-Phasing, and Delaying of Standards

Several approaches are used to establish the time horizon of standards. "Timing" refers to different compliance time requirements for different industries. "Time-phasing" refers to a standard whose requirements become increasingly strict over time. The time horizon of regulatory requirements is often of crucial importance to the degree of compliance innovation which results.

After a standard goes into effect, time-phasing can occur on a firm-by-firm basis via the exemption mechanisms available under every regulatory system. These include variances, abatement agreements, and negotiated compliance schedules. Although these mechanisms have typically been used in the past to mitigate economic difficulties associated with compliance, they could be adapted to encourage firms to develop innovative compliance technology.

Industry often complains that the compliance periods specified in regulations are too short to allow for development of the most effective or innovative response. [123] The validity of this complaint is open to question since most regulations are preceded by several years of government scrutiny, hearings, and public controversy, which provide a signal of the standard to come. Nevertheless, compliance innovations, especially those involving major changes in technology, can take five to ten years to develop. In such cases, extension of the time period between standard promulgation and full compliance could serve as a means of encouraging innovation.

A crucial trade-off problem is raised for regulators in permitting time-phasing and time extensions to comply with regulations. While delay may encourage innovation, it postpones compliance. Thus, unsafe conditions are allowed to exist in the name of fostering innovation. Although delay may produce a cheaper compliance technology that will prevent more harm over the long term, regulators must recognize the current environmental, health, and safety costs it may cause. Also, changing regulatory signals during the implementation of a particular regulation may retard the development of innovative compliance technology by creating a climate of uncertainty and by penalizing those who comply early with the original form of the standard. Lastly, extremely long compliance periods may not present enough stimulus to prompt firms to develop new technology.

Industry-Specific Standards

Regulations do not always take into account the different compliance capabilities of different industrial sectors. When a standard is set, it can be too lax for industries capable of a higher level of compliance, and too harsh for industries where compliance costs threaten financial viability.

Industry-specific standards are a possible solution to this problem. Such standards would not relax a minimum level of protection, but would set different times and means of compliance for each industrial sector. Industry-specific standards may provide a significant benefit to innovation. In cases where standards are made stricter for a specific industrial segment, compliance innovations may result; where they are temporarily more lax, the relaxation may give hard-pressed industries time to develop innovative solutions. In at least one regulatory area, OSHA, the courts have suggested that the agency can consider varying its standards according to the differing compliance capabilities of the target industries. [124]

However, in industries where temporarily lax standards prevail, workers and consumers may perceive unequal treatment as inequitable treatment.

Generic Standards

Regulatory agencies are becoming increasingly aware of the difficulty of promulgating standards on a substance-by-substance basis. This approach is often criticized by environmentalists and others who see it as a ponderously slow method of improving environmental quality, health, or safety; and it is criticized by some industrialists who feel that it contributes to uncertainty, disruptive changes, and excessive paperwork.

An alternative to the substance-by-substance approach is to promulgate a "generic" standard for a class of substances. The proposed OSHA standard for carcinogens is a prototype. The intent of this standard is to identify suspect and proven carcinogens, classify them according to their risk, and prescribe generic kinds of controls for each class. The benefits of this approach, from the perspective of innovation, may be substantial. The generic approach reduces regulatory uncertainty by providing a clear, long-term signal to firms about the kinds of substances likely to be regulated and the nature of the controls required. This may allow for the development of appropriate, long-term technological options in the private sector.

Special Regulatory Consideration for Technologies that Benefit the Environment, Health, and Safety

Regulatory systems for pesticides, food additives, toxic substances, and pharmaceuticals require government approval before new products can be marketed. These products are presumed unsafe until the contrary is proven. (In other regulatory areas such as consumer products, the presumption is just the opposite.)

The effect of such regulatory systems on products that are environmentally sound or that benefit health is a subject of increasing concern. [125] Commentators assert that regulators examine products designed to improve the environment or health with a scrutiny equal, or perhaps greater, to that applicable to products without benefits to the environment or health. This has the perverse effect of penalizing innovation in just the areas where regulatory-based demands are made for new, better products. One example, widely cited as an instance where the government has delayed innovation, is the area of "biological" insect control, which may be environmentally superior to chemical controls.

One policy option for solving this problem is an expedited approval procedure for new products which make an initial showing to the regulators that they are demonstrably superior, on environmental, health, or safety grounds, to existing products. A system of this kind has recently been adopted by the FDA for new drug applications of important health potential, [126], and similar methods could be tried in other agencies. [127] Another mechanism is special exemptions or variances from at least some regulatory requirements for demonstrably superior or innovative technologies. The OSHAct, for instance, has provisions that allow for both variances based on a showing of technologies "as safe and healthful" as the methods presented in a standard, and for "experimental variances" from standards. These have not been widely utilized.

Joint Compliance Activities

There are a variety of policy tools to foster cooperation in regulatory compliance among firms. One is the sharing of testing costs for new products, which has been enacted into the Toxic Substances Control Act (Section 5(h)(2)(b)). Another mechanism sometimes suggested is compulsory licensing of compliance technology to all members of the industry. Joint compliance-related R&D efforts by companies may encourage compliance-related innovation, while decreasing the penalties to ordinary business innovation.

Although they are used abroad, these mechanisms have not been widely used in the United States. In part, this is because of antitrust concerns. One commentator, a former deputy general counsel of EPA, feels that antitrust restrictions need not be an insurmountable barrier. [128] However, attempts to implement this policy option would have to extend beyond agencies concerned with environmental issues to those concerned with antitrust issues.

Existing Tax Incentives for Pollution Control

Two major provisions of the Internal Revenue Code attempt to reduce the cost of complying with pollution control regulations. These grant deductions and credits for investments in pollution control equipment, and allow certain tax exemptions for municipal bond financing of pollution abatement facilities.

Section 169 of the tax code allows more accelerated depreciation for pollution control hardware than for other purchases. Although the deduction applies to both pollution abatement and prevention, it is qualified by the limitation that preventive investments cannot lead to significant changes in the facility. ("Significant" for this purpose means more than a five percent change.) In limiting the prevention deduction to nonsignificant process improvements, Congress was clearly compromising its desire to reward pollution prevention with a reluctance to grant tax windfalls to anyone who modernized to his own benefit in the name of pollution prevention. The investment tax credit may also be claimed for pollution control investments -- but only on one half their value. Because of this limitation on pollution control investments, these supposed tax "benefits" are often not used, since, in many cases, firms find it to their benefit to use only the full investment tax credit. Another limitation restricts the use of these tax benefits to pollution control investments and does not allow them to reduce the cost of worker or consumer protection.

Section 103 of the tax code allows companies to borrow for pollution control facilities, using municipal bond financing at rates lower than those generally prevailing on the money markets because the income to the purchasers of such obligations is tax-free. This provision has been of significant benefit to many companies, but it is limited to facilities for pollution abatement and not for prevention.

The combined effects of the tax provisions are curious. First, as has been indicated above, their impact is smaller than would be expected because they are often not used. Even if the provisions were utilized more frequently, it is not clear that they would encourage compliance. Rather, it seems more probable that they transfer costs to the public and subsidize firms more than they provide an incentive to abate pollution. [129] These tax laws also bias compliance technology. They emphasize air and water pollution control over worker and consumer protection, and they encourage hardware investments exclusively because no benefit is allowed for operating changes. The requirement that changes be nonsignificant biases compliance towards incremental or add-on, rather than major, process changes. These provisions should be critically reconsidered for the difficult trade-off they pose between penalizing innovation by decreasing the cost of incremental technology, and the large-scale subsidy they would represent if modified to encourage more radical technological improvements.

Minimizing Effects on Small Firms and New Entrants

It is commonly asserted that small and new firms are more harshly affected by regulation than larger, more established companies. In the pesticide and drug industries, for example, there is some evidence of displacement of secondary distributors and formulators. [130] However, the differences among sectors may make this a unique phenomenon. If new and small firms are important sources of innovation, and if regulation especially penalizes these firms, there is likely to be some chilling effect on innovation.

In recognition of these problems, Congress and the regulatory agencies have created a variety of programs to deal with the compliance problems of small firms, such as the small business provision in the OSHAct and various OSHA programs. [131] The effectiveness of the programs is open to question. Certainly few are oriented toward assisting innovation. The Toxic Substances Control Act has a provision to allow sharing of testing costs among companies, and this may be an important benefit to small firms' innovation efforts.

An alternative way of addressing the problem is via a currently proposed program to provide technical assistance to small- and medium-sized firms. [132] This program not only would attempt to provide assistance to ordinary business innovation, but also would provide assistance in solving the technical problems associated with regulation, thereby addressing the problem of compliance-related innovation.

Information-Based Strategies

One of the difficulties that inhibits appropriate compliance responses to regulation is a lack of technical information in firms, especially small firms, about the nature of compliance technology. In addition, one of the major difficulties for exposed workers and consumers is that they are often uninformed about the nature (or existence) of hazards to which they are exposed. Thus, they are not motivated to press for the institution of abatement measures.

The problem of the exposed populations is now being dealt with to some extent by the various labeling and disclosure requirements promulgated by the agencies. For example, OSHA is considering a labeling rule applicable to most workplace chemical substances, as well as a rule giving employees access to their medical records. Mandated information flows are the very essence of other regulatory systems, such as those for cosmetics and hazardous materials transportation. Such requirements are likely to increase the demand on firms to develop and/or adopt compliance technologies. Even with current plans for increasing the disclosure of information, many workers and consumers will still be uninformed of the hazards to which they are exposed, because, in some cases, the reforms do not go far enough and because trade secret protection is likely to place a limit on the nature and quantity of information disclosed.

The problem of informational deficiencies in firms is not well addressed by existing and proposed policies. Some of the agencies have consultation and technical information programs to provide such assistance to firms. For example, OSHA has experimented with several different consultation schemes over the past few years, and a similar system exists under the Resource Conservation and Recovery Act. Various technology transfer programs in other nonregulatory agencies provide models for compliance information dissemination in the environmental, health, and safety areas. The regulatory agencies should consider adopting elements of these kinds of programs to provide firms with the technical information that is a prerequisite to the firms' innovating.

Coordination Among Agencies

Because the jurisdictions of various regulatory agencies overlap (e.g., OSHA, EPA, and CPSC are all concerned with the problem of carcinogens), the burden of regulatory requirements on the private sector can be great if there is not close coordination among agencies and if there are not certain common procedures. Different, but duplicative, testing requirements can unnecessarily increase the cost of new product development. Various efforts are now underway to standardize the procedures of these agencies and reduce the burdens which their actions may create. Most notable among these are the Intergovernmental Regulatory Liaison Group (IRLG) and the Regulatory Council. The IRLG is attempting, among other things, to develop a common agency approach to the problem of trade secret data disclosure. Other problems this group might address include the need to develop common agency reporting and monitoring procedures, and to develop and common testing requirements, both of which are the most frequently cited complaints of industry. [133]

Alternatives and Supplements to Regulation

The Nature of Incentive Approaches to Regulation

Standards are only one of the many possible mechanisms which might be used to achieve environmental, health, and safety goals. Incentive schemes have been suggested as an attractive alternative or supplement to existing approaches. Some economists and management scientists believe that the firm is sufficiently rational and does sufficient planning that it can respond very well to alternative programs based on economic incentives. Moreover, it is assumed that firms' responses can be predicted well enough to be able to design adequate incentive programs.

Throughout the 1970s, a growing field of empirical and theoretical literature has described the forms which incentives can take. [134] These financial incentives fall into two broad categories: tax incentives and market rights. Theoretically, financial incentives may be used in lieu of, or in addition to, mandated minimum requirements for environmental, health or safety performance.

Two types of taxes have been proposed. The first is a tax on conditions rather than on results. For example, an alternative to requiring stack gas scrubbers for utility plants is to tax the plant over the time it is operated without scrubbers. In contrast, a tax can be applied in direct proportion to the quantity of pollutant sought to be reduced. For air pollution, this takes the form of an effluent tax, where, for example, the tax might be $.30 per pound of SO_2 of stack emission. For safety, this takes the form of an injury tax, where a tax by severity and injury provides an added incentive to industry to reduce accidents.

Market rights (also called, for example, transferable discharge permits) are similar to taxes, but are transferable. A system of variable taxation can be effectuated by having the government sell the right to pollute, where the aggregate cost of these rights might equal the aggregate tax applied using a tax incentive. Then, depending upon individual differences among firms, one firm may purchase more rights from another firm and pollute more, and others sell their rights and pollute less, depending upon the economics of each individual firm.

The theoretical economic allocative efficiency advantage of incentive approaches is that they lower the total cost of attaining a given level of compliance. This occurs because differences in compliance costs among firms enable some firms to provide more compliance at a lower cost and others, who can only attain compliance at higher costs, to provide less compliance. The value of the cost savings is difficult to estimate, and before the virtue of the incentives approach can be fully appreciated, further study is needed. [136]

As noted by Susan Rose-Ackerman, in a static world of perfect certainty, a tax system, a market rights program, or direct regulation, could all lead to identical compliance results. [136] It is argued that incentive systems achieve the desired level in a more cost-effective manner. However, if the tax does not capture all the costs of pollution or injury <u>and</u> firms do not recognize the hidden costs (e.g. employee absenteeism, reduced productivity of agricultural land, or increased medical costs) of pollution to themselves, their decision to pollute or to pay the tax will be less than correct, and an insufficient amount of compliance will occur. While raising the level of the tax can increase compliance in the aggregate, allocative efficiency may not be achieved if these hidden costs are large and differ significantly from firm to firm.

The tax level could be changed if it does not attain the desired level of environmental quality. However, given the capital intensity of most compliance actions, adjustments to tax changes will not be rapid and may not occur at all if firms adopt a wait-and-see attitude in the face of fluctuating taxes. To avoid this reaction, taxes would have to be carefully set at the beginning with a promise that they will remain unchanged for a fixed period. This creates a temptation to set the tax too high to ensure achieving the announced desired level of environmental quality. For monopolistic industries, these costs could be passed on to the consumer -- the burden of inefficiency would fall on the consumer and not industry. The market rights system could avoid some of these pitfalls since the agency issues no more rights than required to maintain the desired threshold level. Prices adjust to correct for the actual costs of achieving compliance. [137]

Objections to incentives as alternatives to standards are numerous. They primarily center around the unrealistic expectations regarding the ease of implementing incentives and the lack of immediate control. One comprehensive review of the shortcomings of incentives notes that it is a fiction to believe that the tax approach will avoid the political maneuvering process attendant to regulations or the "inevitability of bargaining." [138] Others note that accompanying the lack of immediate control are attendant costs in injuries, illness, and pollution. [139] It is this uncertainty in meeting legislative goals that is one of the most compelling reasons for Congress to prefer standards. In general, successful tax and market rights systems have not arisen because of lack of political support and formidable design difficulties. There is reason to believe that comprehensive design, modeling, and testing would be required to provide successful candidate substitutes for existing standards approaches. [140]

Others have noted that some proponents of tax incentives confuse tax incentives with performance standards, and are actually interested in the flexibility of response that performance standards alone would provide. [141] Further objections to tax incentives are based on arguments about equity. It is not seen as fair to subject workers to more injuries in one industry than another simply on the grounds that injuries in one industry are less costly to prevent than in another. The concept of fairness developed by John Rawls can be extended to rights to health and safety which should not be compromised by economic systems. [142]

Small firms may fare worse than larger firms under an economic incentives approach. Smaller firms have less access to information and capital and may opt to pay the tax rather than to search for a less expensive compliance strategy. Again, fairness becomes an issue.

Some of the objections to economic incentive systems as substitutes for a standards approach are met by a joint approach, i.e., by using them as supplements. Minimum standards can be set, and further compliance induced, by using economic incentives. Giandomenico Majone emphasizes that each situation to be corrected by government is unique and requires the appropriate application of techniques from all available mechanisms. [143]

The effects of standards and incentives differ over time. In response to the standards approach, firms often anticipate levels of required compliance and begin to respond during the period before final promulgation. Following enactment, deadlines require action within relatively short time periods, and firms respond with a flurry of compliance and/or evasive action. In contrast, firms are likely not to begin action in response to a tax incentive or the like until enactment. Further compliance may or may not follow quickly after enactment. If there are economic advantages to corporate planning and longer-range R&D while paying the tax, compliance may be significantly delayed. The difference in business compliance that occurs under the two approaches relates strongly to this difference in time dynamics.

Evaluating the Effect of Alternatives and Supplements on Innovation

Under certain conditions performance standards can be superior to specification standards in encouraging innovation. If alternatives to regulations encourage responses in a manner similar to performance standards, these responses may also be more innovative. However, alternatives to regulations may encourage either more or less innovation than their performance standard counterparts, depending on the conditions.

The motivation for a firm to respond to regulation comes primarily from costs associated with compliance or noncompliance. For the standards approach, there are the costs of compliance and "moral costs" associated with not meeting the standard's deadlines for compliance. These moral costs take the form of (1) actual monetary costs in the form of fines for violation, and (2) stigmatization of the firm, its officers, and its managers for the civil violations. [144] In some cases, the costs are jail sentences for responsible personnel who knowingly violate the law. In the tax or incentives approach, there is a decision to (1) comply or pay a tax, or (2) face the "moral costs" of doing nothing. The "moral costs" of not paying a tax may be less onerous than those of violating a standard. Thus, an incentives approach could encourage more evasion (i.e., no technological response) than a standards approach.

In order to compare the consequences for innovation of the the alternatives of standards and incentives, two contrasting compliance modes are envisioned: a noninnovative and hastily contrived, end-of-pipe solution, accompanied by no real saleability or process cost savings; and, in contrast, a more innovative solution in which the manufacturing process is changed sufficiently so that compliance is intrinsic to its operation. A firm's decision-making and compliance strategy are also envisioned to range between two extremes: a reactive, crisis-driven mode of action; and a carefully thought-out corporate strategy that incorporates considerable R&D and planning. Most responses are likely to be between these two extremes.

The reactive decision mode leads to end-of-pipe solutions with little innovation. The planning mode occurs over a longer period of time, and leads to a highly innovative compliance solution. Standards or alternative and supplementary approaches encourage innovation if they foster the second decision-making mode over the first.

A longer time-frame for compliance may encourage innovative solutions. Thus, standards with short time-frames will result in noninnovative solutions, while tax incentives (and standards with longer time frames) will permit longer time horizons if the value of waiting to develop more innovative responses is worth the tax burden to the firm. However, the incentives approach may not capture the preenactment innovation occurring before the promulgation of a standard, because it encourages no action at all during the preenactment period.

If an incentives approach is perceived as creating more certain signals for firms than the standards approach, more or less innovation could result. Whereas incentives tend to provide a steady, certain signal, standards may result in various degrees of uncertainty. Large uncertainty in standards can immobilize firms and create small incentive to innovate, just as small uncertainty will lead only to the minimum effort necessary to comply. A moderate amount of uncertainty may cause firms to act conservatively and undertake serious development of solutions in order to protect themselves against the future possibility of a stringent standard. Although standards with moderate uncertainty may stimulate compliance, it is not known whether compliance technology would be as innovative as the response fostered via an incentive system.

In summary, the value of alternatives to, and supplements for, regulation depends upon the plausibility of certain theoretical advantages of incentives over the standards approaches. Researchers strongly debate both sides of the issue, and until more definitive work is performed in the area, only educated speculation can be supported.

SUMMARY AND POLICY IMPLICATIONS FOR GOVERNMENT AND SPECIAL INTEREST GROUPS

This final section summarizes the major findings of our investigation of the regulation-innovation relationship and discusses the implications for major actors. Technological innovation is particularly relevant to environmental, health, and safety regulation in three different ways: (1) past technological growth has resulted in problems that created the need for regulation; (2) regulation may affect the future rate and direction of technological innovation, thereby affecting economic growth; and (3) technological innovation is an important pathway to the solution of environmental, health, and safety problems.

The characteristics of regulation and characteristics of the firm and its technology principally determine how regulation affects technological change. Regulation is a complex stimulus. It may have different purposes, control different aspects of development or production, rely on different policy instruments, and have differing legal authority to "force" the development of new technology. Informal government actions, which usually occur well in advance of formal rulemaking, also provide important signals to firms and often result in significant technological change. Uncertainty in the signals given the firm to meet environmental, health, and safety goals -- particularly about the level of, and time frame for, compliance -- may play a crucial part in the firm's response and may either stimulate or retard innovation. This is true for any regulatory stimulus, whether a standard, a tax, or a combination. The uncertainty associated with regulation results from both industry and government action and may be a necessary consequence of the administrative flexibility in the U.S. political system.

Both firm management strategies and firm technology influence the response to regulation. While management strategy tends to be unique to each firm, technological characteristics are fairly uniform within an industrial segment.

It is useful for analytical purposes to separate the impacts of regulation into those affecting: (1) innovation for ordinary or "main business" purposes, and (2) abatement/compliance responses. In the first case, regulation affects a traditional, although slowly evolving, activity; whereas, in the second case, regulation demands technological changes which would not have been previously considered within the ordinary scope of business activity.

Attempts to document the relationship between regulation and innovation present many methodological difficulties. Establishing causality is probably the paramount problem, but other analytical difficulties include: establishing an appropriate measure of innovation, distinguishing between long- and short-term effects, accounting for sector differences, and establishing an appropriate baseline against which to compare measured effects. Most of the existing research suffers from one or more of these difficulties. Thus, the findings in the literature can only be regarded as suggestive, not conclusive.

Regulation may cause changes in main business innovation by affecting profitability. Increased costs have been reported in the pharmaceutical industry, but the unusual character of both regulation and innovation in that sector may make its experience unique. The effect of cost increases on rates of return throughout regulated industry has not been demonstrated. These costs may be passed on. Increased commercial risk may occur as a result of regulation; however, regulation may also decrease risk as compared to, for example, the threat of products liability suits. The number of new products in the pesticide and pharmaceutical industries has been shown to have decreased; however, it is neither clear that the level of significant innovations has declined, nor that the decline is attributable to regulation.

Regulation may increase the number of technically successful innovations that fail because of environmental, health, or safety concerns. On the other hand, regulation may reduce the number of products that would have ultimately failed for environmental, health, or safety reasons by discouraging their development. Even if failures do increase, there will be a compensating effect from increased safety, health, or environmental quality. Moreover, any change in the failure rate is likely to be a transitional, rather than a permanent, effect.

Because regulation can increase market risk, it changes the nature of investment opportunities. Increased risk may deter investment, especially in low-volume products. New applications for demonstrably safe technologies may be preferred to investments in environmentally unproven products and processes. Regulation is also likely to direct resources away from conventional R&D activities into compliance. To the extent that R&D diversion exists, it may tend to reduce main business innovation. There is substantial evidence of a change in corporate R&D, including overall decreases in some industries and a shift from basic to applied research. Whether this results from other factors or from regulation is not clear. Moreover, marginal decreases in R&D have not been shown to lead to a corresponding decrease in innovative output.

Some research has shown that the change in R&D patterns may actually result in more overall innovation, especially in areas "ancillary" to compliance efforts. This phenomenon may occur predominantly in industries which were relatively uninnovative before regulation, but which have responded creatively to regulation. In addition, R&D induced by regulation can often lead to general process improvements. Although these benefits (e.g., greater output, small energy costs) do not usually outweigh the cost of compliance, they offset such costs to some extent. Finally, new organizational structures and skill mixes have been found in firms as a result of regulation. This prompts rechanneling of firm creativity.

Because regulation has different impacts on differently situated firms, it tends to change industry structure. Regulation creates barriers to entry when compliance measures are expensive and subject to economies of scale. On the other hand, many new entrants have solved regulatory problems that established firms were not successfully addressing. Small firms appear to be hit harder by regulation. Firms involved in international trade in areas where foreign standards are less strict than those in the United States may also suffer. On the other hand, because regulation can increase the need to compete and the difficulty of survival in the market, it may lead to more innovation over the long term.

Regulation obviously encourages technological change for compliance purposes. However, these changes will not necessarily be new or novel technologies; indeed, regulation often prompts compliance through new uses or diffusion of existing technologies. In regulated industries (in contrast to the pollution control industry), the adoption of compliance measures may result in health and safety benefits only, with little or no benefit to the firm. The existing evidence indicates that while there have been some creative responses by the regulated industries, compliance technologies have not usually been very novel and were often already developed when the regulation arose. There are significant exceptions to this generalization, however, especially in the case of recent regulations concerned more directly with chemical process technology or product safety.

Although most compliance technologies appear to be developed within the regulated firm for its own use, many compliance technologies are also saleable. The pollution control industry shows that regulation can result in "market pull" innovation. Efforts on the part of regulated firms to sell compliance technology may be seen as "technology push." In some industries, the relationship between suppliers and producers has been altered by regulation, with suppliers often developing innovative compliance technology. It should be recognized that the division of industry into regulated segments and the pollution control industry may not be a real one, especially in the chemical industry. There, the regulated firm and the creator of new compliance technology are often one and the same.

Compliance technologies can produce ancillary benefits to the firm. Compliance technologies may be transferred to non-compliance uses, and vice versa. Although these spin-off benefits may be somewhat unpredictable in individual cases, there is enough evidence to indicate that they are, in the aggregate, a predictable result of regulation.

Firms often reorganize to meet compliance requirements. Environmental affairs departments are an especially common development. The primary purpose of these groups appears to be liaison with the regulators, rather than the development of compliance technology. In addition, environmental affairs groups may function as an internal review mechanism for new technologies developed for main business purposes.

The most important effect of regulation on technological innovation may be its potential for restructuring the nature of industrial production. The current regulatory climate can be seen as a transitional period during which industry responds with compliance measures of limited scope to specific requirements made by the regulators. Over the longer term, industry may adjust to environmental, health, and safety demands with changes in the nature of production that will be more basic and can be accomplished with far less disturbance.

Many of the issues concerning the relationship between regulation and innovation take on a political character. This is an inevitable result of the fact that regulatory decisions must balance many competing concerns. In order to make the research results more relevant to those who participate in the regulatory process, an attempt has been made in the following discussion to frame the major conclusions in terms of action items for specific actors. These actors are the government (both the Congress and the regulatory agencies), industry, and consumer, labor, and environmental groups.

Government Policies

In drafting legislation and regulation, attention should be paid to both innovation in compliance and innovation in product and process technology. Similarly, regulations should work toward both alleviating immediate hazards and promoting longer-term systemic changes in industrial production.

To the extent consistent with the environmental, health, and safety goals established by legislation, regulations should be designed so as not to unduly hamper innovation for ordinary business purposes.

Regulators should consider the technological capabilities and limitations of an industry in order to design more cost-effective regulations and to better achieve the legislative intent. This analysis should be done for each industry separately.

Congress should consider giving regulatory agencies the option of setting health-based standards which "force" new technologies in appropriate circumstances or, alternatively, basing regulations on existing technology. Where mandated, agencies should more aggressively pursue forcing the development of new technology where it is necessary for achieving an environmental, health, or safety goal.

Performance standards have the potential to prompt more innovative solutions, providing they are stringent enough; however, specification standards may also force the introduction of new technologies and encourage their diffusion. Specification standards (especially for new facilities) should be reevaluated periodically against the state of the art. Specification standards can also be desirable because they are more easily enforceable. The agencies should not uncritically accept the idea that performance standards are always preferable.

The time horizon of standards is important in allowing for innovative responses. Although adequate time is necessary to encourage new technology, this gain must be traded off against increased environmental, health, or safety damage. Moreover, too much compliance time may lead only to inaction. Agencies should carefully consider the trade-offs implied by the timing of regulatory requirements.

Generic standards should be seriously considered and encouraged by Congress and the agencies as a means of providing a long-term and definite regulatory signal to firms. Generic standards may also be a more effective way to meet environmental, health, and safety goals than substance-by-substance regulations.

The agencies exercising premarket approval authority over new products should consider giving expedited approvals to new products with demonstrated environmental, health, or safety benefits.

Congress should consider revising the existing tax benefits for pollution control expenditures in order to eliminate their bias toward add-on technology and against consumer or worker protection.

Increased attention should be given to the problems of small firms and new entrants in meeting regulatory goals.

There is a need for more strategies based on the provision of regulatory information, including hazard disclosure and technical information about compliance technologies.

Industry Actions

Few, if any, regulatory requirements appear to have presented insuperable technical difficulties. Industry should, in general, refrain from using this argument in their input to regulatory decision making, because it decreases their credibility in the regulatory debate.

Compliance with regulation presents the possibility for ancillary benefits, including new uses for existing products and general process improvements. These benefits are attainable if firms are aware of their potential and consciously seek them.

Companies should utilize environmental, health, and safety professionals at the design stage of new products and production methods.

Companies can significantly influence the regulatory process for their own benefit by offering constructive suggestions to regulators, especially by offering examples of innovative compliance technology. On the other hand, other firms, the agencies, and the public need to guard against the possibility of regulatory "capture."

Regulations can be anticipated and their adverse effects mitigated by appropriate monitoring of regulatory developments and by adopting in-house procedures such as environmental affairs units to prevent environmental, health, and safety problems. Regulation can result in significant benefits to companies that anticipate regulatory problems and capitalize on the new market needs created by regulation.

Small business should, to the extent possible, reexamine its posture towards regulations in order to ensure that policies reflect its particular concerns, which are not necessarily coincident with those of larger firms.

Joint research and testing is an increasingly viable mechanism to reduce the cost of regulatory requirements. Joint development of compliance technology and sharing of that technology, are likely to be found legally permissible in most instances, but the signals from government could be clarified.

Consumer, Labor, and Environmental Groups

In their participation in the regulatory process, these interest groups should become more cognizant of the technological issues in standard setting.

Consumer, labor, and environmental groups should be more conscious of the fact that their actions -- particularly the institution of law suits, participation in the regulatory process, and collective bargaining activity -- can multiply or leverage the regulatory effort to change technology.

Labor ought to press for a reconsideration of the way technological changes brought about by regulation enter into collective bargaining. Environmental, health, and safety issues ought not to be used as "environmental blackmail" to encourage workers or consumers to reduce their legitimate concerns for these issues, or as an excuse to raise prices or eliminate jobs.

NOTES

1. Barry Commoner, The Closing Circle (New York: Bantam Books, 1971), and Commoner, "The Promise and Perils of Petrochemicals," New York Times Magazine, September 25, 1977.

2. Toxic Substances Control Act, Section 2(b)(3), 15 USC 2602, Public Law 94 469 (1976).

3. Christopher T. Hill et al., A State of the Art Review of the Effects of Regulation on Technological Innovation in the Chemical and Allied Products Industries, vols. 1, 2, 3 (Springfield, Vir.: National Technical Information Service, 1975).

4. "Environmental/Safety Regulation and Technological Change in the U.S. Chemical Industry," Center for Policy Alternatives (CPA), Massachusetts Institute of Technology (Cambridge, Mass., 1979); see also Nicholas A. Ashford and George R. Heaton, Jr., "Effects of Health and Environmental Regulation of Technological Change in the Chemical Industry: Theory and Evidence," in Federal Regulation and Chemical Innovation, ed. C.T. Hill, American Chemical Society Symposium Series no. 109, American Chemical Society (Washington, D.C., 1979).

5. George C. Eads, "Chemicals as a Regulated Industry: Implications for Research and Product Development," in Federal Regulation and Chemical Innovation, ed. C.T. Hill, American Chemical Society Symposium Series, no. 109, American Chemical Society (Washington, D.C., 1979).

6. D. Bruce LaPierre, "Technology-Forcing and Federal Environmental Protection Studies," Iowa Law Review 62 (February 1977).

7. "Environmental/Safety Regulation and Technological Change in the U.S. Chemical Industry" (CPA).

8. William J. Abernathy and James M. Utterback, "Patterns of Industrial Innovation," Technology Review 80 (June/July 1978): 41-47.

9. "Environmental/Safety Regulation and Technological Change in the U.S. Chemical Industry" (CPA).

10. William J. Abernathy, The Productivity Dilemma (Baltimore: Johns Hopkins University Press, 1978).

11. Eads, "Chemicals as a Regulated Industry"; and Glenn Schweitzer, Regulation and Innovation, The Case of Environmental Chemicals, Cornell University (Ithaca, N.Y., February 1978).

12. The reader is referred elsewhere for extensive discussion: "Evaluating Chemical Regulations: Trade-off Analysis and Impact Assessment for Environmental Decision-Making" (Prepared by the Center for Policy Alternatives, Massachusetts Institute of Technology, for the Council on Environmental Quality, Washington, D.C., forthcoming, 1979); and Baruch Fischhoff, "Cost-Benefit Analysis and the Art of Motorcycle Maintenance," Policy Sciences 8 (1977): 177-202.

13. "Evaluating Chemical Regulations" (see footnote 12).

14. Nicholas A. Ashford, Stewart E. Butler, and Eric M. Zolt; "Comment on Drug Regulation and Innovation in the Pharmaceutical Industry" (Prepared for the U.S. Department of Health, Education and Welfare. Review Panel of New Drug Regulation, Washington, D.C., February 10, 1977); and Henry G. Grabowski, Drug Regulation and Innovation: Empirical Evidence and Policy Options, American Enterprise Institute for Public Policy Research (Washington, D.C., September 1976).

15. Schweitzer, Regulation and Innovation, The Case of Environmental Chemicals.

16. Edward Greenberg, Christopher T. Hill and David J. Newburger, Regulation, Market Prices and Process Innovation -- The Case of the Ammonia Industry (Boulder, Colo.: Westview Press, 1979).

17. Ashford, Butler, and Zolt, "Comment on Drug Regulation," and Louis Lasagna, William Wardell, and Ronald Hansen, "Technological Innovation and Government Regulation of Pharmaceuticals in the United States and Great Britain" (Research proposal to the National Science Foundation, Washington, D.C., March 19, 1975).

18. Robert A. Leone, "The Real Costs of Regulation," Harvard Business Review, vol. 55, no. 6 (Nov.-Dec. 1977).

19. Ibid.

20. David Schwartzman, Innovation in the Pharmaceutical Industry (Baltimore: Johns Hopkins University Press, 1976).

21. Ronald W. Hansen, "The Pharmaceutical Development Process: Estimates of Current Development Costs and Times and the Effects of Regulatory Change," in Technological Innovation and Government Regulation of Pharmaceuticals in the U.S. and Great Britain, Louis Lasagna, William Wardell, and Ronald Hansen (Final Report, Contract No. 75 19066-00, to the National Science Foundation, Washington, D.C., August 1978).

22. Ashford, Butler, and Zolt, "Comment on Drug Regulation."

23. A.E. Wechsler et al., Incentives for Research and Development in Pest Control (Prepared by Arthur D. Little, Inc. for the National Bureau of Standards, vols. 1 and 2, National Technical Information Service, Washington, D.C., December 1976).

24. Wayne Boucher et al., Federal Incentives for Innovation, Denver Research Institute, University of Denver (Denver, Colo., January 1976).

25. Hansen, "The Pharmaceutical Development Process."

26. Weschler et al., Incentives for Research and Development in Pest Control; and Schweitzer, Regulation and Innovation, The Case of Environmental Chemicals.

27. Schwartzman, Innovation in the Pharmaceutical Industry.

28. Ibid.

29. Eads, "Chemicals as a Regulated Industry."

30. William Wardell and Jean DiRaddo, "Methodology for Measuring the Effects of Regulation on Pharmaceutical Innovation: Regulatory Disposition, and National Origin of New Chemical Entities in the U.S." in Federal Regulation and Chemical Innovation, ed. C.T. Hill, American Chemical Society Symposium Series, no. 109, American Chemical Society (Washington, D.C., 1979).

31. Boucher et al., Federal Incentives for Innovation.

32. Donald Kennedy, "A Calm Look at 'Drug Lag'," Journal of the American Medical Association, vol. 239, no. 5 (January 30, 1978).

33. Weschler et al., Incentives for Research and Development in Pest Control.

34. Schwartzman, *Innovation in the Pharmaceutical Industry*.

35. Eads, "Chemicals as a Regulated Industry."

36. John Hanley, "The Can-Do Spirit," *Newsweek*, January 8, 1979, p.7.

37. Larry J. Ricci, "Innovative R&D: Gone with the Wind?" *Chemical Engineering*, vol. 85, no. 21 (September 25, 1978): 74-78.

38. Joe C. Iverstine, Jerry L. Kinard, and William S. Slaughter, *The Impact of Environmental Protection Regulations on Research and Development in the Industrial Chemical Industry* (Grant No. PRA 76-21321, Division of Policy Research, National Science Foundation, Washington, D.C., May 1978).

39. T.M. Kelly, "The Influences of Firm Size and Market Structure on the Research Efforts of Large Multiple-Product Firms" (Ph.D. diss., Oklahoma State University, 1970).

40. Schweitzer, *Regulation and Innovation, The Case of Environmental Chemicals*.

41. Edwin Mansfield, "Federal Support for R&D Activities in the Private Sector," in *Priorities and Efficiency in Federal Research and Development*, Joint Economic Committee Compendium (Washington, D.C.: U.S. Government Printing Office, 1976), pp.97-99.

42. Edwin Mansfield et al., *Research and Innovation in the Modern Corporation* (New York: Norton Co., 1971), p.118.

43. Boucher et al., *Federal Incentives for Innovation*.

44. Schweitzer, *Regulation and Innovation, The Case of Environmental Chemicals*.

45. Jacob Schmookler, "The Size of Firm and the Growth of Knowledge," in *Patents, Invention and Economic Change*, eds. Z. Griliches and L. Hurwicz (Cambridge, Mass.: Harvard University Press, 1972).

46. P. Hansel, "Pollution Abatement Research and Development in Industry," 1977.

47. Iverstine, Kinard, and Slaughter, *Impact of Environmental Protection Regulations*.

48. Schwartzman, Innovation in the Pharmaceutical Industry.

49. Morton Corn, "The Impact of Federal Regulation on Engineers," Chemical Engineering Progress, vol. 74, no. 7 (July 1978): 24-27.

50. Iverstine, Kinard, and Slaughter, Impact of Environmental Protection Regulations.

51. Eads, "Chemicals as a Regulated Industry."

52. W. Reddig, "Industry's Preemptive Strike Against Cancer," Fortune, February 13, 1979, pp. 116-119.

53. Schweitzer, Regulation and Innovation, The Case of Environmental Chemicals.

54. L.H. Sarett, "FDA Regulations and Their Influence on Future R&D," Research Management 17 (March 1974): 18-20.

55. "National Support for Science and Technology, An Examination of Foreign Experience" (Working Paper No. 75-12 of the Center for Policy Alternatives, Massachusetts Institute of Technology, Cambridge, Mass., 1975).

56. "The Current International Economic Climate and Policies for Technical Innovation" (Prepared for the Six Countries Programme by the Science Policy Research Unit, University of Sussex, Sussex, England, November 1977, p. 20ff).

57. Thomas Allen et al., "Government Influence on the Process of Innovation in Europe and Japan," Research Policy 7 (April 1978): 41-47.

58. Robert Ricci and Bruce Rubinger, Federal Strategies for Inducing Technological Innovation in the Auto Industry, Transportation Systems Center, Department of Transportation, (Cambridge, Mass., 1978), p. 3-37.

59. "Environmental/Safety Regulation and Technological Change in the U.S. Chemical Industry" (CPA).

60. Iverstine, Kinard, and Slaughter, Impact of Environmental Protection Regulations, p. 86.

61. Boucher et al., Federal Incentives for Innovation, p. 4.4.

62. "Environmental/Safety Regulation and Technological Change in the U.S. Chemical Industry" (CPA).

63. Iverstine, Kinard, and Slaughter, Impact of Environmental Protection Regulations, p. 86.

64. Allen et al., "Government Influence on the Process of Innovation in Europe and Japan."

65. For a review, see Morton I. Kamien and Nancy Schwartz, "Market Structure and Innovation: A Survey," Journal of Economic Literature, March 1975.

66 John Kenneth Galbraith, American Capitalism, rev. ed. (Boston: Houghton Mifflin Co., 1956), pp. 86-87.

67. G.W. Nutter, "Monopoly, Bigness, and Progress," Journal of Political Economy 64 (December 1956): 520-27.

68. John Jewkes, Sawers, and Stillerman, The Sources of Invention (New York: St. Martin's Press, 1959), pp. 158-61.

69. Richard R. Nelson, "The Simple Econometrics of Basic Scientific Research," Journal of Political Economy, 67 (June 1959): 297-306.

70. "Research and Development," Forbes, November 15, 1968, p. 35.

71. E.B. Roberts, "Entrepreneurship and Technology," Research Management, July 1968,: 249-66.

72. Frederick M. Scherer, "Size of Firm, Oligopoly, and Research: A Comment," Canadian Journal of Economics and Political Science, vol. 31, no. 2 (May 1965): 256-66.

73. Schwartzman, Innovation in the Pharmaceutical Industry, p. 101.

74. Edwin Mansfield, Industrial Research and Technological Innovation: An Econometric Analysis, The Cowles Foundation for Research in Economics at Yale University, (New York: Norton Co., 1968).

75. Abernathy and Utterback, "Patterns of Industrial Innovation."

76. Mansfield et al., Research and Innovation in the Modern Corporation; and Schwartzman, Innovation in the Pharmaceutical Industry.

77. Frederick M. Scherer, *Industrial Market Structure and Economic Performance* (Chicago: Rand McNally, 1970), pp. 363-78.

78. Leone, "The Real Costs of Regulation.

79. Nicholas A. Ashford, "The Cost of Not Regulating," *The New York Times*, December 17, 1978.

80. "Environmental/Safety Regulation and Technological Change in the U.S. Chemical Industry" (CPA).

81. Boucher et al., *Federal Incentives for Innovation*; and Iverstine, Kinard, and Slaughter, *The Impact of Environmental Protection Regulations*.

82. *The Impact on Small Business Concerns of Government Regulations that Force Technological Change*, Charleswater Associates (Washington, D.C.: U.S. Government Printing Office, September 1975), pp. 58-9.

83. Eads, "Chemicals as a Regulated Industry."

84. Iverstine, Kinard, and Slaughter, *Impact of Environmental Protection Regulations*.

85. *The Economic Impacts of Federal Pollution Control Programs* (Prepared by the Chase Econometric Associates for the Council on Environmental Quality and the U.S. Environmental Protection Agency (Washington, D.C., January 1975).

86. William J. Abernathy, *The Productivity Dilemma* (Baltimore, Md.: Johns Hopkins University Press, 1978).

87. G. White, "Management Criteria for Technological Innovation," *Technology Review*, February 1978.

88. Albert Rubenstein and John Ettlie, "Analysis of Federal Stimuli to Development of New Technology by Suppliers to Automobile Manufacturers: An Exploratory Study of Barriers and Facilitators" (Submitted to the U.S. Department of Transportation, Washington, D.C., March 1977).

89. "Environmental/Safety Regulation and Technological Change in the U.S. Chemical Industry" (CPA).

90. "Economic Growth with Environmental Quality: Representative Case Studies," Environmental Industries Council (Washington, D.C., 1977).

91. Boucher et al., *Federal Incentives for Innovation*; Iverstine, Kinard, and Slaughter, *Impact of Environmental Protection Regulations*; and "Environmental/Safety Regulation and Technological Change in the U.S. Chemical Industry" (CPA).

92. Iverstine, Kinard, and Slaughter, *Impact of Environmental Protection Regulations*, p. 77ff.

93. Kenneth Leung and Jeffrey Klein, "The Environmental Control Industry" (Submitted to the Council on Environmental Quality, December 1975), p. 104.

94. Murray Cohen and Robert P. Bennett, "Case Studies on Flue Gas Treatment as a Means of Meeting Particulate Emission Regulations," ed. C.T. Hill, in *Federal Regulation and Chemical Innovation*, American Chemical Society Symposium Series, no. 109 (Washington, D.C., 1979).

95. "Economic Growth with Environmental Quality" (Environmental Industries Council).

96. Ibid.

97. "Subsidies, Capital Formation and Technological Change," Charles River Associates (Boston, Mass., 1978).

98. "Environmental/Safety Regulation and Technological Change in the U.S. Chemical Industry" (CPA).

99. Rubenstein and Ettlie, "Federal Stimuli to Development of New Technology by Suppliers to Automobile Manufacturers."

100. "Environmental/Safety Regulation and Technological Change in the U.S. Chemical Industry" (CPA).

101. Boucher et al., *Federal Incentives for Innovation*; Iverstine, Kinard, and Slaughter, *The Impact of Environmental Protection Regulations*; and "Environmental/ Safety Regulation and Technological Change in the U.S. Chemical Industry" (CPA).

102. "Current Developments," *Environment Reporter*, Bureau of National Affairs (BNA), May 26, 1978, p. 101.

103. Boucher et al., *Federal Incentives for Innovation*, p. 53.

104. Iverstine, Kinard, and Slaughter, Impact of Environmental Protection Regulations.

105. "National Support for Science & Technology" (CPA).

106. Iverstine, Kinard, and Slaughter, Impact of Environmoental Protection Regulations.

107. "Environmental/Safety Regulation and Technological Change in the U.S. Chemical Industry" (CPA).

108. "More Firms Set Up Government Relations Units," Chemical Engineering News 57 (July 2, 1979): 18.

109. Eads, Chemicals as a Regulated Industry.

110. "Environmental/Safety Regulation and Technological Change in the U.S. Chemical Industry" (CPA).

111. Iverstine, Kinard, and Slaughter, Impact of Environmental Protection Regulations, p. 73.

112. Boucher et al., Federal Incentives for Innovation, p. 43.

113. Ibid.

114. "Environmental/Safety Regulation and Technological Change in the U.S. Chemical Industry" (CPA).

115. Ibid.

116. Abernathy and Utterback, "Patterns of Industrial Innovation."

117. "Environmental/Safety Regulation and Technological Change in the U.S. Chemical Industry" (CPA).

118. LaPierre, "Technology-Forcing and Environmental Protection Studies."

119. Ibid.

120. Jeffrey Berger and Steven Riskin, "Economic and Technological Feasibility in Regulating Toxic Substances Under the Occupational Safety and Health Act," Ecology Law Quarterly 7 (1978): 285-358.

121. For one example, see "Current Developments" (BNA).

122. See Hercules Incorporated v. The Environmental Protection Agency, 12 Environmental Reporter --Cases 1376, BNA (Washington, D.C., Federal Circuit Court of Appeals, 1978).

123. Boucher et al., Federal Incentives for Innovation, p. 2.37.

124. Industrial Union Department, AFL-CIO v. Hodgson 499 Federal Reporter 2d 467 (Washington, D.C., Federal Circuit Court of Appeals, 1974).

125. W. Tucker, "Of Mites and Men," Harpers, August 1978, pp. 43-58; and Hanley, "The Can-Do spirit."

126. "Kennedy Defends FDA Policies and Actions," Chemical and Engineering News, September 12, 1977.

127. Wechsler et al., Incentives for Research and Development in Pest Control.

128. R.V. Zener, "Antitrust and Pollution Control: An EPA Perspective, University of Pittsburgh Law Review, 36 (1975): 705-14.

129. Reitze and Reitze, "Tax Incentives Don't Stop Pollution," American Bar Association Journal, vol. 57, p. 127.

130. "Chemicals and Health" (Report of the Panel on Chemicals and Health of the President's Science Advisory Committee, Washington, D.C., September 1973).

131. Occupational Safety and Health Administration's Impact on Small Business, Occupational Safety and Health Administration, U.S. Department of Labor, Washington, D.C., 1976).

132. "Government Involvement in the Innovation Process" (A Contractor's Report to the Office of Technology Assessment, OTA-R-73 (Washington, D.C.: U.S. Government Printing Office, August 1978).

133. Boucher et al., Federal Incentives for Innovation, p. 2.33

134. Allen V. Kneese, The Economics of Regional Water Quality Management, Resources for the Future (Washington, D.C., 1964). Kneese and Blair Bower, Managing Water Quality: Economics Technology Institutions, Resources for the Future (Washington, D.C., 1968); Kneese and Charles P. Schultze, Pollution, Prices and Public Policy (Washington, D.C., 1974); Susan Rose-Ackerman, "Market Models for Water Pollution Control: Their Strengths and Weaknesses," Public Policy 25 (Summer 1977).

135. It appears almost paradoxical that the quantification of a desired rate of injuries and illness in the form of an injury tax may in fact reduce the capability of the government to achieve sufficient health and safety. This arises from the differences that will be noted between the level of the injury tax, when calculated on a cost per employee basis, and the wage differential for hazardous pay, calculated on a cost per employee basis. The injury tax will be substantially higher than the wage differential and will, if set properly, represent the dollar amount required to internalize health and safety costs to the industry. The tax will be higher than the wage differential for two reasons: (1) it includes the accident costs to workers <u>and</u> to the employer, and (2) workers may value their own lives less than the society or government does. The reduced capability to achieve a desired level of health and safety will arise from the difficulty to justify the tax level when compared with the wage differential in the political arena. A dollar for dollar justification would be possible only if and when all benefit streams from greater health and safety and the value of increased equity could be monetized.

136. Rose-Ackerman, "Market Models for Water Pollution Control."

137. However, it can be argued that the market rights approach may be even more uncertain. The situation could arise where the government issued a fixed number of rights and many firms make the decision to buy rights later and not comply. A point would be reached where there would be no compliance <u>and</u> no rights available for purchase. Presumably, the price of the rights should be made sufficiently high to preclude this market failure, but this problem and others such as the prospect of hoarding rights make the proper design of these alternatives imperative.

138. Giandamenico Majone, "Choice Among Policy Instruments for Pollution Control," <u>Policy Analysis</u>, 1976.

139. Nicholas A. Ashford, <u>Crisis in the Workplace: Occupational Disease and Injury</u>, A Report to the Ford Foundation (Cambridge, Mass.: MIT Press, 1976).

140. For cases of serious hazards, the incentives approach is probably inapplicable:

...The discharge of highly toxic substances would still have to be prohibited by law and regulation. Schedules of effluent and emission charges that truly minimized the cost of pollution control would be too complex for practical application; the consequent simplified schedules would eventually introduce some inefficiencies into the system. Kneese and Schultze, Pollution, Prices and Public Policy, p. 107.

141. Russell Settle and Burton Weisbrod, "Governmentally-Imposed Standards: Some Normative Aspects" (Discussion Paper #439-77, Institute for Research on Poverty, University of Wisconsin: Madison, Wis., September 1977).

142. John Rawls, "Concepts of Distributional Equity: Some Reasons for the Maximum Criterion," American Economic Review, vol. 64, no. 2 (1972): 141.

143. Majone, "Policy Instruments for Pollution Control."

144. The OSHAct requires the posting of a citation for violating a standard. This is considered more onerous by the firm than the small fines generally collected. Ashford, Crisis in the Workplace.

6 Labor, Productivity, and Technological Innovation: From Automation Scare to Productivity Decline

Clinton C. Bourdon

INTRODUCTION

A little over a decade ago, U.S. society was beset by fears of automation. Visionaries predicted that the rapid growth of highly mechanized, automatic control processes would so rapidly displace manufacturing workers that social stability would be imperiled by mass unemployment. The U.S. Commission on Technology, Automation and Economic Progress was formed to discuss the role of technology and automation in economic progress. After many hearings and the preparation of voluminous staff reports, it tried to assuage the fears of progress by concluding piously that "technology has, on balance, surely been a great blessing to mankind -- despite the fact that some of the benefits have been offset by costs." [1] While the fears of Americans were centered around the issue of potential technological unemployment, European discussions of automation in the early 1960s emphasized the potential social and economic gains. Since European economic growth had been threatened by labor shortages, the prospect of large-scale, labor-saving automation was welcomed as an aid to expansion. [2]

Times have changed. Now U.S. officials and economists are concerned about a slowdown in the rate of productivity growth, perhaps brought on by a decline in investment in capital equipment and a slowing of product and process innovation in major industries. This comes only twelve years after the Commission on Technology, Automation, and Economic Progress had to work to reassure troubled citizens that although "the pace of technological change had increased in recent decades and may increase in the future, ...a sharp break in the continuity of technical progress has not occurred, nor is it likely to occur in the next decade." Meanwhile, in Europe, public attention has shifted away from concern over labor shortages -- temporarily solved by importing low wage "guest" workers, to fears of the social costs of massive unemployment due to the decline of both traditional basic industries such as steel, shipbuilding, and apparel, and of newer sectors such as electronics and aircraft.

In the course of a little over a decade, Western industrial societies have gone through both extremes of the fears of progress: from concerns over too much technical change to worries over too little. Yet, credit for much of these wide swings in public mood (or, at least, in official statements of concern) might best be attributed to the substantial cyclical fluctuations in the economy during the past decade. At the national level, the last ten years have brought booms and busts unparalleled in the post-World War II era, while also carrying unprecedented shocks in the international costs and availability of such basic commodities as oil and food. In the midst of these continuing disruptions, academics and public officials would be wise not to draw implications for long-term secular trends from what may be short-term cyclical adjustments or random economic shocks. In particular, it should be obvious by now that it is very difficult to predict either the rate and direction of technological change or its impact on the labor force, never mind both. Attempts to do so, especially if dramatized by fears of rampant robots, at one extreme, or zero productivity growth, at the other, are thereby doubly suspect.

The realities of the interaction between labor and technological change are much more mundane and, as yet, too poorly understood to carry the freight of melodrama. If by "labor" it is meant (as it should be and is here) both direct production workers and managers or professional corporate staff, i.e., all workers in the modern corporation, there is no single story that can be told of their impact on technology or of technology on them. Although general macroeconomic trends can be delineated, relating technological change to the general increase in wages, incomes, and leisure in modern society, the real impact of changing technology can only be understood at the level of a firm, an industrial sector, or a specific labor market. Only at these finer levels of microanalysis can the causes, costs, and benefits of technological change be identified concretely.

Yet microanalysis or case studies in and of themselves may not be very helpful in providing a basis for valid generalization. Most frequently, for example, the first questions asked about the interrelation of labor and technology are ones related either to the degree of resistance of workers to technological change or to the deleterious impact of mechanization or automation on the work force. Although there is no dearth of case studies of the impact of technological change on workers, it is still virtually impossible to give any simple answer to these types of questions. Because of the sheer diversity of industries and technologies in the modern economy, it is very difficult to say what is typical about the process of technological change. Moreover, the academic analyses and popular accounts in the literature focus on what may be relatively rare or unusual events while neglecting the ordinary and usual processes of change. For example, everyone has heard of the Luddites and of Lordstown -- both celebrated stories of worker resistance to technology. Few people realize, however, that neither of these cases was typical of their times, nor was the issue of technology solely the cause of the protest or conflict.

To reorient the study of labor and technological change away from dramatic incidences of worker resistance, different questions must be asked and more information provided about the actual processes of technological evolution. First and foremost, more research is needed on the sources of variations in productivity. At present, we know very little about the causes of the substantial differences in interfirm, intersectoral, and intertemporal levels and rates of growth in labor and total factor productivity. Without any detailed knowledge on this, it is difficult to identify and analyze systematically the role of workers and managers in promoting or retarding technological change. Second, the combination of public attention to automation scares and to worker (usually union) opposition to technology has focused concern on the major dislocations that can result from abrupt or rapid technological change. This concern has led to heated debate on the benefits of technology or over the proper level of adjustment assistance. While such debate is not without merit, widespread public concern over these issues may be misplaced. If the normal or typical mode of technological change is one of gradual or incremental process change, neither the resulting costs or benefits will be so great or visible as to necessitate major public programs for specific adjustment. In general, public policy might be better concerned with engendering an economic climate which contributes to, and supports the process of, gradual evolution rather than focusing largely on abrupt or large-scale technological changes which are exceptions to it. Third, it is unlikely that any type of public policy can provide a final resolution to the major problems of adjustment which _are_ caused by technological change. In the course of events, some firms and workers will bear the costs of dislocation, while others will benefit. Disagreement over who should bear what level of costs is a source of endemic conflict in any pluralistic, democratic society. Learning to manage such social conflicts, without recourse to either inefficient levels of subsidies or callous disregard of economic distress, is central to public policy. However, the current lack in the United States of any consistent level of adjustment assistance, or any major creative public committment to readjustment, forces affected groups to lobby individually for aid or protection. In this process, the politically strong and well-organized are more likely to succeed than the weak. But, since a positive outcome is rarely certain for even the strongest interest groups, there is presently no climate of general social security which may engender acceptance of change.

TRENDS IN PRODUCTIVITY AND TECHNOLOGY

> The ribbon-loom was invented in Germany. The Italian abbe Lancellotti, in a work that appeared in Venice in 1637, but was written in 1623, says this:
>
> > "Anthony Muller of Danzig saw about fifty years ago in that town a very ingenious machine, which weaves four to six pieces at once. But the mayor of that town became apprehensive that this invention might throw a large number of workmen onto the streets, and therefore had the invention suppressed and the inventor secretly strangled or drowned."
>
> Karl Marx
> Das Kapital

Because of the adjustment costs imposed on particular groups in society, there has always been at least some resistance to technological change. Yet, while direct opposition has occurred sporadically since the time of Marx's illustration, the pace of technological change has continued virtually unabated. The beneficial results of this trend are well known: higher real incomes, greater leisure time, and longer life expectancy for nearly all members of industrial society.

The current slowdown in the rate of productivity growth in the United States threatens to erode these gains. During the last decade, the rate of change in productivity has been 40 percent less than the historical trend; this drop has been matched by an equivalent decline in the rate of real earnings. There is no widely accepted explanation for this apparently abrupt change in the rate of productivity growth. It is unlikely, however, that labor (in the sense of direct production workers) is responsible for any substantial part of the decrease. Unorganized workers are usually powerless to resist the degree or type of technological change -- although they may informally affect the rate and manner in which technology is introduced in firms -- while the relative power of organized workers, as reflected in their membership in major unions and in the competitive strengths of the basic industries, is generally conceded to have weakened during this period. Although isolated examples of resistance to technology may still be found in some unionized sectors of the U.S. economy, there is no evidence that such resistance is either more widespread or more effective than was in the past.

Nonetheless, the current concern with the slow growth of productivity in the United States provides an opportunity to reassess labor's role in the process. In the following pages, the role and impact of labor on productivity and technological change is examined through the existing literature on macroeconomic and sectoral trends in productivity and on the basis of case studies of technology change in selected industries. It should be noted at the outset, however, that the existing economic literature on the study of the interrelationship of labor, productivity change, and technological innovation may not be sufficient to draw any general lessons for public policy or private labor-management relations. Despite the recent studies of growth accounting and of the "sources of growth," the economics literature, particularly at the macroeconomic level, has been so distinguished by the usual high levels of abstraction as to make it virtually useless in understanding the concrete process of technological innovation and change. For example, there is still some disagreement in the economics literature over the degree to which productivity advances have been "embodied" in specific improvements in capital and labor and to what extent they are "disembodied." But, clearly, the answer to this question carries tremendous policy implications in understanding the relative impact that investment in machinery, or in labor skills, or in innovations in capital equipment have in raising productivity.

A brief review of the macroeconomic and sectoral characteristics of productivity change and its impact on labor follows. After that, a series of case studies characterizes the micro-nature of productivity and technological evolution in different industries. Implicitly, three issues are discussed.

1. How is the specific nature of technological change best characterized? If, as may be the case, incremental process change may be both more ubiquitous and more significant in raising productivity than major (discontinuous) product or process changes, how are these incremental developments affected by the work force, and what impact do they have on workers?

2. What are the distributive implications of rapid technological change? If, in some cases, the employment, wage, and skill impacts are so great as to make worker resistance both understandable and effective, how can the changes still occur?

o What role do managers and professionals play, as corporate employees, in affecting the rate of productivity change? What incentives presently exist (both explicitly and implicitly), and how might these be changed if different outcomes are desired?

Aggregate Trends in Productivity and Welfare

One way to relate aggregate changes in productivity to impacts on the labor force is to focus on real wage trends and the relationship between real wage and productivity changes. Assuming perfectly competitive markets and other restrictive conditions, such as full employment, labor will be paid its marginal product as its real wage. As the marginal product grows over time, due to technological change either embodied in new equipment or disembodied in "advances in knowledge," the real wage should increase commensurately. This steady increase in the real wage, coupled with a less than equivalent decline in the amount of labor employed, has resulted in a constant or slightly rising share of labor in national product. [3]

While perhaps abstruse, this type of analysis was the long-run model behind the famous wage guideposts of the early sixties, first propounded by the Council of Economic Advisors of the Kennedy and Johnson administrations and now resurrected in the 1978-1979 voluntary guidelines. The theoretical construct was combined with a pragmatic short-run macroeconomic analysis of inflation which saw excessive demand as the main reason behind the generalized price increases. Whenever purchasing power exceeded productive capacity, it could generate inflationary pressures in the economy. This naturally implied that whenever wages (an important but not the only source of purchasing power) grew faster than productivity, i.e., productive capacity per capita, demand-pull inflation would follow. Therefore, current economic stability required that the growth of wages should not exceed that of productivity. In fact, the proponents of the wage guideposts suggested this as a long-run rule, because a constant relationship had been observed historically, and because it meant that labor's share of national income would at least remain constant.

Indeed, long-run trends in real wages and productivity show that they have followed very similar paths of growth in the United States (see Fig. 6.1). A more recent and precise comparison can be made by considering the average annual growth rates for five year periods between 1950 and 1975, and the actual annual rates for the period 1976-1977, which reflect shorter-run variations. These are presented in Table 6.1.

The long-term figures between 1950 and 1960 reflect that while the rate of growth in real wages tended to decline steadily over the whole period, trends in productivity, while also declining, manifested higher fluctuations -- with a brief acceleration of growth in the first half of the 1960s. The impression that real wages tend to fluctuate less sharply than productivity is confirmed by the figures in the second half of Table 6.1: when productivity rose sharply, real wages also rose, but less dramatically (for example, between 1970 and 1971; 1974-1975; and 1975-1976); on the other hand, when productivity

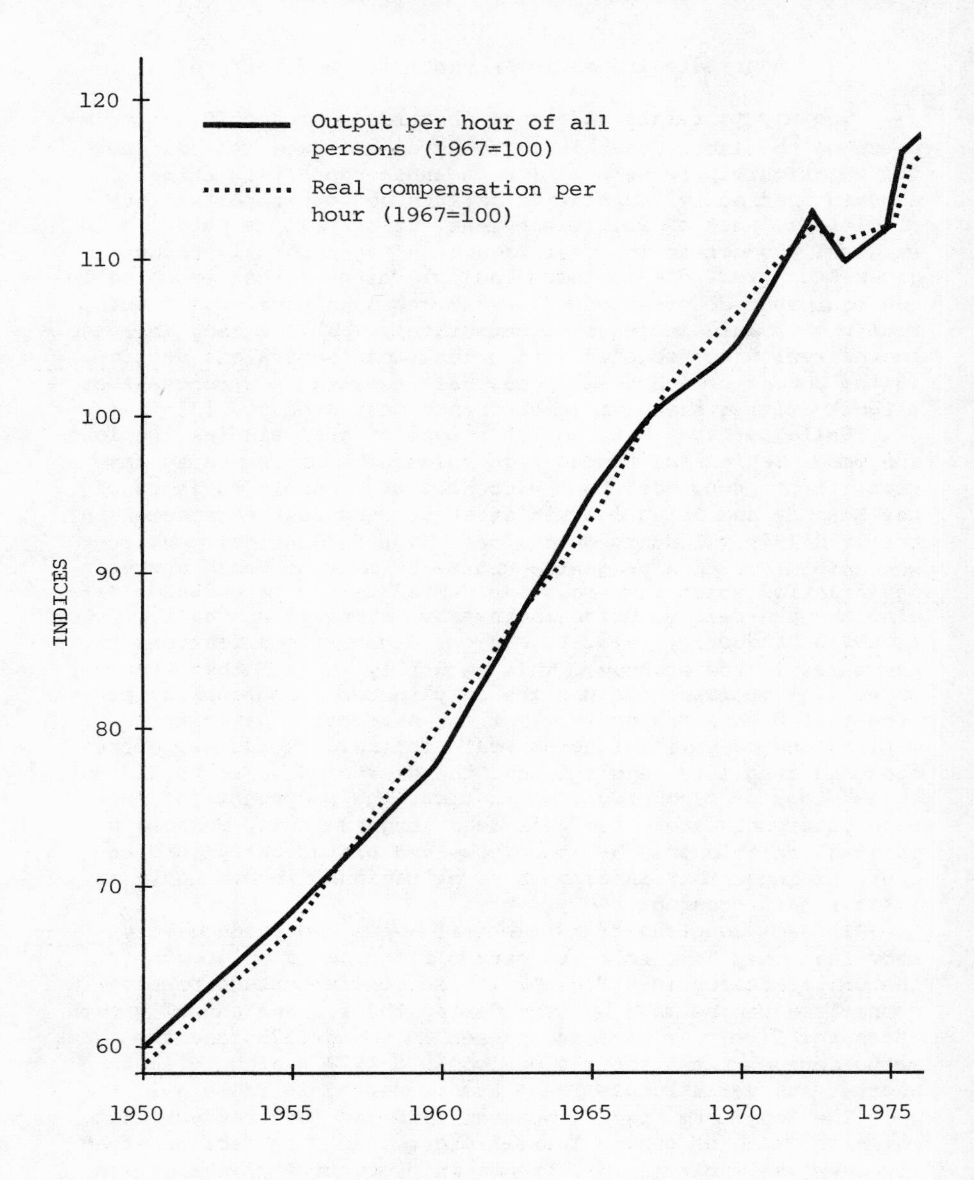

FIGURE 6.1

EVOLUTION OF PRODUCTIVITY AND REAL WAGES IN THE UNITED STATES, 1950-1977

Source: Bureau of Labor Statistics, Monthly Labor Review, July 1978.

TABLE 6.1

RATES OF GROWTH OF LABOR PRODUCTIVITY AND REAL COMPENSATION IN THE UNITED STATES, 1950-1977

Year	Growth Rates (Percent Per Year)	
	Labor Productivity	Real Compensation per Hour
1950-55	3.00	3.49
1955-60	2.45	3.28
1960-65	3.93	3.06
1965-70	1.99	2.52
1970-75	1.36	1.07
1967	2.3	2.7
1968	3.3	3.3
1969	0.3	1.5
1970	0.7	1.1
1971	3.2	2.2
1972	2.9	2.3
1973	1.9	1.8
1974	-2.8	-1.4
1975	1.8	0.5
1976	4.2	3.1
1977	2.6	2.3
1967-1977	1.86	1.76

Source: U.S. Bureau of Labor Statistics, Monthly Labor Review, July 1978.

growth slowed down substantially, so did the growth of real wages but in a less marked way (as between 1968 and 1969; 1972-1973; 1973-1974; and 1976-1977). In 1974, for the first time, both productivity and real wages declined (as reflected by the negative growth rates). In sum, there has been a tendency towards a considerably slower growth of both productivity and real wages in the United States since 1960, and the fluctuations of growth in real wages, although roughly following those of productivity, tend to be less marked than the latter both upwards and downwards.

Another trend, in addition to that of real wages, which is important in evaluating the welfare impact of technological change is that of increasing leisure time. On the average, the employed person in the private sector of the economy worked 36.2 hours per week in 1977, while in 1890, the average worked per week was 61.9 hours. While the shortened workweek has accounted for most of the gains in leisure during a person's work life, there have also been substantial increases in paid holidays, vacation, and sick leave. The total of these has added an average of three more weeks of leisure per year. Even more important in creating potential leisure, however, has been the shorter time people spend in the work force. Since the turn of the century, the average worker has entered employment later and left it sooner. One study estimates that a worker has gained an extra nine years of leisure with these changes. [4]

It is also notable that a parallel to the recent slowdown in productivity has been a decrease in the rate of decline in hours worked. Fig. 6.2 shows the trend in hours worked for the last 100 years; the decrease in the last 25 years has been markedly less than in some previous periods. This lack of decline in hours worked in the last decade is particularly true in manufacturing (see Table 6.2) where average weekly hours have remained virtually constant at around forty per week, while hours in other sectors have continued to decline. (Indeed, the sectoral disparity in these trends may be one of the motivations for some current union bargaining goals of a thirty-five hour week.)

These macroeconomic theories and time series tell a very simple story: aggregate increases in productivity have been distributed to workers in the form of real wage increases and of leisure. The shares of capital and labor in national income have remained at least constant, or changed slightly in labor's favor, over time. Consequently, the neoclassical story implies that workers as a whole have little to fear and much to gain from continued increases in productivity, however these are achieved. The negative reaction or resistance by some workers to technological change is, therefore, seen as either a purely distributive problem (different workers or skill groups may gain or lose from some changes), or as a problem due to microeconomic

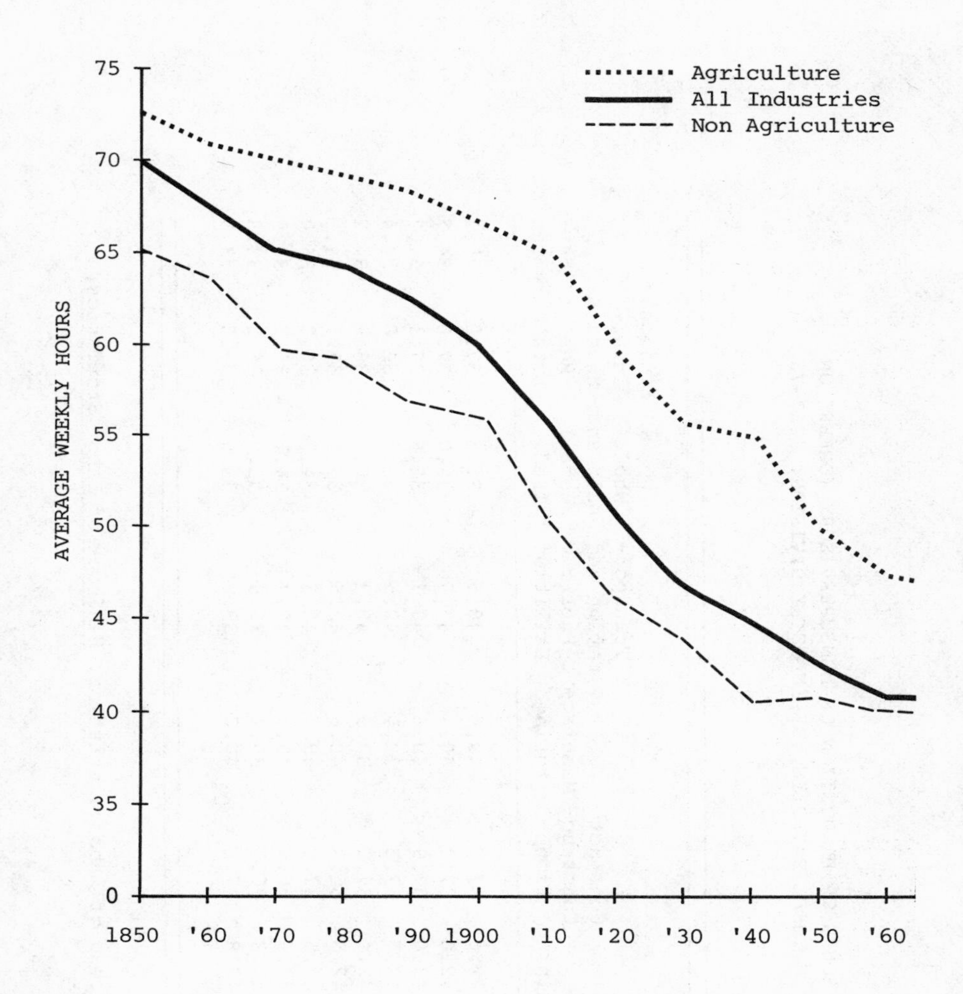

FIGURE 6.2

AVERAGE NUMBER OF HOURS WORKED PER WEEK IN THE UNITED STATES, 1850-1963

Source: U.S. Bureau of Labor Statistics, presented by Ewan Clague in Hours of Work, Hearings before the Select Subcommittee on Labor of the Committee on Education and Labor, House of Representatives, 88th Congress, Part I, p. 96.

TABLE 6.2

AVERAGE WEEKLY HOURS OF PRODUCTION OF NONSUPERVISORY WORKERS ON PRIVATE NONAGRICULTURAL PAYROLLS, BY INDUSTRY DIVISION, 1967-76

Year	Total Private	Mining	Contract Construction	Manufacturing	Transportation and Public Utilities	Wholesale and Retail Trade	Finance, Insurance, & Real Estate	Services
1967	38.0	42.6	37.7	40.6	40.5	36.5	37.0	35.1
1968	37.8	42.6	37.3	40.7	40.6	36.0	37.0	34.7
1969	37.7	43.0	37.9	40.6	40.7	35.6	37.1	34.7
1970	37.1	42.7	37.3	39.8	40.5	35.3	36.8	34.4
1971	37.0	42.4	37.2	39.9	40.2	35.1	36.9	34.2
1972	37.1	42.5	36.9	40.6	40.5	35.1	37.1	34.1
1973	37.1	42.5	37.0	40.7	40.6	34.7	36.9	34.0
1974	36.6	42.4	36.9	40.0	40.2	34.1	36.7	33.9
1975	36.1	42.3	36.6	39.4	39.6	33.8	36.5	33.8
1976	36.2	42.8	37.1	40.0	39.9	33.6	36.6	33.5

Source: U.S. Bureau of Labor Statistics, Handbook of Labor Statistics, 1977.

distortions in markets (such as wage rigidity or labor immobility, which results in unemployment). But, according to the story, since society as a whole benefits, the gainers could always compensate the losers, and all would be better off. The key policy issues, then, at the end of the neoclassical economic rationale of growth, are ones of compensation (for graceful but deserving losers) or market efficiency measures such as deregulation or tariff reductions (for recalcitrants). Simply put, the resulting neoclassical view of adjustment policy comes down to either buying out or outgunning those who resist change.

While there is much to recommend this neoclassical macroeconomic approach to growth, it has some clear limitations. Three are of particular interest here. First, the basic assumptions of the model may not be very realistic. The world might be better characterized if more attention were paid in the model to monopoly power of business and labor; to imperfections in factor markets which hinder mobility; to economies of scale in production; and to institutional distortions in the distribution of work opportunity and income. Since taking any of these elements into account in an aggregate model makes it nearly intractable, macroeconomists have tended to rely on the usual simple assumptions of a competitive economy. [5] It follows that the realism and saliency of their results suffer.

Second, no strong conclusions about general social welfare can be drawn from the standard model of neoclassical growth and technological change. Since utility is derived from the nature of work itself, as well as from other social interactions, the simple increase in real wages and leisure need not be taken as a cardinal measure of social gains and benefit. Indeed, Marx predicted glumly that capitalism might achieve "the better remuneration of slaves"; and there is now no lack of opposition to those technological changes in work or society which promise higher output at a cost to the quality of work life.

Third, the macroeconomic approach tells very little about the extent and causes of variation in rates of productivity between productive sectors of an economy. However, some economists have tried to explain the current productivity slowdown as being caused by both shifts in output among sectors and changes in the rate of productivity growth in particular sectors. This more disaggregated approach is clearly useful in understanding the sources of productivity change. [6]

Despite these caveats, the macromodel of growth does call attention, in the context of a general equilibrium analysis, to the overall gains to labor in the process of growth. Although highly abstract, it also represents the most rigorous theoretical analysis of the sources of growth. The sectoral analysis and case studies which follow, while more realistic and concrete, usually suffer from the alternate sins of a partial and purely descriptive approach.

Sectoral Variations in Productivity

While the aggregate rate of labor productivity growth appears to have been declining steadily since 1950, and particularly since 1965, an analysis of different industries reveals that the slowdown of the last decade could be subdivided in two distinct phases. The first was roughly between 1965 and 1970, when a general slowdown can be observed for all sectors. During the second phase, after 1970, many industries kept the same slow pace of growth, but a few grew more rapidly than the average (notably, hosiery, wet corn milling, malt beverages, synthetic fibers, aluminum rolling and drawing, telephone communications, and less markedly, but still well above the national average, the motor vehicle industry), while others actually had a decrease in productivity (coal mining, copper rolling and drawing, iron mining, hydraulic cement, retail foodstores, cereal breakfast foods, and blended and prepared flour). [7]

These three patterns of labor productivity growth by industry are exemplified best by: the motor vehicle industry, which shows nearly continous high growth; the footwear industry, with continuous slow growth; and the coal mining industry, with a high growth before 1965, a sharp slowdown around 1965, and a sharp decline after 1968 (see Fig. 6.3). Since the general slowdown in productivity growth can be attributed to specific sectors (and it should be remembered that most industries are in the slow growth group after 1970), more attention should be concentrated on those sectors with declining productivity. They are certainly largely responsible for the fact that average annual growth was slower in the aggregate in 1970-1975 than in 1965-1970.

Measures of the rate of growth of total factor productivity by industry also show considerable variation. Data compiled and estimated by Dale Jorgenson and Frank Gollop (see Table 6.3) show that the rate of growth in the highest industry was five hundred times that in the slowest industry with a positive growth rate. [8] Five other industries had negative rates of growth during the entire period under study, (1947-1973). In addition, approximately one-half of the major sectors studied had lower rates of growth during 1966 and 1973 than during 1953 and 1957. Many of these were sectors such as services, retail trade, and finance, insurance, and real estate, which have grown faster in both employment and output than the rest of the economy. This sectoral shift from relatively rapidly growing sectors to slower growing ones, in addition to the lower recent rates of growth in particular sectors, is another of the major causes of the aggregate slowing of productivity increase. [9]

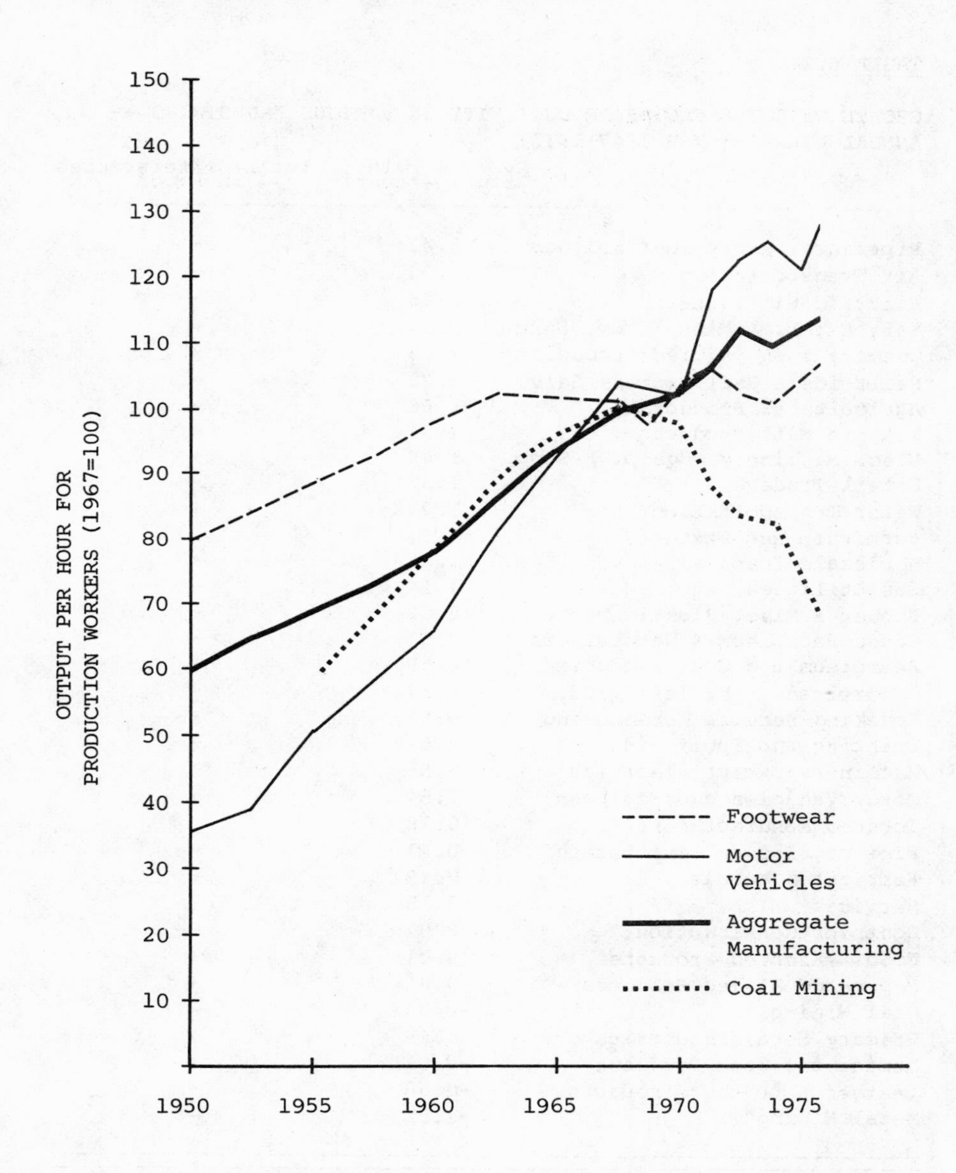

FIGURE 6.3

LABOR PRODUCTIVITY GROWTH FOR VARIOUS SECTORS

Source: U.S. Bureau of Labor Statistics, Productivity Indexes for Selected Industries, 1976 edition.

TABLE 6.3

GROWTH OF TOTAL FACTOR PRODUCTIVITY IN VARIOUS INDUSTRIES -- ANNUAL AVERAGES FOR 1947-1973

Industry	Aver. Growth Rate (percent)	Int'l Differerences in Rate*
Pipelines, except Natural Gas	5.42	+
Air Transportation	3.82	-
Electric Utilities	2.66	--
Tel. & Tel. & Misc. Comm. Serv.	2.33	-
Chemicals and Allied Products	2.33	+
Railroads & Rail Express Serv.	2.20	+
Agricultural Production	1.68	-
Textile Mill Products	1.54	+
Elec. Machinery, Equip. & Supp.	1.48	+
Retail Trade	1.32	-
Water Transportation	1.22	+
Furniture and Fixtures	1.21	-
Wholesale Trade	1.21	+
Gas Utilities	1.14	-
Rubber & Misc. Plastic Prod.	0.99	+
Crude Petroleum & Natural Gas	0.98	-
Petroleum and Coal Products	0.69	+
Apparel and Fab. Tex. Prod.	0.67	+
Trucking Serv. & Warehousing	0.67	--
Printing and Publishing	0.57	-
Machinery Except Electrical	0.55	+
Motor Vehicles and Equipment	0.54	+
Tobacco Manufacturers	0.28	-
Finance, Ins. & Real Estate	0.21	--
Fabricated Metals	0.19	+
Services	0.15	--
Contract Construction	0.07	--
Food & Kindred Products	0.01	+
Paper and Allied Products	0.01	+
Coal Mining	-0.02	-
Primary Metal Industries	-0.49	+
Lumber and Wood Products	-0.61	+
Leather & Leather Products	-0.68	+
Metal Mining	-1.55	-

* Differences in rate of productivity change between the periods 1953-1957 and 1966-1973. (+) indicates higher rate in later period; (-) indicates a lower rate; (--) indicates a negative rate in later period.

Source: Dale Jorgenson and Frank Gollop, "U.S. Productivity Growth by Industry, 1947-73," mimeographed, University of Wisconsin (Madison, Wis., 1977).

Labor's Impact on Productivity

Unfortunately, economists have had very little to say about the explanation of these interindustry and intertemporal variations in the rate of labor and total factor productivity growth. Perhaps the sole exception is John Kendrick's work on postwar productivity trends, in which he extends earlier work by Terleckyj to identify industry characteristics which are significantly correlated with rates of change in total factor productivity. [10] Specifically, he examines the relationship between the rate of change in total factor productivity and nine characteristics of the industries studied including: rate of change in output, rate of change in real capital, variability of output changes, average education per employee, ratio of R&D to sales, average hours worked, concentration ratio, rate of change in concentration, and unionization ratio. While simple correlation coefficients between the characteristics reveal a few strong patterns of association, such as those between the rate of change in productivity and the rate of change in output, they do not manifest many statistically significant interrelationships. For the most part, the simple correlation between the rate of change of productivity and the other nine variables is positive but small. Multivariate regression of productivity growth on some or all of the nine variables did not help to clarify the interrelationships either. Only a few variables were significant and positive (such as R&D expenditures and average education of employees); while the rate of change of output was, somewhat circularly, found to be consistently important.

Because the explanation of these intersectoral variations in productivity change is so primitive, it is difficult to assess the impact of either production workers or unions on the rate of change. In a recent article in Fortune, however, Kendrick is quoted as listing five factors which affect productivity increases in a given industry. Fortune says, "Kendrick's fifth rule is that because of rigid work rules, labor unions generally retard gains in productivity." [11] However, Kendrick's academic analysis shows a negative, but statistically insignificant, impact of the degree of unionization on productivity, and, in his book, Postwar Productivity Trends in the U.S., 1948-69, he is more cautious in his comments:

> The negative influence of the degree of unionization tells us that apparently the influence of unions with regard to the rate of innovation and changes in economic efficiency has outweighed possible positive union influences mentioned earlier. Even this result is not unambiguous, of course, in view of the possible correlations between unionization ratios and other parameters. [12]

The "possible positive union influences" referred to is the fact that the (hypothetically) higher rates of wage increase in unionized industries could spur a higher rate of substitution of capital for labor and thus a faster growth in labor productivity, if not of total factor productivity in those industries. Kendrick concludes, "Only quantitative analysis can indicate what has been the net effect of unionization on productivity." Recent research by Kim Clark has shown that across plants and firms in the cement industry, unionization contributes to higher productivity. [13]

While there is no conclusive evidence as to the impacts of unionization on productivity at the sectoral level, it is interesting to note that the deleterious impact of unions on productivity may not be as great as sometimes is assumed. In Table 6.3, five of the six industries with the highest rate of factor productivity growth are highly unionized. At the least, this indicates that it is possible to have high rates of productivity increase in union environments -- whether these have occurred in spite of the unions or not. (Some specific examples of union contribution to, and support of, productivity increases and technological change in particular industries will be presented.)

Besides the issue of unionization, it is important to note the positive association between education per employee and the rate of productivity change. To the extent that employee abilities and learning on the job contribute to higher productivity, greater rates of productivity change may be realized through higher levels of education. Thus, the apparent correlation between the decline in current levels of educational achievement in the United States and the slowing of productivity growth may not be spurious. In fact, yet another explanation for the recent productivity decline is a change in the character of the labor force which is becoming more heavily weighted with younger and less educated workers. [14]

Impact of Technological Change on Labor

At the sectoral level, there are three ways to assess the impact of a change in productivity or technology on labor. First, change in output per worker over time can be correlated with employment changes. Fig. 6.4 presents a scattergram of the relationship between a measure of output per worker and employment in several subsectors of manufacturing; the pattern reflects a mildly negative relationship over the time period 1950-1975. However, using other data and a different timespan, 1948-1966, Kendrick reports a "mild positive correlation between rates of change in [total factor] productivity and in labor input for thirty two-digit industries." [15] His measure of labor input, however, is man-hours weighted by average hourly

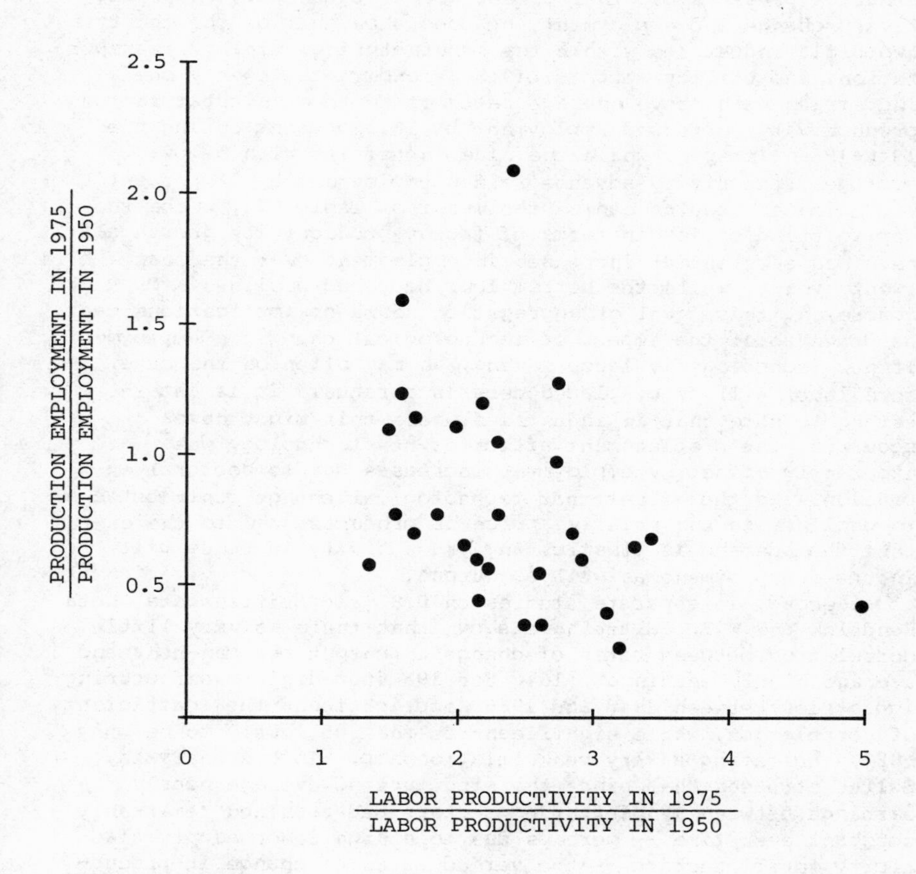

FIGURE 6.4

RELATIONSHIP BETWEEN CHANGES IN INDUSTRY EMPLOYMENT AND CHANGES IN LABOR PRODUCTIVITY FROM 1950 TO 1975 FOR SELECTED 3-DIGIT MANUFACTURING INDUSTRIES

Source: U.S. Bureau of Labor Statistics, Productivity Indexes for Selected Industries, 1976 edition.

compensation, so reductions in man-hours actually worked could be offset by compensation increases in some industries. While Kendrick presents no other direct correlations between productivity change and employment, he does show that of the thirty two-digit industries within the manufacturing, mining, transportation, and utility sectors of the economy, the twenty-one industries with above average rates of advance in total factor productivity increased employment by 16.5 percent during the 1948-1966 timespan, while the nine industries with below-average productivity advance raised employment by 11 percent.

Similar results can be derived from Table 6.3, where the top four industries in terms of factor productivity growth all have had substantial increases in employment over the last twenty years, while the bottom four have had declines. Of course, at this level of aggregation no major implications can be drawn about the impact of technological change on employment. If new technology is labor saving, as may often be the case, some labor will be displaced ceteris paribus. It is interesting to note that in industries where this might have occurred, the displacement effect of new technology was apparently offset by employment increases due to sectoral expansion. To the extent that technological change contributed to declines in the relative price of products, and to the extent that demand is elastic, any productivity increase will increase employment as well as output.

Second, in separate studies on U.S. and British data, both Kendrick and W.E. Salter have shown that there is very little correlation between rates of change in output per man-hour and average hourly earnings. [16] For 395 four-digit manufacturing industries between 1948 and 1966 Kendrick found the coefficient of correlation, while significant at the .05 level, to be only .022, indicating a very weak relationship. In his analysis, Salter stressed that since the structure of average hourly earnings between manufacturing sectors had remained remarkably constant over time -- perhaps due to a high labor supply elasticity to all sectors -- the varied rates of change in productivity between sectors could not be reflected in wages without seriously distorting those differentials. Fig. 6.5 presents a scattergram relating labor productivity change to hourly compensation of production workers for a time period longer than that studied by either Salter or Kendrick. These data continue to support the apparent lack of relation between wages and productivity at a sectoral level.

Third, one of the great fears stemming from the automation scare was that even if technology did not cause higher levels of unemployment in the aggregate, it might lead to "structural unemployment." That is, changes in technology would so change the mix of skills demanded in the economy that certain groups, particularly those of low skill, would be virtually unemployable, while others, the highly skilled, might be in short

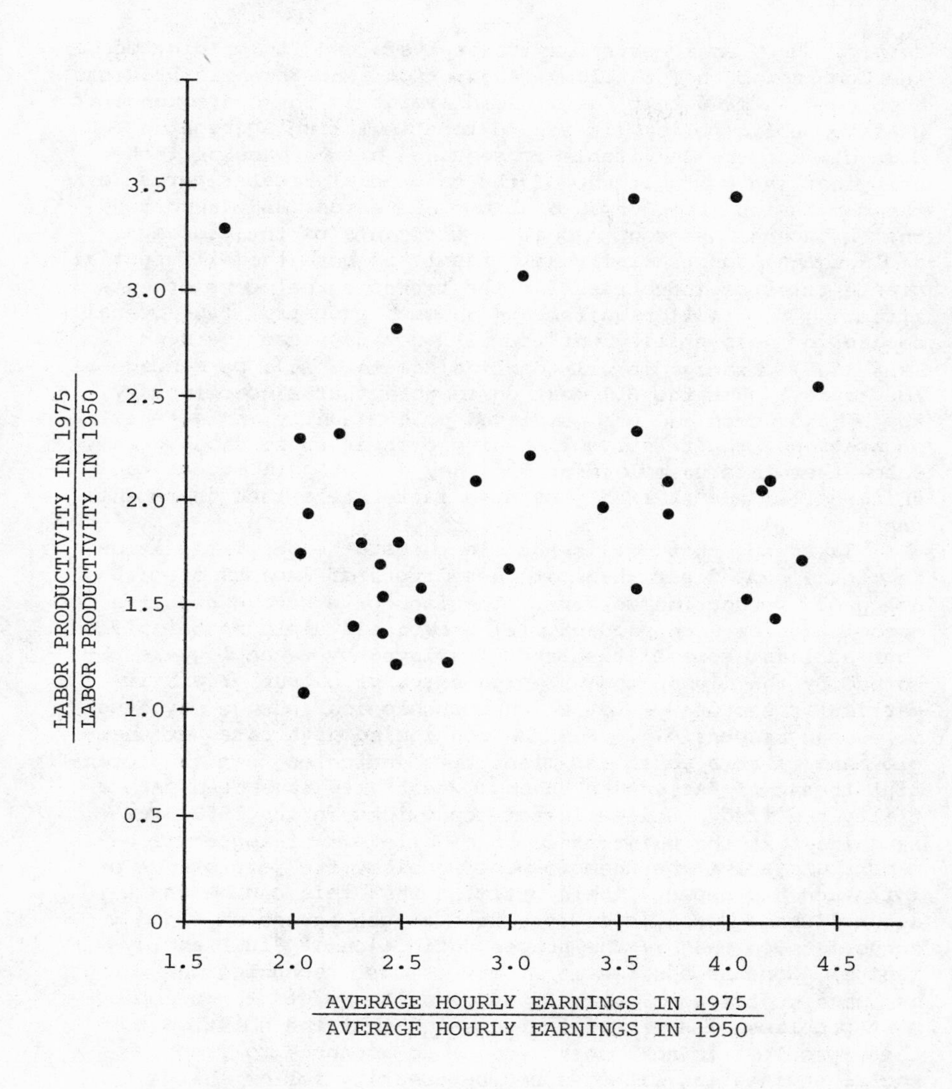

FIGURE 6.5

RELATIONSHIP BETWEEN CHANGES IN LABOR PRODUCTIVITY AND CHANGES IN AVERAGE HOURLY EARNINGS FROM 1950 TO 1975 FOR SELECTED 4-DIGIT MANUFACTURING INDUSTRIES

Source: U.S.Bureau of Labor Statistics, Productivity Indexes for Selected Industries, 1976 edition

supply. But, as a result of its analysis and its predictions, the Commission on Technology, Automation, and Economic Progress concluded in 1966 that "high unemployment is the consequence of passive public policy (in regard to stimulating aggregate demand), not the inevitable consequence of the pace of technological change." In one of the major analytical reports for the commission, the Bureau of Labor Standards (BLS) reported that although the occupational requirements of the economy would change substantially as a result of both the differential growth rates of industries and the technological developments affecting the skill requirements in each industry, "the overall demand for less-skilled workers will not decrease (between 1964-1975) although it will decline somewhat as a percentage of the total." But the BLS went on to note that since minority, youth, and women had been employed predominantly in low-skill occupations, the relative lack of growth in these jobs would force them into unemployment "if they do not gain access to white-collar and skilled jobs at a faster rate than in recent years..." [17]

Taken together, it is not clear what the net implication is of the sum of all three of these types of impacts of technology on production workers. The lack of a strong negative association between productivity growth and employment implies that at least some of the labor displaced by technology is absorbed by the higher than average rates of output growth in particular sectors -- for which technological change may itself have been responsible. But the continuing high rates of unemployment of both youth and minorities indicates that the potential threat of "structural unemployment" may have been partially realized. While liberal economists in the 1960s were sanguine that the maintenance of high levels of aggregate demand could solve the unemployment problem, the persistence of inflation has reduced their optimism that this can be easily accomplished. (Now it is even hard to get agreement among economists on what is the proper definition of "full employment.") Nonetheless, while it may be that technological change may have contributed, even in some small way, to the unemployment problems of some skill groups <u>and</u> that the solution to these problems is no longer seen as so amenable to proper social engineering, it does not necessarily follow that a reduction in the rate of change would provide greater benefits. It simply reinforces the need for effective training programs and for the enforcement of equal opportunity statutes to mitigate the impact of technological change on particular skill groups.

MICROANALYSIS OF PRODUCTIVITY CHANGE

Scattered throughout the literature on labor relations and on economic history are numerous case studies of technological change in particular firms or industries. The topics covered by these studies range from the dieselization of the railroads in the late 1930s and 1940s to the introduction of numerically controlled machine tools in metal-working in the 1960s to the rise of cold-type technology in printing in the 1970s. Unfortunately, although there are many interesting stories to be told in each of these and other industries, there is no coherent summary of their implications for the understanding of the interaction between technology and labor. This is particularly unfortunate because a microfocus on particular firms and industries provides not only an illustration of the specific costs and benefits of technological change but also permits the identification of major factors in managing change successfully. It is important to recognize that the costs and benefits of particular technological changes are not necessarily given or determined by the technology itself; rather, their size and distribution may be affected substantially by the manner in which they are introduced and managed. Thus, many cases of extreme resistance to technology by workers may actually be less a reaction to the process change itself than to the manner in which the change was undertaken.

For example, James Bright's Automation and Management, which is the only example of a combination of case studies and general analysis in the literature, shows that in twelve of the thirteen examples of the introduction of automatic machinery in different industries, "management claimed that workers were generally receptive to new machinery and, in many cases, were highly enthusiastic. Reasons for this largely stem from the advantages that automation offered the worker in the form of improved working conditions." Bright also calls attention to the fact that automation did not have a substantial negative impact on either the employment or the wage of workers in the firms he studied, largely because most of the changes affected only a small part of the processes in the firm or plant. If management worked to inform workers of the possible changes, to protect them against displacement, and to train them to adapt to the new machinery, no substantial resistance was encountered. Bright concluded his analysis by commenting that the workers accepted automation in the cases studied because they realized its benefits. He noted, "worker gains can be recalled from these 13 plants. Automation may offer the workingman:

1. Easier work, physically and mentally.
2. More interesting work through a variety of job activity.

3. A more satisfying job through a sense of responsibility.
4. Pleasanter job surroundings.
5. Higher pay.
6. Greater job security against both the firm's competition and the effects of old age.
7. Greater continuity of employment.
8. A safer job.
9. The pride of running a fine machine.
10. Opportunity to learn more of the total process and machinery.
11. Experience that may be even more valuable in the future." [18]

Bright's work dates from 1958, and there is no current, comparable research to test his hypotheses and conclusions. Yet, on the basis of an eclectic collection of more recent case studies, the following observations can be made on the micro-impacts of technological change.

Much technological change in production processes is incremental and thus does not always cause abrupt or substantial changes in employment, skills, or wages. For example, Stephen Hollander's study of the chemical industry showed that "half the reduction in the cost of producing rayon...has been the result of gradual process improvements which could not be identified as formal projects or changes." [19] This process of incremental change, which William Abernathy and James Utterback have described as being characteristic of much of industrial innovation, is relatively easily handled by management and labor. [20] Perhaps because of this, only 8 percent of labor contracts in the late 1960s had provisions explicitly related to automation. For the most part, the impacts of technological change were subsumed under the usual bargaining categories for job classifications and wage levels, with management retaining the right in most contracts for complete control over the choice of technology and operations of the production process. [21]

Many unions support technological change as a means of ensuring high wages and good working conditions. While union views on automation are not homogeneous, varying substantially between and within unions and over time, they are by no means uniformly negative. One review of union attitudes found differences in opinion at various levels of union leadership and membership. National leaders, in their role of "statesmen," rarely opposed automation. Local leadership was much more hostile; one survey found that 60 percent "agree that automation is rarely in the workingman's interests." While rank-and-file union members were generally more favorable, another survey concluded that, "The worker who does not hold a position with his local union has a positive attitude toward the general impact of automation." [22]

Some unions have actively promoted mechanization. During the 1950s, John L. Lewis led the United Mine Workers (UMW) in supporting a mechanization and modernization campaign which helped double the output per worker between 1947 and 1959. When asked his reaction to the modernization policy, Lewis responded:

> The United Mine Workers not only cooperates with the operators on that -- we invented the policy. We've encouraged the leading companies in the coal industry to resort to modernization in order to increase the living standards of the miner and improve his working conditions... We are not trying to keep men in the mines just to retain jobs. It will be a millennium if men do not have to work underground but can all work in God's sunshine...The UMWA has never undertaken to oppose modernization or progress from the standpoint of compelling the retention of uneconomic employment in the industry...We never have advocated the retention of more men than the industry needs. [23]

As another example, after 1957, the International Longshoremen's and Warehousemen's Union representing dock workers on the West Coast changed their policy to support containerization of cargoes and mechanization of dock work in order to obtain more control over the process and to obtain compensation for some affected workers. The implementation of the resulting Mechanization and Modernization Agreement permitted the gradual attrition of displaced workers through retirement so that present manning levels on the docks are efficient. (In contrast, agreements to provide income security for East Coast longshoremen in exchange for acceptance of containerization have led to considerable overmanning and excessive payments of benefits.) [24]

A variety of other positive statements and actions about automation and technological change can be gathered from different union leaders and industries over the years. A prominent union leader can be quoted as saying in Congressional testimony that "organized labor welcomes the new advances in technology. We want to see the new equipment introduced as promptly as possible so that we and our heirs can receive the benefits of lowered operating costs and higher productivity." [27] Given this view, it is usual to find organized labor, particularly in industrial unions, bargaining for a variety of measures to cushion any displacement impact of automation rather than opposing change directly. For example, the Communications

Workers of America have recently begun incorporating extensive income security measures in their contracts (such as earnings protection plans, expanded supplemental unemployment benefit coverage, and early retirement benefits) in response to the quickening pace of labor-displacing technological change in communications. While the CWA has not resisted the technology, AT&T has provided the benefits -- after bargaining, of course. Yet it should not be overlooked that the level and coverage of displacement benefits is a continuing source of sometimes bitter disputes in major industrial sectors. The belated settlement in late 1978 of some of the job security provisions in the 1977 United Steelworkers contract with the major steel companies is symptomatic of the deep conflicts that can arise over these types of benefits.

Workers disagree over the distribution of the costs and benefits of technological change. In the two cases in which the coal and longshore unions supported mechanization, some union members benefited through higher earnings while others were unemployed. In fact, while the number of miners employed fell from 419,000 in 1947 to 180,000 in 1959, the UMW never paid supplementary unemployment benefits from the royalty fund on tons of coal mined. In 1960, the UMW convention under Lewis' direction faced down a dissident faction composed of unemployed miners who were seeking to extend medical and hospital care benefits. Lewis said, "...the Trustees of the Fund have determined that the Fund cannot continue indefinitely to pay out more money to the beneficiaries than it receives in royalty payments from the industry. That is pretty simple arithmetic for anybody to understand." Similarly, in the West Coast Longshore agreement only some of the dock workers were eligible for payments from the Mechanization and Modernization (MM) fund -- these were the "A men," who were full union members. The "B men" and "casuals," who worked sporadically and only after most of the "A men" were employed, were not covered by the M&M benefits. In both of these cases, the only benefits the displaced workers received were publicly funded: unemployment insurance or general relief.

Another interesting example of these kinds of intraunion disputes has been the struggle within the American Federation of Musicians (AFM) for control over the Music Performance Trust Fund. The AFM, after the Lea Act declared illegal many of its outright featherbedding activities, negotiated the creation of the fund with major record companies. The fund was financed by royalty payments on musical records and broadcast transcriptions. The income of the fund was to be used:

> ...solely to employ instrumental musicians... in rendering musical services to the public on occasions when no admission is charged and when it will increase the public appreciation of music. [It may not be used to] provide welfare or other benefits." [25]

As the recording industry grew, the fund's resources increased enormously. Moreover, some musicians such as those in the Los Angeles and New York areas found themselves contributing over 75 percent of the fund's revenue (due to concentration of music and recording in those areas) while receiving only 10 percent of the benefits. They responded by forming a rival union in the early 1960s, the Musicians Guild, which sought to redirect the royalty payments away from the fund and toward their own wages and benefits. After some litigation and various inter-union political battles, the recording musicians ended up with 50 percent of the royalty payments. [26]

Thus, even in cases where unions have accepted technological changes in exchange for higher wages or privately funded unemployment benefits, some workers still bear the costs of displacement. Apparently, for good or ill, union solidarity and brotherhood is often limited by simple pragmatism: the willingness to sacrifice some to improve the lot of others.

<u>Worker participation can aid and promote technological change and productivity increases</u>. Proponents of various forms of worker participation plans often point to their substantial impact on productivity. There is no doubt that many experiments in worker participation have reduced labor costs and led to many incremental improvements in production. A 1975 National Science Foundation survey of fifty-seven field studies of worker participation experiences in the United States found that four out of five reported productivity increases. A 1977 study by Raymond Katzell of New York University of 103 U.S. work experiments (including work enrichment as well as participation) confirmed these findings. [27] Over thirty years of experience with combined participation and profit-sharing plans, such as the Scanlon plan, have shown that considerable unit cost-savings can be realized through "cooperative management." [28]

It is surprising that these findings have not led to wider adoption of this form of management. Although Scanlon plans and other work enrichment and participation schemes number in the hundreds and are found in plants of such corporations as General Motors, Cummins Engine, and TRW, they are still the exception and not the rule in American management. Government attempts to promote wider use of these types of plans have not been extensive or significant. Thus, wider diffusion of these apparently successful methods of raising productivity is still a problem.

The attention to explicit and formal mechanisms of worker participation should not disguise the tremendous amount of informal participation and consultation which occurs in American industry. In his study of the adoption of new metalworking technologies, for example, Harvey Belitsky found that:

> Some American firms resemble "the best managed firms" of Japan, in which plant engineers elicit the suggestions of even semi-skilled workers regarding materials, machine adjustment, and other factors which can lead to higher productivity." [29]

American management works in a more participative mode than it is generally given credit for, while American workers often contribute to productivity gains without formal incentive plans. For example, in Hollander's study of DuPont's rayon mills, he found that the largest number of process innovations came from production personnel, rather than from the R&D laboratory, and these innovations had the largest economic impact. [30]

<u>Management is still interested in automation and technological change, but there are competitive alternatives</u>. Despite the effect labor can have in promoting or retarding the pace of technological change, all decisions on process changes are still made by middle and top corporate management. There is no direct evidence that American managers are currently any more averse to technological improvements in production than they were in the past. What has changed, however, is the general climate towards investment and risk. The general slow growth in the U.S. economy during the 1970s, coupled with the series of external shocks -- ranging from oil prices to regulatory reform -- has made management much more risk-adverse in making production decisions. The recent decision of Zenith to move its production of modules for color televisions out of the United States is one example. Zenith had pioneered in automating U.S. production and was, in 1976, the only major U.S. manufacturer which produced color TV components in the United States. In 1975 and 1976, Zenith had to decide whether to continue to mechanize U.S. production, through automatic component insertion of electronic components onto modules and in automatic testing of assemblies, or follow everyone else and relocate production in another country. For the first time, Zenith opted for relocation, not because continued automation was estimated to be ultimately infeasible and noncompetitive, but because (1) the relative costs and risks of relocating production outside the United States were now much lower than in the past; and (2) due to intense and immediate price competition, the payback time of the automated equipment was too long. [31]

Despite these pressures on management which retard the rate of adoption of new technology, there does not appear to be a decline in the rate of, or interest in, process innovation itself. One survey of American and foreign scientists and engineers by the International Institute for Production Engineering Research (CIRP) found that "computer-automated facilities will be a full-blown reality before the end of this century." Their ultimate objective, the automated factory, will be approached through such "viable, economic steps as integrated manufacturing software systems, group technology, cellular manufacturing, computer control and multistation manufacturing systems." [32]

The CIRP survey admits, however, that the key to making this new level of automation feasible is an increase in the ability to produce relatively nonstandard items in batch lots. At present, most automated machinery is simply not flexible enough to produce anything but very standardized items in high volume. And the demand for these must be very stable and predictable before the high fixed costs of automated equipment can be tolerated. The relative price of equipment versus labor is also a factor in this choice, of course. But despite perceptions of rapidly increasing labor costs, it is not clear what the overall relative factor price trends are. For example, the wholesale price index of metal-working machinery and equipment increased by 83 percent between 1967 and 1976, while average hourly earnings in the sector which uses this equipment increased by 74 percent. These trends, if representative, also would help explain the relative slowness of the adoption of new equipment.

The major implication of these case studies is simply that workers can and have aided and contributed to the process of technological change. The most important conditions for this contribution appear to be, however, that management informs workers of the changes, encourages their participation in implementation, and accepts at least some of the costs of job security, retraining, and other disruptions to the work force. For the most part, management has acted in this way -- an approach Bright describes as being simply humane and in accord with common sense. Workers have been willing, in general, to cooperate, not only because the costs of incremental technological change are usually low, but also because the burden of adjustment for major changes is often limited (in some cases by the workers themselves) to a relative few. Indeed, the most celebrated cases of worker resistance to technological change may be, at bottom, less pure opposition to technology than a reaction to a company's and even society's resistance to accepting and bearing enough of the cost.

PUBLIC POLICY AND ADJUSTMENT ASSISTANCE

There is no adjustment assistance designed specifically for workers displaced by technological change or, indeed, by any other form of private domestic market change or industry relocation. Workers not covered by private programs have to rely on general public assistance such as unemployment assistance or programs under the Comprehensive Employment and Training Act provisions for retraining. It is not at issue here whether these policies are either effective mechanisms or adequately funded. Extensive discussions and debates are available in the economics and manpower literature over such issues as the best forms of providing public layoff benefits or operating private interplant transfer schemes after layoffs. All of these discussions are simply symptomatic of continuous social bargaining over the distribution of the costs of change. Since any level of social benefits can be attacked for either being too great, by acting as a disincentive to work and adjustment, or, conversely, for being too small, so as to be ineffective in mitigating the costs of adjustment, it is unlikely that any particular level of benefits will be accepted as desirable for very long.

In the ongoing political struggle over optimal levels of adjustment assistance for technological change as well as trade adjustment and plant closing, the following themes will continue to emerge:

(1) General versus specific benefits: The United States now manifests two levels and systems of public benefits for displaced and unemployed workers (trade and defense-related on the one hand, and general unemployment insurance and CETA on the other) and two levels and systems of private benefits (substantial income and employment security in the major union sectors and in large nonunion companies like IBM, Xerox and Eastman-Kodak, coupled with little or no benefits in smaller union or nonunion firms). Specific readjustment benefits, such as those for displaced workers under the 1974 Trade Act or those found in major union contracts, tend to be very costly; therefore, they are very limited in coverage. Political pressures seeking to expand the size and coverage of these specific benefits will intensify while other political forces may bring a reduction in budgets for general programs such as CETA. These opposing trends will only increase existing social inequities.

(2) Union versus management control over change: The current emphasis among U.S. trade unions concerned with worker displacement is on prior notification of impending layoffs or relocation decisions. The U.S. unions have been impressed by the private and public control over layoffs which exists in Europe. [33] Attempts to legislate similar controls over management actions in the United States will be a major goal at the state and national level. Management, although it lives

with these measures in major European countries, will undoubtedly resist their introduction in the United States.

(3) Reactive versus creative adjustment measures: U.S. policy has always been purely reactive in nature in promoting readjustment: it provides what some call "burial insurance." While even this sort of insurance is useful, it is too cynical a policy if the victim has been unfairly killed, or, worse, is still half alive. Recently, there have been programs developed to save the competitively viable parts of industries decimated by import competition. For example, efforts are now underway in the apparel and shoe industries to experiment with new technologies and new marketing strategies to save declining industries. These programs represent a combination of government funding with new institutional forms of labor-management cooperation. They seek not to prevent readjustment (though they have protested against dumping and other unfair trade practices), but to actively promote it by working to save some parts of an industry while other segments decline. Although these more "interventionist" policies are unique in the United States, they are commonplace in Europe.

(4) Risk versus security: Perhaps the main paradox behind the current thinking about the slowdown in productivity growth and the apparent decrease in technological innovation is a contradictory concern over the role of risk. Some commentators see the presumed decline of innovation in the United States and in many industrial societies as due to an unwillingness by firms to be risk takers. Yet, managers of these very firms explain their reluctance to invest in new machinery and in R&D as stemming from a sense that the world is too risky to permit such investments. Another paradox lies in the contradictory attitudes of workers and managers toward the risks of technological change. Workers accept technological change if their jobs or earnings are made secure, while managers promote it in order to make their jobs and salaries secure. In general, risk no doubt promotes change: it provides both a stimulus and an incentive. Yet some stability is needed in the general economic and social climate in order to implement and invest in new technologies and management techniques. Where this climate of stability and security exists at the level of the firm, as in Japan, considerable change can still occur. Its absence in some firms in the United States, coupled with a general economic climate of uncertainty, explains much of the resistance to technological change and the decline in innovation. Were the United States to develop a policy to promote innovation, it would be wise to find the illusive, but optimal, combination of risk and security -- and not err in the direction of risk. Thus, the challenge to adjustment policies is to provide some security during the process of change, while not providing a disincentive to change itself. To date, neither the design nor the management of U.S. adjustment policies have met these objectives.

NOTES

1. Technology and the American Economy, U.S. Commission on Technology, Automation, and Economic Progress, (Washington, 1966), p. xiii.

2. Sol Barkin in Manpower Aspects of Automation and Technical Change, Organization for Economic Cooperation and Development (Paris, 1966), p. 94.

3. The precise explanation of what might be called the relative constancy of relative shares has been the subject of much dispute. See "Macrodistribution Theory," Chapter 16, in, Income Distribution Theory, Martin Bronfenbrenner (Chicago: Aldine, 1971) and C.E. Ferguson, "Neoclassical Theory of Technical Progress and Relative Factor Shares," Southern Economic Journal, vol. 34, no. 4 (1967): 490-504.

4. The Employment Impact of Technological Change, U.S. Commission on Technology, Automation and Economic Progress (Washington, D.C., February 1966).

5. Larry Blair, "Welfare Theory, Technological Change, and Worker Well-Being," (mimeographed, Oak Ridge Associated Universities, Knoxville, Tenn., 1975). A paper by Thurow incorporates some of these distortions and finds labor's actual earnings to be significantly below its marginal product. See Lester Thurow, "Disequilibrium and the Marginal Productivity of Capital and Labor," Review of Economics and Statistics, vol. 50, no. 1 (February 1968): 23-51.

6. See William Nordhaus, "The Recent Productivity Slowdown," in Brookings Papers on Economic Activity, Brookings Institution (Washington, D.C., 1972): 3.

7. Data are from Productivity Indices for Selected Industries, 1976 Edition, U.S. Bureau of Labor Statistics (Washington, D.C., 1977).

8. Dale Jorgenson and Frank Gollop, U.S. Productivity Growth By Industry, 1947-73, mimeographed, University of Wisconsin (Madison, Wis., 1977).

9. Nordhaus, " The Recent Productivity Slowdown."

10. John Kendrick, Postwar Productivity Trends in the U.S., 1948-69, National Bureau of Economic Research (Cambridge, Mass., 1973).

11. Edward Meadows, "A Close-up Look at the Productivity Lag," *Fortune*, December 4, 1978, pp. 82-90.

12. Kendrick, *Postwar Productivity Trends*.

13. Kim B. Clark, "Unionization, Management Adjustment, and Productivity" (Working Paper No. 332, National Bureau of Economic Research, Cambridge, Mass., April 1979).

14. George L. Perry, "Labor Force Structure, Potential Output, and Productivity," *Brookings Papers on Economic Activity*, 1971: 3.

15. Meadows, "A Close-up Look at the Productivity Lag."

16. Kendrick, *Postwar Productivity Trends*; and W.E. Salter, *Productivity and Technical Change*, (Cambridge University Press, 1969).

17. "American Industrial and Occupational Requirements 1964-75," in Appendix, "The Outlook for Technological Change and Employment," in *Technology and the Economy*, vol. 1 (Washington, 1966).

18. James R. Bright, *Automation and Management*, Harvard Business School, Harvard University, (Cambridge, Mass., 1958).

19. Stephen Hollander, *The Sources of Increased Efficiency*, (Cambridge, Mass.: MIT Press, 1965).

20. William Abernathy and James Utterback, "Patterns of Industrial Innovation," *Technology Review*, 80 (June/July 1978):.

21. Robert A. Swift, *The NLRB and Management Decision-Making*, University of Pennsylvania, (Philadelphia, Penn., 1974) and Lawrence Stession, *The Encyclopedia of Collective Bargaining Clauses*, (New York, 1977).

22. Julius Rezler, *Automation and Industrial Labor*, (New York: Random House, 1968).

23. Thomas Kennedy, *Automation and Displaced Workers*, Harvard Business School, Harvard University (Cambridge, Mass., 1962).

24. Ibid., and Clinton C. Bourdon, Issues and Trends in Income and Employment Security, Report to the National Commission on Productivity and the Quality of Working Life (Washington, D.C., March 1978).

25. Kennedy, Automation and Displaced Workers.

26. Ibid.

27. Bruce Stokes, Worker Participation - Productivity and the Quality of Work Life, (Worldwatch Paper 25, Worldwatch Institute, Washington, D.C., December 1978).

28. James Driscoll, "The Scanlon Plan: Implication for Union Management Cooperation," (mimeograph, Sloan School, Massachusetts Institute of Technology (Cambridge, Mass., October 1978).

29. Harvey Belitsky, New Technologies and Training in Metalworking, (National Center for Productivity and Quality of Working Life, 1978).

30. Hollander, The Sources of Increased Efficiency.

31. Clinton C. Bourdon, "Note on Zenith's Offshore Decision," mimeographed , Harvard Business School, Harvard University (Cambridge, Mass., 1978).

32. Eugene Merchant, "Future Trends in Manufacturing - Toward the Year 2000," Annals of the CIRP, vol. 25, no. 2 (1976).

33. Sheldon Friedman, Factors Affecting the Receptivity of the Labor Force to Technological Innovation, (MIT Center for Policy Alternatives, Massachusetts Institute of Technology (Cambridge, Mass., July 1975).

7 Direct Government Funding of Research and Development: Intended and Unintended Effects on Industrial Innovation

Paul Horwitz

INTRODUCTION

In FY79 the federal government will spend over $27 billion, or over 5 percent of the total budget, on the direct support of research and development activities. [1] This substantial sum of money amounts to $125 for every man, woman, and child in the United States. More R&D is financed by the federal government in this country than by all other sources combined.

For this discussion, R&D is defined as the systematic search for new knowledge of the natural world, together with the process by which such advances in knowledge become embodied in one-of-a-kind devices of potential practical utility.

The literature on federal R&D funding is vast and is growing faster than the funding itself. This chapter presents the most important data on the current situation in this and other countries, describes what is known or can reasonably be inferred concerning the effects of such funding on technological innovation, and draws certain broad conclusions in reference to the various policy options that have been suggested. Although the government supports R&D for a wide variety of reasons, this discussion focuses on the connections between that support and industrial innovation. Some of these connections are quite unexpected and may involve indirect consequences of federal policies, procedures, and practices adopted with little or no consideration for their effect on innovation.

Also, briefly considered is the role played by demonstration projects, which are increasingly supported by federal funds. Both the role of the government in funding such projects and effectiveness of these projects in promoting or accelerating the commercialization of new technologies have been examined in several recent publications. [2]

Two broad themes running through this chapter are important to enumerate at the outset. They are:

1. R&D is neither a necessary nor a sufficient condition for technological innovation.

2. R&D is not a fungible commodity: both its desired effects and its ability to achieve them may depend critically on the context in which it is funded and performed.

Many of the points in this chapter stem more or less directly from the first statement. It is curious, therefore, to find that while most technology analysts and planners have come to accept it without question, there is still a body of opinion, both within the scientific community and, for example, in the Congress, which appears to treat the terms "R&D" and "innovation" as practically synonymous. While the output of an R&D project may correspond roughly to the completion of the creation phase of the innovation process, to equate R&D and innovation ignores the other steps in the process and discounts the fact that creative design can be the product of activities unrelated to R&D.

Since the innovation process includes many steps that do not involve R&D, actions of the federal government which affect the ease with which these steps can be accomplished, or the probability that they will be undertaken at all, also affect the innovation process. This leads to the conclusion that many government activities, seemingly unrelated to technological innovation, may have indirect, and often unintended, effects on it.

For example, during the last fifteen years, government policies, programs, and procedures affecting venture capital formation, regulation of firms, ease of entry or exit, labor training and mobility, and markets for new products have variously been identified as having significant effects on innovation. [3] Most of these effects are indirect, in that the respective government agencies are not purposefully supporting (or retarding) innovation per se. Most often these indirect effects are also <u>unintended</u>. Thus, a concern for investor protection at the Securities and Exchange Commission has given rise to regulations which affect the entrepreneur's ease of access to sources of venture capital. [4] As another illustration, too detailed specification of the characteristics of a new product to be procured by the government hinders innovation on the part of prospective suppliers, while a broader set of performance specifications has led to innovative responses. [5]

But if government actions in such widely disparate areas have had indirect effects on innovation, it should not be surprising to find similar indirect effects from the support of

R&D. Thus, in addition to its intended effect in bringing forth specific innovations, R&D support by the government may have indirect, and often unintended, effects on technological innovation.

In particular, a policy or program intended to bring about technological innovation through direct support of the early stages in the process may be either strengthened or significantly weakened by its unintended and indirect effects on the later stages. This aspect of R&D support has received little attention in analyses of science and technology policy. In discussions of government support for R&D, the emphasis is usually placed on the overall level of such support, for example, as a fraction of GNP, or as compared to other government programs or the amount R&D received last year. Such an emphasis can be misleading since it conceals a tacit assumption that R&D activities do not differ significantly from one discipline or industrial sector to another. It also ignores the importance of such factors as the motivation of the researcher, the institutional environment in which the R&D is performed, and the nature of the commercial and industrial system that produces and delivers the final innovation.

In view of the importance of these factors, it is useful to distinguish between types of R&D on the basis of three characteristics: the identity of the customer (e.g., government, public utility, industry, consumer), both for R&D and for innovation; the nature of the stated objective of the R&D (e.g., increase in knowledge, proof of principle, commercial demonstration); and the institutional character of the performer of the R&D (e.g. university, government or non-profit laboratory, large or small firm, private inventor).

The next section describes recent trends in federal funding of R&D, and the third section compares these trends with similar data from other countries. The fourth section is devoted to examining the main policy considerations which are usually adduced in this country and elsewhere to justify a role for the government in the support of R&D. The fifth section presents a taxonomy of different kinds of R&D, and uses it to structure a discussion of the direct and indirect effects of R&D support on industrial innovation. Much of this discussion is qualitative and is based on specific examples rather than proceeding from macroeconomic statistics that are generally too aggregated for our purposes. The last section summarizes the findings and addresses a number of broad policy issues arising out of them.

R&D FUNDING POLICY IN THE UNITED STATES

The federal R&D budget is not the outcome of an overall resource allocation plan resulting from a centralized decision making process. Rather, it is the result of a series of unrelated disbursements which are broken out as individual items from the budgets of many different agencies in the government and assembled by the Office of Management and Budget (OMB) into an Appendix to the President's Budget Message to the Congress.

The process by which the money is requested by the different agencies, the funds authorized and appropriated by the Congress, and the disbursements finally made is a somewhat arcane one involving many separately identifiable steps on the part of the funding agency, the OMB, and the Congress. In recent years the American Association for the Advancement of Science has commissioned an annual analysis of the resultant "R&D budget." [6] As a result of this excellent document, it is probable that more is known about how and where R&D money is spent in the United States than in any other country.

Although overall funding priorities for R&D are generally set in the White House, the decisions affecting R&D outlays within an individual government agency are usually the outcome of a complex interplay between the agency, the OMB, and relevant committees of the Congress. This makes coordination of such decisions, particularly when they come under the jurisdiction of different groups, both difficult and unlikely.

Another point to bear in mind is that by no means are all the R&D activities in the country supported by the federal government. On the contrary, as Fig. 7.1 shows, government R&D outlays account for only slightly more than one half the R&D funding in the United States, a ratio that has remained remarkably constant over the past decade. Fig. 7.1 also shows that overall R&D financing, both federal and nonfederal, has remained very nearly constant during this period, once the effects of inflation are taken into account. (It is argued that the GNP implicit price deflator used by the National Science Foundation and others to reduce all dollar amounts to a given year is unsuitable for R&D expenses, and that the appropriate deflator is actually higher, [7] but the points to be made here do not depend on such subtleties.)

The impression given by Fig. 7.1 of a relatively unchanging U.S. policy with respect to R&D is altered by Fig. 7.2, which displays the distribution of federal R&D funds. It is clear, in fact, that a major shift has been taking place over the past decade in the relative priorities given to various program areas. Space activities have declined as a percent of the total federal R&D budget; civilian R&D has increased; and defense has held fairly constant. This trend toward civilian R&D is unprecedented in this country: for the first time, 40 percent of federal R&D outlays are directed at the civilian economy.

This shift is more fundamental in one sense than the earlier one associated with the rise of space-oriented R&D during the 1960s. The National Aeronautics and Space

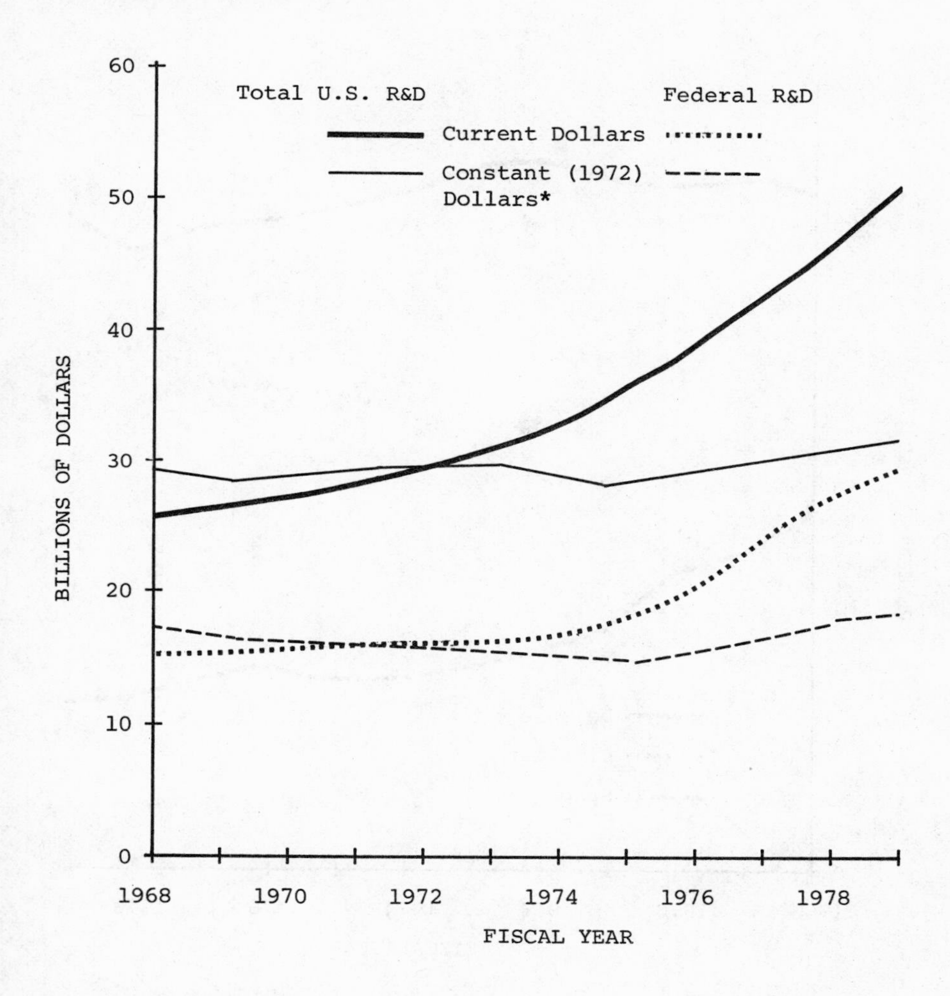

FIGURE 7.1

TRENDS IN TOTAL U.S. AND FEDERAL R&D EXPENDITURES

*Based on the GNP implicit price deflator with an estimate for 1978 and 1979.

Source: "Science and Technology Report," October 1978, Committee on Science and Technology, U.S. House of Representatives, 95th Congress (Washington, D.C.: U.S. Government Printing Office).

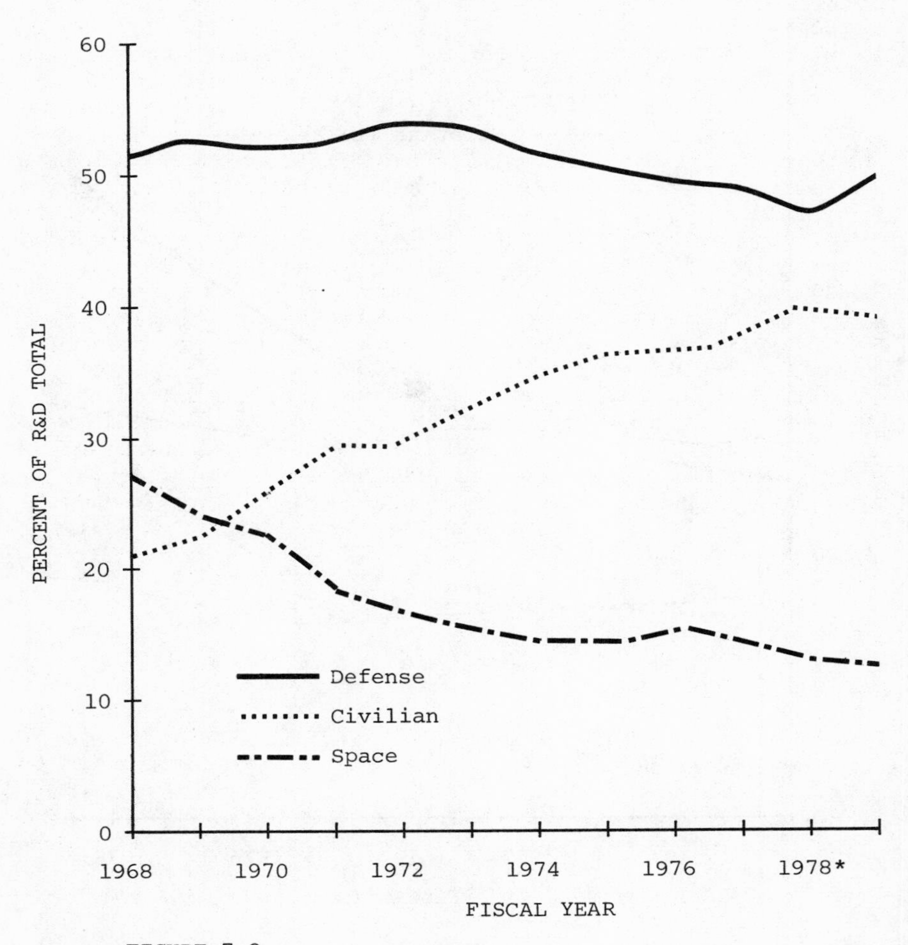

FIGURE 7.2

TRENDS IN THE DISTRIBUTION OF FEDERAL R&D OBLIGATIONS BY MAJOR PROGRAM AREA

*1977 to 1979 estimated from President's 1979 budget.

Source: "Science and Technology Report," October 1978, Committee on Science and Technology, U.S. House of Representatives, 95th Congress (Washington, D.C.: U.S. Government Printing Office).

Administration was the ultimate user of almost all the products that it was underwriting, much as the Department of Defense had traditionally been. This is not the case with most of the civilian R&D which has overtaken the declining space budget in recent years. The innovations which are expected to flow from R&D in this category are usually commercial products to be sold to customers, whether public or private, quite distinct from the federal agency that funded the inception of the innovations (although the money used to buy them may in certain cases be provided, directly or indirectly, by that agency). In this case, the funding agency is the judge of the quality and usefulness of the R&D, but the ultimate utility of the resulting innovation, if any, will be determined by an end-user distinct from the federal government, and possessing quite different needs, desires, resources, and constraints. The difference between these two kinds of R&D is therefore a fundamental one.

Who are the funders and who are the intended users of federally funded civilian R&D? Fig. 7.3 provides an answer to the former question: the two agencies with the largest budgets in the civilian technology area are the Department of Energy and the Department of Health, Education, and Welfare, the latter largely because it includes the National Institutes of Health. The National Science Foundation, the primary agency charged with the support of basic research in the sciences, is a distant third.

Fig. 7.4 displays the allocation of R&D resources by function. It will probably come as no surprise that there have been substantial increases in R&D since 1969 in the areas of energy and environmental protection. Less well known, perhaps, is the reflection in the R&D statistics of the increased attention given in recent years to such areas of public concern as transportation, food, and natural resources. It is clear from these data that there is relatively little support in the United States for innovation for general economic purposes such as increasing manufacturing productivity. This is in sharp contrast with the situation prevailing in most other industrialized nations.

R&D FUNDING POLICY IN FOREIGN COUNTRIES

Fig. 7.5 displays government funding of R&D as a fraction of the total budget for seventeen industrial nations ranging in size from the United States to Ireland. Two nations, the United States and France, stand out as spending more, proportionately, on R&D than any other country. The United States spends 7 percent of its total federal budget, while France spends close to 8 percent.

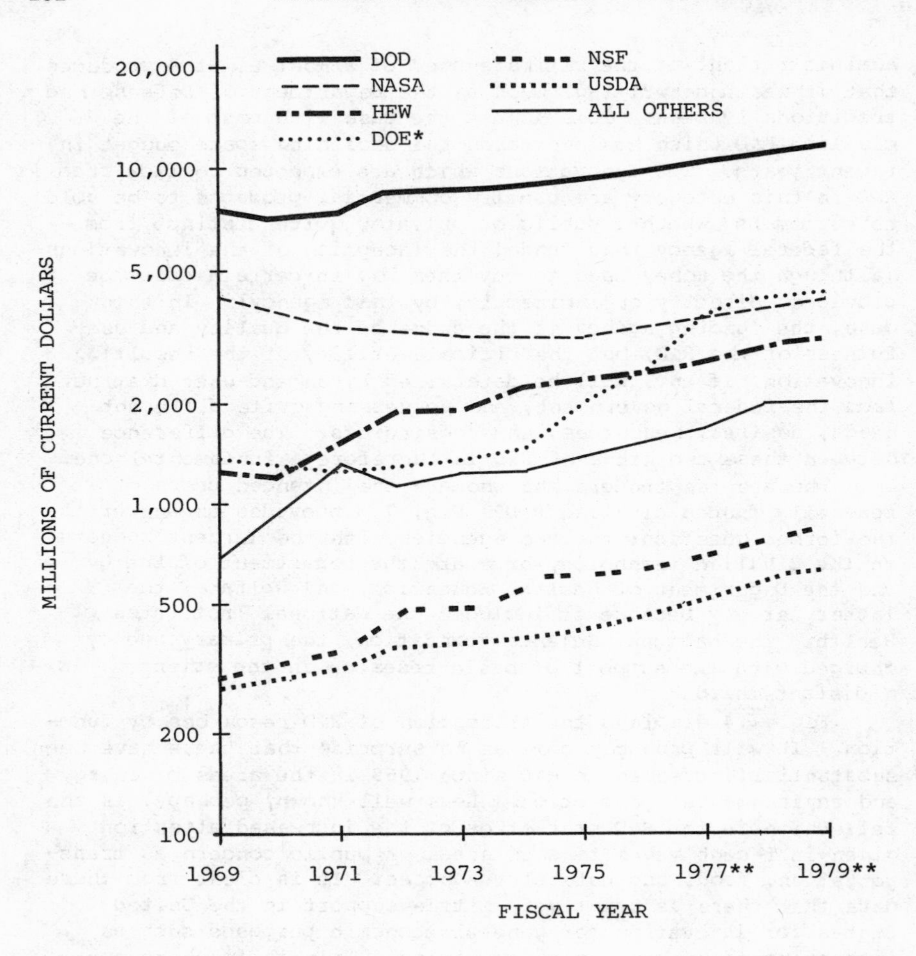

FIGURE 7.3

TRENDS IN R&D OBLIGATIONS OF FEDERAL AGENCIES LEADING IN R&D PROGRAMS

*From 1969 to 1974, AEC data were used; from 1974 to 1976, ERDA data were used.

**1977 to 1979 estimated from President's 1979 budget.

Source: "Science and Technology Report," October 1978, Committee on Science and Technology, U.S. House of Representatives, 95th Congress (Washington, D.C.: U.S. Government Printing Office).

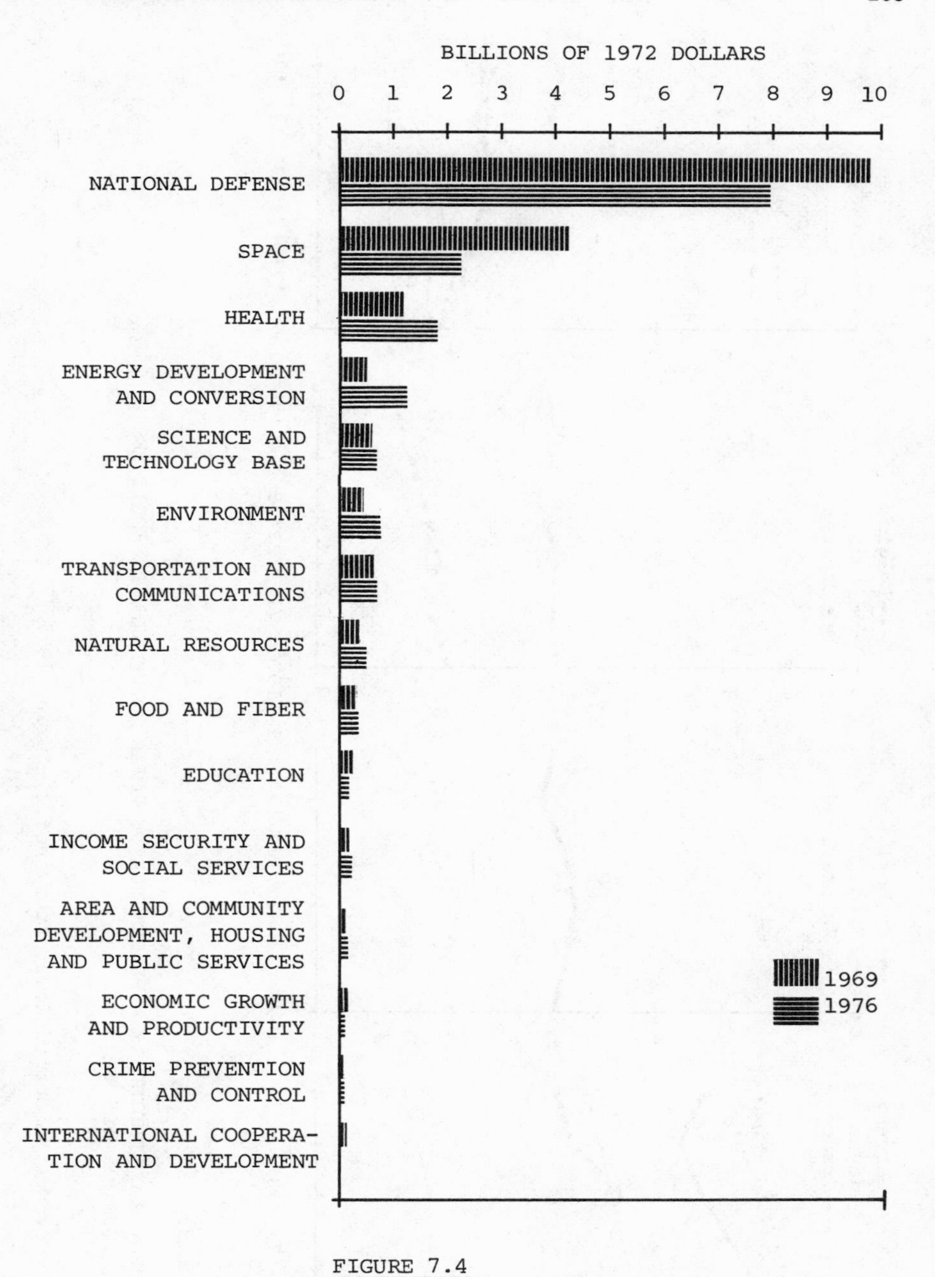

FIGURE 7.4

FEDERAL OBLIGATIONS FOR R&D BY FUNCTION

Source: Science Indicators 1976, National Science Board, National Science Foundation (Washington, D.C., 1977).

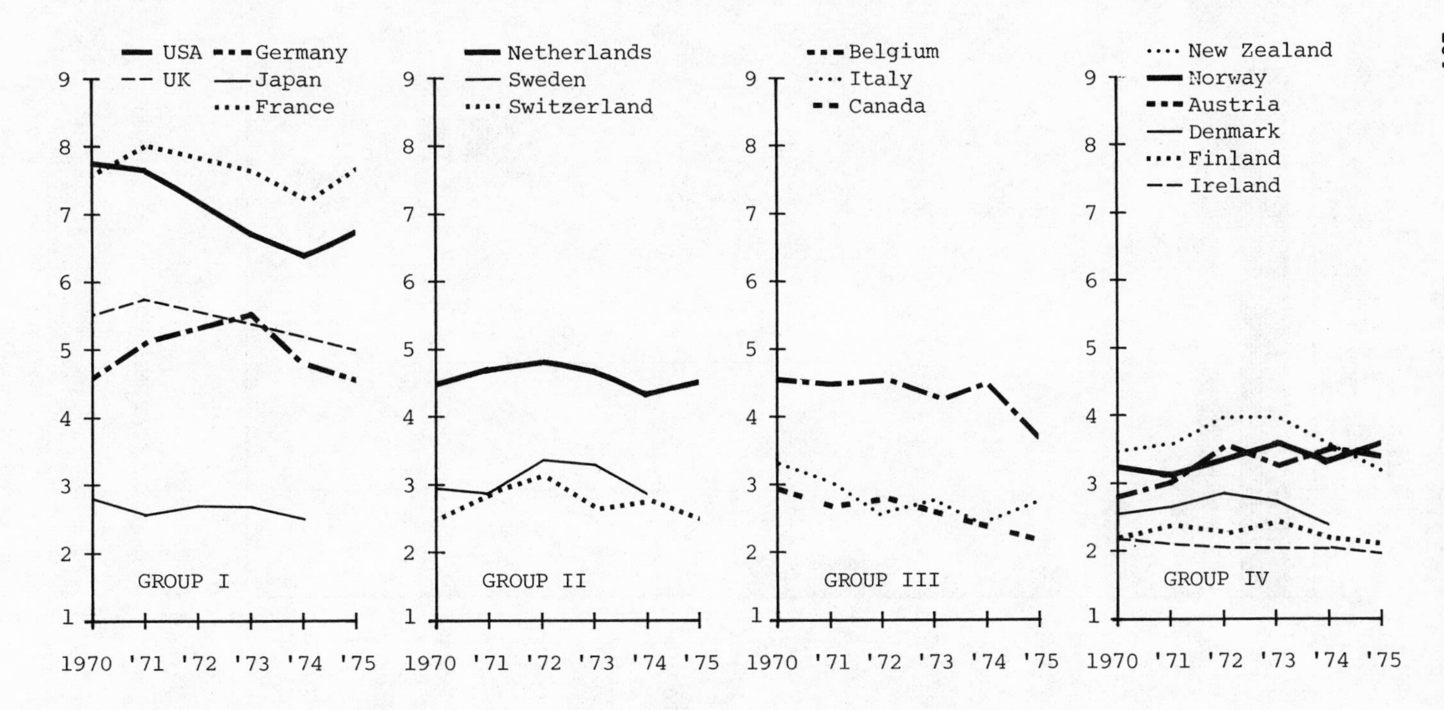

FIGURE 7.5

GOVERNMENT R&D FUNDING AS A PERCENTAGE OF TOTAL GOVERNMENT EXPENDITURES IN FOUR COUNTRY GROUPS

Source: Science Resources Newsletter, no. 2, Spring 1977

This is a striking fact, particularly in light of the relatively poor showing of France in technology development and innovation [8] when compared, for example, to Japan which spends only 3 percent of its budget on R&D. Perhaps equally surprising is the fact that Great Britain and Germany each devote about 5 percent of their budget to R&D, yet it would be hard to find two countries less alike in terms of industrial innovation. It is clear from these examples that while government sponsorship of R&D may have something to do with innovation, it does not ensure it.

The R&D sponsored by these countries is applied to a variety of social goals. Fig. 7.6 shows an estimate of this distribution between seven areas, for six countries. Once again, the United States and France look similar, this time in their emphasis on R&D for national defense. They are joined in this respect by Great Britain, the only country besides the United States which spends more than half its R&D on defense. Germany and Japan, on the other hand, are distinguished by having the greatest concentration on research directed at the advancement of knowledge. The United States is unique in devoting fewer resources to economic development than any other of the countries shown.

It can be shown using the data from Fig. 7.6 and data on productivity growth that there is a strong positive correlation between economic growth and the fraction of R&D devoted to the category "advancement of knowledge," while a similarly strong, but negative, correlation exists between growth and emphasis on defense-related R&D. [9] This correlation does not necessarily imply, however, that the royal road to economic prosperity is via the diversion of R&D funds from defense to basic research. The data do not give a clear picture of just what is included under the rubric of "advancement of knowledge," and there is some evidence that it may mean different things in different countries. In particular, there is a very significant difference between knowledge which is useful for science and that which is useful for technology. Inasmuch as "advancement of knowledge" can refer to activities as diverse as radio astronomy and corrosion research, it would seem a rather vague measure of support for innovation.

It should be noted that every nation studies the science and technology policies of every other, hoping to learn something of value to itself. The European nations, in particular, have in many cases consciously imitated the science and technology policies of the United States. They have been impressed by the success of the United States in encouraging the development of technology and have feared that this success could have an adverse effect on their balance of trade. Thus, starting in the 1960s, many nations increased their R&D spending as a way of "keeping up with the Joneses" in many of the areas (e.g., aircraft, computers, and electronics) in which the United

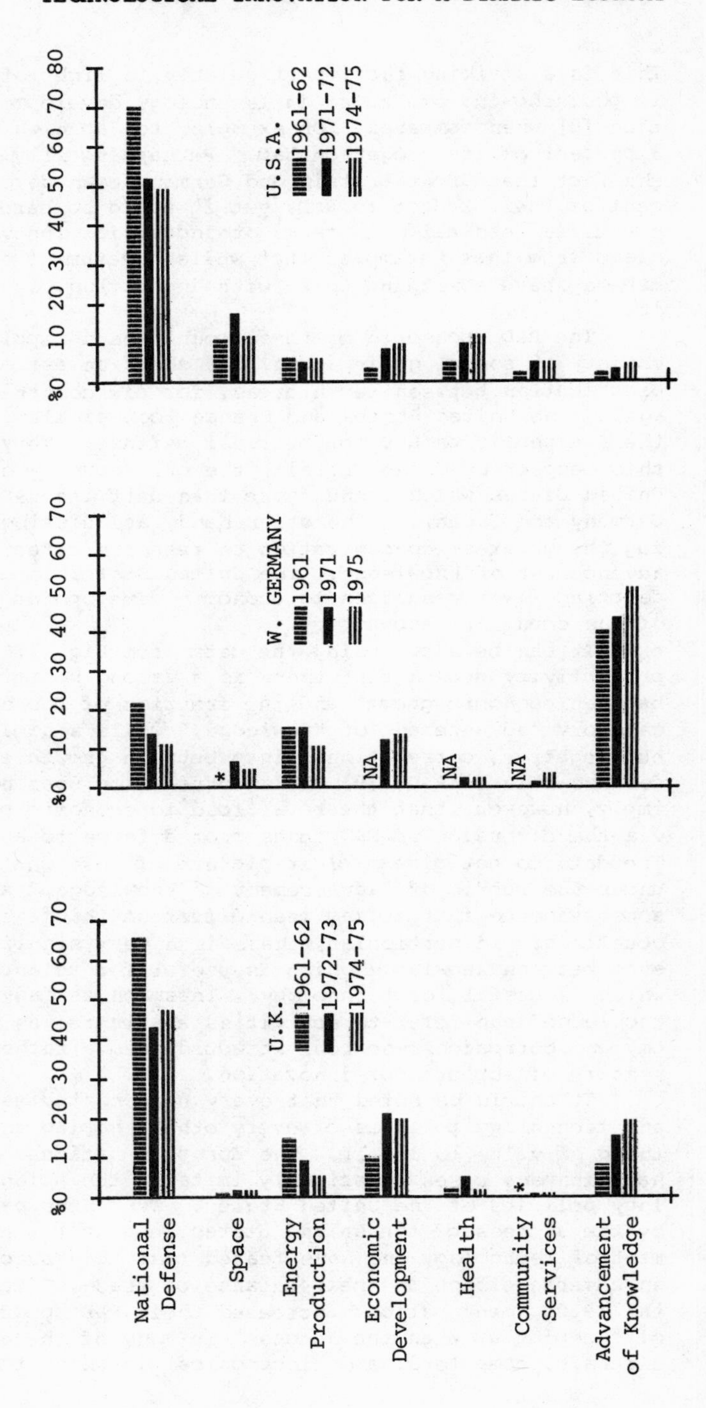
U.S.A.
1961-62
1971-72
1974-75
%0 10 20 30 40 50 60 70 80
W. GERMANY
1961
1971
1975
%0 10 20 30 40 50 60 70
*
NA
NA
NA
U.K.
1961-62
1972-73
1974-75
%0 10 20 30 40 50 60 70
*
National Defense
Space
Energy Production
Economic Development
Health
Community Services
Advancement of Knowledge

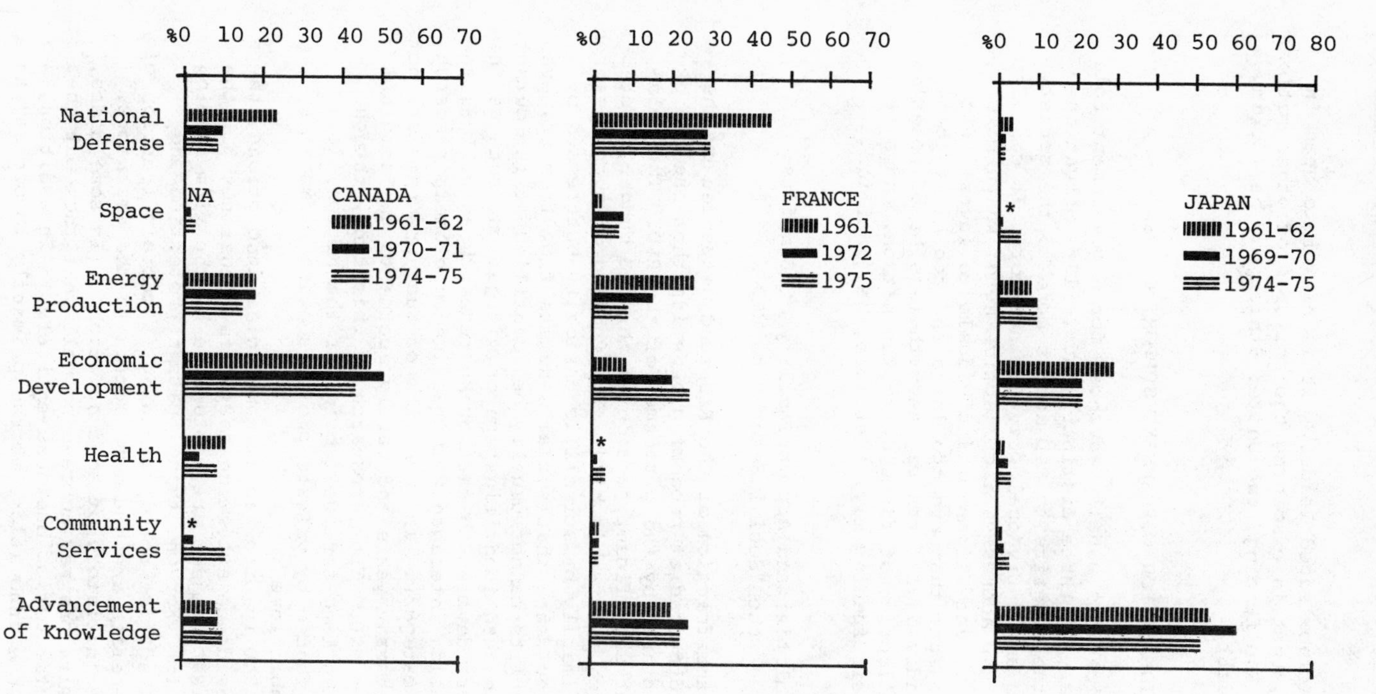

FIGURE 7.6

ESTIMATED DISTRIBUTION OF GOVERNMENT R&D EXPENDITURES AMONG SELECTED AREAS BY COUNTRY

* Less than 5%

Source: Science Indicators 1976, National Science Board, National Science Foundation (Washington, D.C., 1977), p.56.

States was the acknowledged leader. It is useful to bear in mind, therefore, that in examining the policies of other nations with respect to R&D funding, the United States may be looking at traces of itself.

RATIONALE FOR GOVERNMENT SUPPORT OF R&D

The justifications usually advanced for a government role in R&D funding are of three distinct kinds. [10] Situations exist in which the benefits to R&D accrue more to the society at large than to any individual private investor. It can also happen that the R&D addresses a collective need, which makes it impractical for private firms or individuals to invest in it. Finally, R&D projects that are too large and too risky to be undertaken by individual firms may nevertheless be of potential benefit to a collection of firms or to the nation, and a federal role in their funding may, therefore, be justified.

Externalities and Appropriability of Benefits: the "Social Return" to R&D

A significant fraction of the R&D funds expended by the government are directed at producing benefits that are not expected to be captured by the performer of the R&D. The relatively small but significant portion of the R&D budget devoted to basic research is of this kind. The funding of such research is motivated primarily by scientific curiosity buttressed by the historical evidence that social benefits flow from it, even though such benefits cannot usually be identified at the outset. R&D devoted to hazard identification or mitigation, such as the increasing effort devoted to earthquake research as well as the recent $100 million interagency initiative on climate research, is in the same category. The benefits of such programs are not usually felt for many years and rarely accrue to the original performer of the R&D. Thus, research of this kind, though important for the long-term well being of the nation, is unlikely to be funded by private profit-seeking firms at a socially desirable level.

Even when R&D is directed at commercial innovations, there is increasing evidence attesting to the fact that not all the benefits of an R&D project are capturable by the firm funding and performing it. For one thing, patent protection and technological lead time notwithstanding, a certain fraction of the economic benefits are bound to accrue to the imitators, rather than the initiator, of an innovation. (In some cases, in fact, the imitators may succeed so well that they put the original innovator out of business entirely.) In addition, benefits from some innovations accrue primarily to society as a whole, rather than to the innovating firm. For example, the health benefits resulting from the adoption of a new scrubber technology on fossil fuel fired power plants may not be reflected in profits to the innovator sufficient to justify the investment necessary to develop the technology. This results, in part, from the fact that the costs of air pollution from unscrubbed plants are not borne in full measure by the polluting firm. The presence of such externalities, to the extent that they are not internalized in the form of higher prices, for example, through regulation, constitutes one of the justifications for federal support of civilian R&D.

Economists have devised various means of measuring the "social returns" from R&D and have found not only that they are very high, but that they are significantly higher than the "private returns" to the investing firms. For example, by a careful case-by-case investigation of seventeen specific innovations, Mansfield and his co-workers found a median of 56 percent per year for the social return to R&D and 25 percent per year for the private return. [11] Terleckyj, using a somewhat different method, found returns of 80 percent and 30 percent, respectively, to R&D funded by the firm itself. [12] The disparity between the two results is easily accounted for by differences in method, but the point is clear: the returns to R&D were quite high in both studies, and the social returns were significantly higher than those to the innovator. Thus, there would appear to be strong justification for a federal role in the funding of R&D.

The straightforward reporting of the results of measurements such as these has a tendency to disguise both the difficulties in making the measurements and their inherent ambiguities. Even the rate of return of R&D to the investing firm can be difficult to judge accurately, since it involves a judgment of exactly where an invention originates, [13] a careful accounting of the cost of unsuccessful R&D, and the contribution of R&D performed outside the organization. In estimating the social returns to R&D, attention must be paid in addition to effects on displaced products and to interindustry flows of technology. Terleckyj's results, in particular, show a strong effect of R&D performed in one industry on the productivity gains of another.

An interesting feature of Terleckyj's work is that he found no effect on the productivity increase of an industry or a firm resulting from government supported R&D. This is not too surprising since most government financed R&D is not intended to improve productivity, but rather to produce specific innovations for use by the government. Since the productivity of the government is arbitrarily fixed at unity in Terleckyj's study, the introduction of such public goods, no matter how innovative, has no measurable effect.

Indivisibilities and Economies of Scale: Public Goods

Certain functions of the government are undertaken because it would be impractical or inefficient for smaller groups to perform them or because their benefits necessarily accrue to the nation as a whole. Thus, not only is it impractical (not to mention dangerous!) for individuals or small groups to raise armies to protect themselves, but the protection afforded every American citizen by the armed forces of the United States necessarily accrues to them, whether they want it or not. Therefore, it is considered reasonable to require that they help to defray the cost of raising, maintaining, and equipping a national military force.

Military hardware is thus a public good, bought by the federal government for a public purpose. Modern military equipment encompasses a great deal of innovation, and the R&D required to produce this innovation is a public good in the same sense and is appropriately paid for by the government. The justification for a federal role in this instance resides in the indivisibility of the protection from foreign enemies which is the ultimate purpose of the R&D.

A similar justification may be made in certain cases on grounds of efficiency. For example, certain kinds of data are sufficiently important to a wide variety of firms that each of them is motivated to perform whatever R&D is necessary to obtain them. Rather than have individual firms duplicate one another's work, it is useful to have a centrally financed facility where such work can be done, and the results made available to everyone. Much of the industrially oriented work done at the National Bureau of Standards is of this kind.

Attitudes Toward Risk Taking

A private firm, no matter how profitable it may be, is often forced, by what Klein in Chapter 3 describes as "Type I uncertainty," to be more risk averse than is optimal for the long term prosperity of the nation at large. The firm may feel that it cannot afford to place itself in a position in which it may sink or swim based on the outcome of a long-term high-risk R&D project, even if the rewards to the success of the project are considerable. The more prudent course to follow is to invest very modestly in such a product, primarily to keep tabs on the general advance in the state-of-the-art, and to avoid, as far as possible, any technological surprises which may be initiated by the competition. This is the way most firms are managed, particularly if they are producing a standardized product line using highly automated and capital intensive process machinery. Yet, from the point of view of society, what is fiscally responsible for an individual firm can lead to stagnation, inflexibility, and reluctance to innovate in the industry of which it is a part. This may ultimately result in greater economic hardship and peril for every firm in that industry if external conditions change and they are unable to adapt.

By this reasoning, a federal role in the support of R&D might be indicated, if the government is willing to assume greater risks than the private sector. Two comments are in order however. First, the criterion that the government should assume risks runs directly counter to the desire for greater accountability in the conduct of public affairs. [14] Second, there are alternatives to government sponsorship of R&D which may be more effective in stimulating innovation.

The inherent conflict between desirable risk taking and the need for accountability is fundamentally unresolvable. The above discussion warns agains judging an R&D program merely on the basis of its successful projects. Too high a success ratio, in fact, might even be construed as evidence of too strong a tendency to place bets on a sure thing, and too little willingness to take the risks necessary for the long range prosperity and stability of the nation. But one man's great adventure is another man's boondoggle. In the present political climate it is difficult for the head of a federal R&D program to argue, for example, that the funding of a substantial number of projects which failed to achieve their hoped-for objectives is evidence that the program has been successful in supporting projects too risky for the private sector. It is nevertheless true that under these circumstances the agency may have been behaving in an appropriate fashion in assuring that promising, but risky and unproven, ideas be accorded a reasonable examination.

Faced with an aversion to risks on the part of firms, the government has a second set of options available to it. A more promising alternative to federal support for high-risk R&D might be to arrange that the firms in question be motivated to do the R&D themselves. As Ashford, Heaton, and Priest point out in Chapter 5, one possible way of doing this is through creative use of regulation. New regulations, properly timed and applied, can have the effect of changing the rules of the game. By forcing all firms to make technological changes which they would not otherwise be motivated to make, new regulations create an atmosphere in which innovation can take place. An example is the stratified-charge engine brought out by Honda in response to emission control requirements. Similarly, proposed requirements limiting emissions of solvents in conventional inks have given rise to increased interest in inks cured by ultra-violet light within the graphic arts industry.

Another option open to the government is simply to promote competition, and especially to promote new entries in oligopolistic situations. Since such newcomers typically rely on technological advances to find an initial market niche, their presence, or the threat of it, tends to act as a powerful incentive for innovation on the part of the existing firms, as Klein points out in Chapter 3.

The two alternative kinds of government action described above are not mutually exclusive. On the contrary, they are complementary. For example, a policy of offering funding primarily or exclusively to would-be new entrants to conduct R&D on alternatives to existing technology might have a very strong "multiplier effect" in eliciting a response from firms already in the field. Similarly, a program of regulation coupled with support for R&D might prove more effective in promoting innovative responses to social needs and desires than either program alone.

EFFECTS OF R&D FUNDING ON TECHNOLOGICAL INNOVATION

It is necessary at the outset to make a clear distinction here between a _program_ which is set up to support R&D of a particular kind for a specified purpose, and an individual R&D _project_ which may be funded under such a program. The importance of this distinction stems from the different criteria that are appropriate for evaluating the success of programs and projects.

From the point of view of its effect on innovation, it is reasonable to judge an R&D _project_ primarily on the basis of the utility of the results obtained to the intended (or perhaps some other) end-user. To apply the same set of criteria to the evaluation of an R&D _program_, however, would be to judge such programs on the basis of the fraction of their projects which succeed, i.e., lead to commercially useful innovations. This, as previously argued, would militate against precisely the sort of high-risk, long-shot projects which in the aggregate can contribute immensely to the vitality and long-term stability of the country, but for which the private returns are too small and too risky for firms to invest their own resources on R&D.

Categories of R&D

In examining the connection between government sponsorship of R&D and technological innovation, distinctions must be made between different kinds of R&D performed for different purposes. It is useful to divide government activities in support of R&D along lines which take into account the motivations of the funder and of the performer. As illustrated in Fig. 7.7, the major such division is between research aimed at increasing the total store of human knowledge about a broad scientific or technological area, i.e., basic research, and R&D designed to produce a specific innovation or group of innovations to serve a particular function, i.e., applied R&D. Within the basic research category, the distinction between science research and technology research is based on whether the primary goal of the acquisition of new knowledge is, or is not, the facilitation of the innovation process. The distinction between these two subcategories is not a standard one and is illustrated below by a few examples.

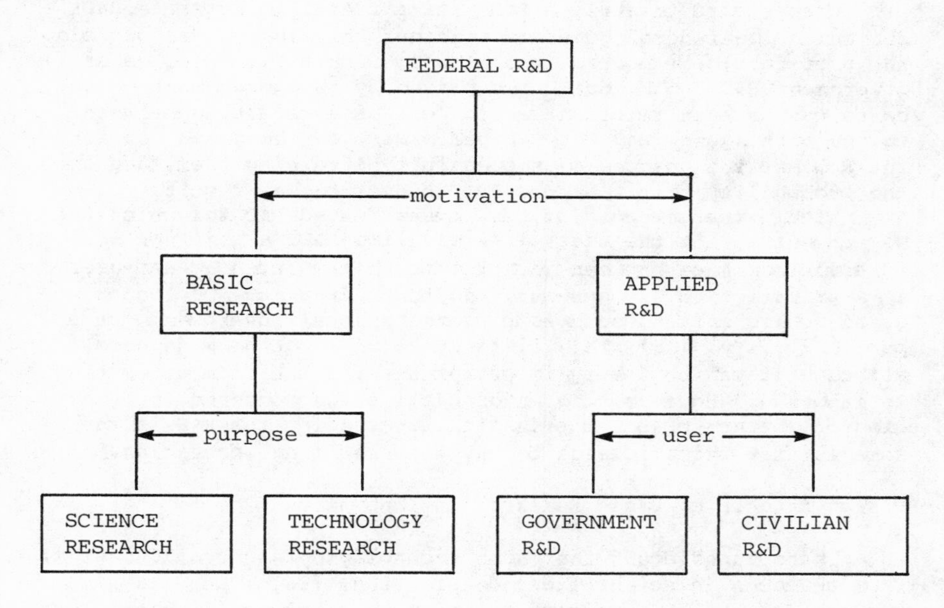

FIGURE 7.7

A TAXONOMY OF FEDERAL R&D EXPENDITURES

Research aimed at describing and understanding the interactions of elementary particles is considered science research, even though it can lead to technological innovation, for example, through its use of sophisticated instrumentation. Advances made in accelerator design for use in high energy physics have, in fact, been incorporated in commercial innovations. In addition, there is the very real possibility that such research may eventually open up wholly new opportunities for revolutionary innovation through a better understanding of the fundamental nature of matter. For example, exotic states of matter composed entirely of heavy quarks have been proposed as a highly compact medium for energy storage. The critical point, however, is that such innovations, important though they may be, are basically serendipitous, and are not the main motivating force either for the individual doing the research or for the sponsoring agency.

An example of technology research would be a fundamental investigation of the chemistry and physics of corrosion processes. Such work, although it may have considerable intrinsic scientific merit, is motivated in large part by the practical effects it may and is expected to have in the short term, on a wide range of new products and processes.

With regard to applied R&D, it is useful to separate R&D for which the federal government is not only the funder but also the user for the expected innovation. Such R&D is denoted as government R&D. The dual role played by the government in such cases can have an important effect on the interaction between the support agency and the R&D performer, on the goals set for the R&D and the chances of successfully attaining them, and on the probability that a successful innovation may result.

Within the category of government funded R&D for which the government is not the user, i.e. civilian R&D, a distinction is sometimes made between work for which the intended end-user is a private firm or consumer, and that aimed at a public or quasi-public market, such as a state or local government, or a public utility. [15] This distinction will not be made here, although it can be a very important one for the firm attempting to innovate. However, the major distinction concerns the question of whether or not the institution funding the R&D is the same as that which intends to buy the resulting innovation.

Direct Effects of Basic Research on Innovation

There is no doubt that some technological innovations stem from advances in scientific understanding. [16] There is also no question that the process by which this happens is often indirect, serendipitous, and time consuming. The person doing the research is rarely the same person who recognizes the opportunity it presents (very often when combined with other research results); puts it in the context of a perceived user need; develops the technology necessary to put it to use; produces a prototype; and finally manufactures, distributes, sells, and services an innovative product. Typically, in fact, no single individual performs all these separate tasks. Rather, each task may be accomplished by quite different people, operating in different institutional environments, and rarely in communication with one another.

A great deal of attention has been given in the past to a process known as "technology transfer," by which research results funded by the government are supposed to be communicated to an industrial user who can apply them to problems of the "real world," and produce innovations based on them. Arising out of a concern that this technology transfer was not taking place as rapidly or as smoothly as it might, proposals have been made for intervention by the federal government aimed at improving the communication links between the various individuals and institutions involved in the innovation process. In this way, it is hoped that the flow of scientific research results into new, technology-based products will be accelerated.

More recent work into the nature of the innovation process has cast doubt on the simple linear model on which this idea is based. The difficulties encountered in an attempt to apply basic research results to innovation have proven to be more fundamental than merely problems of poor communication between different groups, and in any case, government efforts to improve such communications have not been notably successful. [17]

It would appear that communication may not be the only barrier preventing the rapid adoption of basic research results by industry. The applications of such research are generally not apparent at first sight, and too early an effort to force such application can result in disastrous failure for the firm attempting it. It is questionable, in fact, whether the government has the resources to guess any better than the entrepreneur which particular applications of scientific advances are the ones whose time has come. Thus, rather than concentrating on solving the technology transfer problem, the government might better turn its attention to the quite different question of how to improve its support for basic technology, as distinct from science.

Technology research includes work undertaken to develop new knowledge useful to a broad range of potential innovations. Included among these are incremental improvements in existing technology, which have been shown to have a very important effect on industrial productivity. [18] The motivation for this work is significantly different from that for science research, although there is, obviously, an appreciable area of overlap.

An example of technology research is that undertaken by the National Advisory Committee on Aeronautics during the 1930s into the basic technology of aircraft design and construction. The benefits of this program, while they did not accrue exclusively to any single aircraft manufacturer, resulted in the growth and prosperity of a highly innovative aircraft industry. It is noteworthy that, in contrast to the situation in the United Kingdom, the U.S. Government did not directly support the design of new commercial aircraft, but limited its assistance to the support of generic R&D of utility to the industry as a whole. The Department of Defense has, of course, supported the design of military aircraft, some of which have been adapted to commercial uses. Although it came close with the SST, [19] the U.S. Government has never assumed the burden of bringing a new commercial aircraft into production. The example of the British/French Concorde, which, though a technical success, is too high-priced and too expensive to operate commercially without massive government subsidies, supports the wisdom of this approach. [20]

Another example of technology research is to be found in the programs of the West German Fraunhofer Society for the Advancement of Applied Research, cited by Hollomon in Chapter 8. The Society receives joint funding from the government and industry, and has the objective of transferring new knowledge to industry. Another German initiative, the "new technologies" program, started in 1969, finances the exploratory development in industry of potential new commercial products and processes in such areas as high speed ground transport and biomedical technology. [21] No equivalent programs exist in the United States, although there are individual cases in which the government has supported the development of technology basic to a specific industry. However, such actions have been taken on an idiosyncratic basis, and no overall policy guidelines applicable to a variety of cases appear to exist.

The case of the American nuclear power industry illustrates a consequence of the lack of attention by the government to the support of technology research. Initial R&D on a wide variety of reactor designs was undertaken during the 1950s primarily by the government, both in national laboratories owned by the Atomic Energy Commission (AEC) and in privately owned industrial facilities. However, pressure on the AEC, particularly from coal industry interests, to eliminate the "subsidization of private industry" led to a decision in 1962 to discontinue most of these projects, and to concentrate government sponsored R&D on particular types of advanced convertors and reactors. [22]

In retrospect, however, it is clear that insufficient attention was paid at the time to a federal program of support for technology research in two specific areas of vital importance to the industry as a whole, but in which private firms had insufficient incentive to invest, namely reactor safety and the management of the fuel cycle, with particular emphasis on waste disposal. Failure to solve the technological problems in these areas in a timely and socially acceptable way has done incalculable harm to the nuclear power industry; yet it is difficult to see how individual firms in that industry could have been expected to expend their own resources to the extent needed to solve problems of interest and importance to the entire industry. A case can be made, therefore, either for a government role in the support of R&D on problems of safety and the fuel cycle, or for some alternative action, e.g., regulation aimed at inducing firms in the industry to carry out such research themselves. The role of the government in this case is justified by the presence of both indivisibilities (the research conducted by one firm would simply have to be duplicated by the others) and externalities (the costs to the public of hazards associated with reactor operation and with disposal of radioactive waste products are insufficiently internalized by regulation).

The problems associated with certain industrial operations essential to the nuclear industry appear to offer an unusual opportunity for government support for a new technology of broad potential usefulness. Such tasks as fuel reprocessing and reactor decommissioning require a flexibility of approach attainable at present only with human labor. However, they must take place in circumstances where high levels of radioactivity present an unacceptable health risk and preclude the use of manpower. The arguments made above concerning indivisibilities and externalities justify an effort on the part of the federal government to solve this problem, which is of crucial importance to the nuclear industry. Federal assistance for the development of automated, flexible machinery, both by R&D support and through procurement of early prototypes, would have very important effects not only on the nuclear industry, but on a broad range of manufacturing industries.

The importance of government support for basic technology can hardly be overemphasized. Programs of this kind exist in virtually every industrialized nation in the world, as Hollomon discusses in Chapter 8. In the United States, however, this type of R&D has a tendency to fall between two stools. It is too diffuse and uncertain in its effects to be undertaken by private firms, but being motivated by a desire to advance technology, as opposed to science, it is alien to many scientists. In particular, the National Science Foundation, which is the lead agency for basic research, is almost wholly oriented toward the university science departments and is not in a good position to evaluate and fund research proposals based on their technological promise. In recognition of this problem, a recent National Science Foundation program, Small Business Innovation Applied to National Needs, makes explicit use of private venture firms for the evaluation and eventual support of R&D projects of potential commercial significance.

Direct Effects on Innovation of Applied Research

This section deals with the related questions of how to evaluate as well as how to affect the success of those federally funded R&D programs designed to result in specific innovations. An important dichotomy exists between those innovations for which the federal government is the sole or major customer, government R&D, and those intended for nonfederal adoption and use, civilian R&D.

The present apparatus for R&D funding by the government has grown up largely in response to a perceived need for an increased output of innovations for the government's own use, mainly in the areas of national security and space exploration. More recently, however, in response to a variety of perceived social needs ranging from a secure supply of energy to an improved system of law enforcement, there has been an increasing tendency to procure R&D for the purpose of accelerating the development of technological innovations whose ultimate user may be an industrial firm, a public utility, a division of state or local government, or even a private individual. It is evident that the motivation, justification, and evaluation of R&D programs are, and should be, quite different in the two cases.

In the case of government R&D the primary criterion for judgment is clearly whether the funding of R&D has resulted in innovations which work, i.e., serve the needs of the sponsoring agency. Other considerations obviously include such factors as cost and timeliness of delivery of the innovations, as well as the effectiveness of the program in supporting a technical base capable of producing the innovations which will be needed in the future. Conspicuously absent from this list of criteria are any considerations of the relevance to its overall mission of the agency needs which prompted the R&D, though, clearly, if these have been misrepresented or misunderstood, the innovation may not adequately serve the social or national goal to which it is directed. Thus, an innovation may be deemed a "success" if it meets the needs set down by the sponsor, but fail nevertheless to contribute as expected to the societal goal toward which it is directed. The successful development of a weapons system which, upon deployment, proves ineffective or unnecessary is an example of this possibility.

Contrast this with the situation regarding civilian R&D. First of all, there is the added complication of avoiding as far as possible the displacement of private funds. Ideally, this entails a careful scrutiny of individual projects and results in the government undertaking ventures which are too risky to be attractive to the private sector or ones for which the expected social return is significantly greater than the private return, or where the presence of indivisibilities makes a federal role appropriate. Furthermore, each R&D project funded by a program in civilian technology will require a detailed appreciation of the needs and requirements of the specific market at which the innovation is directed. Since the success or failure of such an individual project will be judged substantially on the degree to which the innovation is adopted and used in the private sector, such factors as a misreading of the market or lack of appreciation of institutional, financial, or legal barriers to its adoption can be critical in deciding the outcome.

It seems clear that the procedures which have been developed over time for the procurement of R&D for the government's own use are inappropriate for the relatively new area of civilian R&D. [23] In particular, too little attention is given under current procedures to the problems of marketing the intended innovation. Marketing can be relatively straightforward when the funding agency is able to specify its own needs. But, as any innovative firm can attest, it is quite a different task to understand someone else's problem with the accuracy and completeness necessary to define and constrain the scope of the required R&D effort.

Unfortunately, the marketing problem becomes harder to solve as the government spends more R&D funds on a project. Government expenditures in support of a given innovation tend to distort the market for it, and the greater the level of R&D support the greater such distortion will be. It is extremely difficult to sort out and identify this distortion and, thereby, to evaluate the "real" demand for the innovation under development. In a market economy, government support for an innovation is expected to continue for a limited period only, and the innovation must eventually find its own commercial niche. The question of the appropriate time to "cut the umbilical cord" can be very difficult to answer and involves, among other things, the evaluation of the objective market demand for the innovation. It is made doubly difficult by the political problems involved with the building up of a constituency which favors continuation of the project.

The problem of government sponsorship of civilian R&D has been successfully approached in the case of agricultural research by the Extension Service of the Department of Agriculture. This example has been widely studied by science policy analysts, and suggestions have been made that it be adapted to fit the needs of research in technological areas of importance to wide sectors of industry. The policy of subsidizing the development of new products directly in firms has been employed in France, Germany, the United Kingdom, Sweden, and Japan; but it has been attempted in the United States only in special cases (see Chapter 8). Although such subsidies would constitute a shift in the traditional relationship between government and industry in the United States, there would appear to be a persuasive argument in favor making such a shift in those cases where they are justifiable under the criteria outlined. In any event, there is general agreement that some mechanism for providing adequate feedback between the performer and the user of the research is necessary, both in order to ensure the relevance of the research projects undertaken and to provide the close coupling necessary for the adoption of successful innovations.

The question of how far the government should go down the road toward the commercialization of particular innovations is one which has received a great deal of attention. [24] The general impression one gets from a study of specific case histories [25] in a number of different countries is that, in areas where they are not the customer, governments have generally been unsuccessful in supporting R&D projects designed to produce specific innovations. Very few such projects, in fact, have resulted in commercially successful products. There is also evidence that R&D projects which have been carried too far in the direction of commercialization may have diverted resources from other promising projects and may have stifled competition by concentrating resources among a small number of large firms, or in government owned research facilities. If the purpose of federal sponsorship of civilian R&D is to hasten the development of innovative solutions to societal problems within the framework of the market economy, it would seem appropriate to place a greater emphasis on the effects of such sponsorship in creating an atmosphere in which such innovation can flourish, rather than concentrating one's efforts on bringing specific innovations to the marketplace.

Indirect Effects of Basic Research on Innovation

Most basic research is performed in universities, and the most important indirect effects of government funding for this research are on the system of higher education and, through it, on the half million scientists and engineers engaged in research in the United States.

Corresponding to the lack of federal support for technology research, noted above, there is a comparable lack of interest in such research in most American universities. Industrial research managers have repeatedly commented that new Ph.D.'s in the sciences are generally too narrowly trained in their disciplines and frequently require extensive on-the-job training to enable them to work on the interdisciplinary problems of greatest interest to their industrial employers. In engineering departments, the "scientification" of engineering education has resulted, in the opinion of some, in a reduced emphasis on practical problem solving in favor of a deeper understanding of the fundamental underpinings of the discipline. [26] This trend undoubtedly reflects, if it is not caused by, the current imbalance in government R&D support between science and technology.

An attempt to address this imbalance and to make research more "relevant" has resulted in the current policy shift toward "directed" scientific research. Recent attacks in the Congress on scientific projects with unusual sounding titles or subjects are symptomatic of a concern that federally funded scientific activities be of direct benefit to the taxpayer. However, in a recent survey of educators and researchers, one of the most persistent criticisms levelled at the government was the trend toward overmanagement of scientific research, as well as the emphasis on its direct and immediate usefulness. [27] The trend toward overmanagement of research programs has resulted in the requirement of detailed schedules for projected progress, as well as a significant increase in reporting requirements, both of which have a tendency to restrict the freedom allowed the researcher to pursue his work wherever it leads. From the point of view expressed above, much of the difficulty stems from a failure to distinguish between science research and technology research.

The uneasiness being expressed both by the attackers and by the defenders of such projects stems in large part from a failure to distinguish adequately between the support of science and the support of technology. The two are quite distinct. They deserve recognition and support for different reasons and should be evaluated on different grounds. The evaporation of the "era of good feeling" enjoyed by scientific researchers a decade or more ago may stem in part from the failure to take adequate account of these differences, and from the application to scientific work of criteria more appropriate to technology research.

The gap that has grown up between universities and industry over the past few decades may well be another result of government R&D funding patterns. In the years before government support for basic research in universities became a way of life, there was considerable interaction between university science and engineering departments and specific technology-intensive firms. Such firms supported university research projects, although not at today's levels, and benefitted from the results. [28] This pattern has changed dramatically as funds have become available from the federal government, and with this change has come a diversion of interests and a lack of understanding between the two groups. Any federal policy directed at increasing support for basic technology should take into account the possibility, as a by-product, of effecting an improvement in university-industry relations. The recent NSF program to promote joint university-industry basic research projects is a useful step in this direction.

Indirect Effects of Applied R&D on Innovation

An important indirect effect flows from the observation that most innovations sponsored by the government have been for its own use, and that most of these have been in the areas of defense and space. There is an undeniable "opportunity cost" associated with this allocation of R&D support. Resources which might have gone into the civilian sector of the economy have been diverted into the public sector.

To say this is not to ignore the obvious fact that the jobs created in the defense and space industries have been almost entirely civilian jobs, nor to deny that there has been an appreciable spin off of commercial products modelled on innovations created originally as public goods. On the contrary, the defense and space expenditures of the federal government, particularly in the form of procurement of innovative products, have had some very important effects on the civilian economy. [29] The fact remains, however, that the potential effect on industrial innovation of a large fraction of the federal R&D funds has been diluted by their diversion to the development of products of little or no commercial significance, and that opportunities to use them in ways that would more effectively have stimulated innovation have thereby been lost.

This discussion is not meant to argue that it has been wrong to allocate funds as they have been. There were and are compelling reasons to do this, unrelated to the economic effects. But it is useful to recognize the opportunity costs associated with any resource allocation, and it raises the question whether the negative correlation between defense R&D expenditures and economic growth noted above has a causal basis.

One of the most obvious changes that has taken place in government supported R&D since the Second World War is the rise of "Big Science." Spurred by the successes of highly visible and successful government sponsored projects such as the space program and the Manhattan Project, federal R&D planning has in many instances adopted the large, integrated, project oriented systems approach to such technical problems as the commercialization of solar energy or the improvement of mass transit systems. Without examining individual programs in detail, it is possible to draw some general conclusions concerning the indirect effects of this approach on the innovative ability of technology-based firms.

In directing a large program like the Apollo project, a major problem facing the program manager is to ensure that all the disparate parts of the R&D team are coordinated with one another and are working harmoniously toward a common goal. One of the products of the Apollo project, in fact, was the management system which it developed for doing this sort of extremely complex planning job -- an output which may turn out in the long run to be one of the most important innovations to emerge from the program.

But there is a danger that the lessons of the space program may have been learned too well, and attempts to apply them to the solution of other kinds of societal needs, where neither the problem to be solved nor the nature of the desirable solution are well defined, will fail. [30] In this case, the kind of planning and coordination necessary and useful for running a complex project with a specific goal can be counter-productive and can result in over-management and inflexibility.

Since a single government agency typically will direct the entire R&D effort aimed at a given societal goal, there is a danger that the planning may become monolithic and concentrate prematurely on a single technological choice, disregarding the development of potential alternatives. Pressures to "get on with the job" as well as efforts to cut costs can make the simultaneous pursuit of alternative solutions unattractive, even relatively early in a project. As the work proceeds and grows in scope, powerful constituencies will arise behind the chosen technology, making it difficult to abandon or significantly modify it in the face of technical problems. Instead, when these arise, a mood of indomitable perseverance is more likely to influence the decision making process, particularly in the absence of attractive technical alternatives. The decision to concentrate on a single solution, made originally in an effort to reduce the cost of the project, thus becomes a self-fulfilling prophecy, ensuring that the initial choice will be the one developed, and greatly increasing the risk of failure.

The same process which results in a premature focusing of the technical effort is often mirrored in a concentration on one or, at most, a small number of research institutions to carry on the work. This is particularly true when the research institution is actually a part of the funding agency itself, as is the case, for example, with the former Atomic Energy Commission laboratories, some of which have been given very significant roles in the development of civilian energy technology. Such laboratories are frequently given the dual role of advising the parent agency on research policy and then performing the research they themselves have advocated. They may also be asked to comment on and evaluate research proposals received from other sources. The imbalance created by the competition between an "inside" laboratory and an "outsider" with an innovative idea is an unintended effect of government support for R&D which, by raising barriers to new entrants, may militate against the intended effect of fostering a supportive environment for innovation. [31] It is particularly counterproductive in view of the lack of ties between the government owned laboratories and either the manufacturer or the user of a commercial innovation.

The same comment applies, of course, to competition for R&D contracts between individual firms. It is undoubtedly less risky for a funding agency to contract to have an R&D project performed by a firm with which it has dealt before, and which has a good track record, than to take a chance on a company of which it knows very little. Even if the better known firm is unable to perform successfully, the project manager can hardly be held at fault, whereas he may well be called upon to justify his faith in the less well known company. The tendency to "play it safe" in such a situation is a natural one, but the result of such an unwritten "grandfather clause" is to place a considerable barrier in the way of smaller firms, and particularly of new entrants, which have been shown in numerous studies to be important sources of innovations. [32] Thus, short-term risk taking may be necessary to ensure a diversity of solutions and thus promote long-run success.

So long as the government is the customer for the innovations stemming from the R&D it sponsors, specialization in which one segment of industry performs the R&D and another produces the resultant finished product can work efficiently. Since it is up to the federal agency which funded the R&D to see to it that the eventual innovation is designed, produced, tested, manufactured, and deployed, it can ensure that all these steps be undertaken, if necessary, by some firm totally unrelated to the one that did the original R&D. The situation is altered, however, if the innovation is intended for a user other than the funding agency.

The necessary responsiveness of the researcher to the needs and constraints of the end-user of an innovation is likely to be called forth by the profit motive, which presupposes the existence of a buyer-seller relationship between the two parties. This, in turn, argues for the use of industrial firms familiar with the intended market as the performers of R&D for civilian use. But the firms with the closest connections to the government funding agency will most likely be precisely those for which the government, not the user, is the traditional customer, and for which the R&D itself, not the innovation, is the accustomed end product. Thus, the traditional mode of funding R&D for the government's own use can set up a significant barrier to the successful transfer of its benefits to a nongovernment user.

An unintended effect of government sponsorship of R&D has been the establishment of a class of technologically advanced firms which are organized and motivated in such a way that they are quite unlikely to produce commercially successful innovations. [33] Significant government expenditures for R&D particularly since the Second World War, have created a business opportunity which in turn has given rise to a segment of industry whose only product is the R&D itself, and whose only customer is the government. This unique situation has some interesting implications for industrial innovation.

It is true that certain firms, for example, in the electronics industry, have been very successful in spinning off technological innovations originally designed with government R&D support into products for the civilian economy. However, this has happened primarily when R&D financing has been coupled to a program of procurement of the finished product, and then only when the firm performing the R&D also produced the final innovation. For such firms R&D was not conceived of as the final product, and they were, therefore, motivated at an early stage in their planning to look beyond the government to the private sector as a market for the innovation. The role of the government in providing an early market niche through procurement of an innovative product was crucial in facilitating the utilization of the R&D work which it sponsored.

POLICY IMPLICATIONS

This section uses the material presented above concerning the effects of government support for R&D to formulate guidelines for increasing its effectiveness in promoting innovation. It does not attempt to answer the difficult questions of how much and what kind of innovation is socially desirable. These questions are important, however, and should be borne in mind since they lead directly to the issue of federal control of the pace and direction of innovative activity in industry.

The following points summarize the findings of the body of this chapter:

- Direct support of R&D is not the only, and may not be the best, way for the federal government to promote innovation.
- Rather than attempting to produce innovations itself, the government should affect their production indirectly.
- It can influence the ability and willingness of private firms to innovate by enhancing their ability to do so, increasing their desire to do so, or increasing their fear of not doing so.

In many cases, the most immediate substitute for R&D funding can be a strategy of facilitating and encouraging the licensing of technology from other countries. For one thing, this course of action can be considerably cheaper than doing the R&D domestically, primarily because by licensing, the United States can buy the results of successful R&D, without having to finance all the unsuccessful projects. It is a truism that a policy of relying exclusively on licensing for the development of new technology, if employed by all countries and carried to the logical extreme, would result in world-wide technological stagnation. Nevertheless the selective "borrowing" of foreign technology can be a very valuable substitute to an expensive R&D program of uncertain outcome. This procedure has been followed very successfully, for example, by Japan, which between 1966 and 1975 spent approximately $1.5 billion in royalties and fees in the United States alone.

Another often suggested substitute to the present project-by-project evaluation procedure for R&D support would be to institute a general tax credit for R&D expenses. Since at present R&D expenses are an allowable deduction, this proposal would amount to a rough doubling in the amount of money made available to R&D through the tax code. The objection to this idea primarily involves the lack of effective controls on the use of the money and the fact that in many cases it would displace private funds. In other words, it increases the total resources available to the firm, but does nothing to affect the firm's allocation of those resources between its various functions.

In order to affect such resource allocation decisions within firms, the government has available a number of alternatives which have nothing to do with R&D, but rather with the financial, competitive, and regulatory environment in which those decisions are made. It is beyond the scope of this chapter to discuss them in detail, but clearly changes in the capital gains tax or in SEC regulation of qualified stock options can have a substantial influence on the capital available for innovative activities, particularly to new firms. In addition, patent protection, or the lack of it, can have a powerful effect on a firm's desire to invest in new product or process development. There is evidence indicating that government policy on the granting of rights to inventions made with federal R&D funds can be an important factor in determining whether such inventions will become innovations. [34] For example, the Department of Energy's statutory policy to deny such rights except under special circumstances runs counter to the expressed purpose of the R&D expenditures of that agency.

Finally, some actions of the government can force a firm to invest in innovation because the risk of not innovating has become greater than the risk of trying to and possibly failing. As has been pointed out, R&D subsidies can act in this way if they tend to promote credible competition in an otherwise oligopolistic situation. Other forms of action include enforcement of antitrust legislation, as well as regulation and procurement of innovative products if they are handled in such a way as to create a market for new ideas.

In the context, then, of these alternatives, if the decision is made to go ahead with federal R&D funding to promote innovations, what are some of the do's and don't's which emerge from this analysis?

- DO emphasize support for basic technology research.

The United States does fund some work of this kind, for example through the Applied Science and Research Applications (ASRA) Directorate at the National Science Foundation, but there is no coherence or centralized purpose to its support. This is in sharp contrast to the situation in virtually every other industrialized nation.

- DO maintain close contact with the user.

The example of the agricultural extension service is an overused one and cannot be copied slavishly, but it is useful to bear in mind.

- DO concentrate on promoting competition between firms, and, particularly, on facilitating new entries.

The most useful tools for this purpose are the unsolicited proposal and the sole-source grant. It is imperative to institute and oversee procedures for processing the former efficiently and fairly, and to avoid appropriating a new idea and advertising it through a request for proposals.

- Above all, DO sponsor risky projects. Leave the financing of sure things to the banks.

As regards the DON'T's:

- DON'T try to support the later stages of the innovation process.

The role of government R&D in civilian technology development should be to increase the opportunities for innovation, not to decide which particular innovations are needed or to produce them. Governments have been quite unsuccessful when they have attempted to do this.

- DON'T put all your eggs in one basket.

Investigate a diversity of technical approaches even though this can entail duplication of effort. This technique has been used very successfully by the Advanced Research Projects Agency of the Department of Defense for the support of government R&D.

Just as diversity of approaches is most effective for solving a technical problem, it is equally important for the government to apply a diversity of techniques simultaneously to the problem of stimulating innovation. This chapter has stressed the policy of R&D subsidy, but, it has also pointed out that there are many other options available to the government. There is strong evidence that a symbiotic effect exists between these various approaches, and that two or three of them operating simultaneously are much more likely to be effective than any one alone. To take full advantage of this symbiotic effect, however, may require planning and coordination at the highest levels of the government, and a broad interagency approach to the problem.

NOTES

1. Willis H. Shapley and Don I. Phillips, "Research and Development -- Report III," American Association for the Advancement of Science (Washington, D.C., 1978).

2. Walter S. Baer, Leland L. Johnson and Edward W. Merrow, "Analysis of Federally Funded Demonstration Projects: Final Report," The Rand Corporation, (Santa Monica, Calif., April 1976); U.S. Office of Technology Assessment, The Role of Demonstrations in Federal R&D Policy (Washington, D.C.: Government Printing Office, July 1978); and "Government Support for the Commercialization of New Energy Technologies: An Analysis and Exploration of the Issues," MIT Energy Laboratory Policy Study Group (Cambridge, Mass., November 1976).

3. "Government Involvement in the Innovation Process" (Report of the Center for Policy Alternatives (CPA), Massachusetts Institute of Technology, to the U.S. Office of Technology Assessment, Washington, D.C., 1978).

4. "Report of the Small Business Administration Task Force on Venture and Equity Capital for Small Business" U.S. Small Business Administration (Washington, D.C., January 1977).

5. James M. Utterback and Albert E. Murray, "The Influence of Defense Procurement and Sponsorship of R&D on the Development of the Civilian Electronics Industry," Center for Policy Alternatives Report CPA 77-5, Massachusetts Institute of Technology (Cambridge, Mass., 1977).

6. Shapley and Phillips, "Research and Development -- Report III."

7. M.A. Shankerman, "A Research and Development Price Index and Real R&D Expenditures, 1957-1975," in "Essays in the Economics of Technical Change (Ph.D. diss., Harvard University, 1978).

8. Keith Pavitt, "Governmental Support for Industrial Research and Development in France: Theory and Practice," Minerva 14 (1976): 330-35.

9. C.T. Hill, unpublished material, Center for Policy Alternatives, Massachusetts Institute of Technology, (Cambridge, Mass).

10. J. Herbert Hollomon et al., Energy Research and Development (Cambridge, Mass.: Ballinger Publishing Co., 1975).

11. E. Mansfield, J. Rapoport, A. Romeo, S. Wagner, and G. Beardsley, "Social and Private Rates of Return from Industrial Innovations," Quarterly Journal of Economics, May 1977.

12. Nestor E. Terleckyj, "Output of Industrial Research and Development Measured as Increments to Production of Economic Sectors" (Paper delivered at the Fifteenth Conference of the International Association for Research in Income and Wealth, York, England, 1977).

13. Eric A. von Hippel, "The Dominant Role of the User in Semi-conductor and Electronic Subassembly Process Innovation," IEEE Transactions on Engineering Management (May 1977): 212-39.

14. Bruce L. R. Smith and D. C. Hague, The Dilemma of Accountability in Modern Government (New York: St. Martin's Press, 1971).

15. "Government Involvement in the Innovation Process," (CPA).

16. Interactions of Science and Technology in the Innovative Process: Some Case Studies, Battelle Columbus Laboratories (Columbus, Ohio, March 1973).

17. Federal Technology Transfer: An Analysis of Current Program Characteristics and Practices, National Science Foundation (Washington, D.C., 1975).

18. William J. Abernathy and James H. Utterback, "Patterns of Industrial Innovation, Technology Review 80 (1978): 40-47.

19. Mel Horwitch, "The American SST Experience -- The Transformation of Multi-faceted Technological Enterprises" (Working paper for the American Association for the Advancement of Science's Symposium on Macroengineering, Washington, D.C., February 13, 1978).

20. Pavitt, "Governmental Support for Industrial Research and Development in France."

21. O. Keck, "West German Science Policy since the Early 1960's: Trends and Objectives," Research Policy, 5 (1976): 116-57.

22. "Federal Funding of Civilian R&D, Volume 2: Case Studies" (Final report from Arthur D. Little, Inc., to the Experimental Technology Incentives Program, U.S. Department of Commerce, Washington, D.C., 1976).

22. Applications of R&D in the Civil Sector: The Opportunity Provided by the Federal Grant and Cooperative Agreement Act of 1977, U.S. Office of Technology Assessment (Washington, D.C., 1978).

24. George Eads, "U.S. Support for Civilian Technology: Economic Theory and Political Practice," Research Policy 3 (1974): 2.

25. George Eads and Richard R. Nelson, "Governmental Support of Advanced Civilian Technology -- Power Reactors and the SST," Public Policy 19 (1971): 405; J. Zysman, "Between the Market and the State: Dilemmas of French Policy for the Electronics Industry," Research Policy 3 (1975): 312; and A. Gardner, "Launching Aid," Proceedings of the Conference on Industrial Subsidies, U.K. Department of Industry ,1975.

26. C. A. Anderson and C. H. Kimzey, "National Productivity: The Responsibility of Engineering Educators," Engineering Education 69 (1978): 188.

27. "Science at the Bicentennial: A Report from the Research Community," National Science Board (Washington, D.C., 1976).

28. David F. Noble, America by Design: Science and Technology and the Rise of Corporate Capitalism (New York: Alfred A. Knopf, 1977).

29. Utterback and Murray, "The Influence of Defense Procurement and Sponsorship on R&D.

30. Richard R. Nelson, The Moon and the Ghetto (New York: W. W. Norton & Co., 1977).

31. "Government Involvement in the Innovation Process," (CPA).

32. William K. Scheirer, "Small Firms and Federal R&D" (Report to the U.S. Office of Federal Procurement Policy, Washington, D.C., 1978).

33. Paul Horwitz, "Public Funds and Private Technology," in Technology at the Turning Point (San Francisco: San Francisco Press, 1977).

34. Government Patent Policy Study (Boston, Mass.: Harbridge House, 1968).

8 Policies and Programs of Governments Directed Toward Industrial Innovation

J. Herbert Hollomon

INTRODUCTION

The governments of industrialized nations directly or indirectly affect the process of industrial or commercial innovation. Government taxation, monetary, expenditure, and regulatory policies help establish the economic climate in which the innovation process occurs. In addition, all governments have undertaken programs to influence specific stages of this process or to support the development of particular techniques to serve as the basis for commercial developments of national significance. [1]

Because of its complex and interacting nature, no single comprehensive theory specifies the type and extent of government action that will affect the innovation process. More importantly, the effects of a specific government action or program are impossible to assess. For example, all industrial nations have a patent system designed to increase the rewards to the entrepreneur-innovator. Clearly, rewards like those afforded by the patent system are a necessary condition for insuring that the innovation process takes place. Yet, in spite of a large number of studies, there is no definitive analysis of the effectiveness of patent systems in fostering innovation.

In studying the effects of R&D on the innovation process, some investigators have attempted to determine the profitability of R&D to the firm or to relate the total R&D in a sector of industry to its growth, productivity, or profitability. Growing firms with high rates of productivity improvements often support R&D, but some firms grow and become productive without it. Does R&D induce growth and profitability, or, conversely, do profitable and growing firms support R&D? Surely, R&D contributes directly or indirectly to some innovations in some firms in some stages of the product life cycle. Since, according to neoclassical economics, all firms will underinvest in R&D for new products and processes, government subsidies are believed necessary to augment their investments. [2] But how much subsidy, in what sectors, and through what means should governments spend public funds for this purpose?

There are three approaches that can be used to examine government policies to stimulate innovation. One approach is to use the knowledge of the nature of the innovation process and ask what government policies are likely to affect the various stages of innovation occurring in different firms during the several stages of the product life cycle. For example, what government policies and programs affect the start-up of new firms? Or, how is the knowledge of changing techniques provided to or generated by firms when their products are mature, and their product and process improvements occur incrementally?

A second approach provided by neoclassical economic theory for determining, or at least rationalizing, the role of governments in the innovation process in market economies. [3] Arguments based on this theory justify government intervention to stimulate industrial innovation, as illustrated in the following examples.

Governments provide public goods and services, (e.g., defense and security forces, monetary systems, and means of exchange) that benefit their societies generally, but cannot be easily provided by each citizen. If new technology and advanced products or systems are required to improve the goods that the government buys for public purposes, then the government must support the development of these products or systems.

In a market economy, individual firms, even with the strongest patent protection, may not be able to appropriate the full benefits of industrial innovations and may underinvest in them as a result. Governments, therefore, invest in activities believed to contribute to innovation, including education, training, and basic science. The recognition of the inability of a firm to capture the full benefits of its innovations has led many governments to encourage innovation by reducing costs to the firm either by direct subsidy, as in France, [4] or by tax credits for R&D, as in Canada. [5]

Industrial activities create costs to society that are not included in the internal calculus of the firm; these are called "externalities." Air and water have been considered free goods. Innovation stimulated by government regulation or subsidy may help reduce such externalities as air pollution or the morbidity and mortality of workers.

A third approach to understanding the role of government in the innovation process is to determine what governments <u>do</u> and, through an organized examination of their various activities, to derive lessons for the United States. [6] Since a completely acceptable and logical rationale for government policy is not possible, a comparative description of the action of governments should provide guides that will be useful in determining the appropriate role for the U.S. Government.

Direct transfer of a government action or policy from one nation directly to another is not possible without considering the differences in their political, social, and economic environments. However, when the same type of government intervention is used successfully in very different societies, there is a strong argument that it should be applicable to the United States.

All three approaches to the analysis of the role of governments in the innovation process are considered in this chapter. However, the emphasis is on the third approach: a comparison of the roles various competitor countries' governments actually play in the innovation process. The discussion begins with a description of programs that affect innovation generally, is followed by specific government actions that are likely to affect one or more stages of the process, and concludes with a brief discussion of government programs that attempt to ameliorate the adverse effects of innovation. Selected activities in the United Kingdom, France, West Germany, Japan, the United States, and occasionally other countries are used to illustrate the range and scope of government involvement and to provide suggestions for U.S. policy.

ACTIONS OF GOVERNMENTS THAT AFFECT INNOVATION GENERALLY

Two of the oldest and most prevalent government actions that affect innovation are the establishment of a patent system and of a system of uniform weights and measures. More recently, nearly all governments have assisted in the adoption of industrial standards and contributed financial and other support to science and to the development of industrial technology. Each of these kinds of support affect the overall climate for industrial innovation generally.

The Patent System

The system of patents of all industrial countries derived from the "letters patent," which provided a commercial monopoly to individuals or companies to encourage the exploitation of resources or the development of markets. Patents are often thought to be primarily a stimulus to invention, but they play an even more important role as a stimulus for innovation. A patent prevents the use of the invention by other than the inventor, his assignee, or licensee. The patent increases the possible return to the innovator and the appropriability of the benefits from his investment, thereby stimulating innovation. However, since the patent precludes others from practicing the invention, it may discourage the diffusion of the new.

The strength of the patent system is different in different countries. The system in the United Kingdom does not protect process invention. The French system is essentially a system of registration with little examination of prior art. In West Germany, the employee may not sign a patent agreement with his employer, and the employed inventor has guaranteed rights and control of his invention. While this system may not have significantly influenced the innovation process, it has minimized questions about the government ownership of patents resulting from government funded work. The Common Market countries have now agreed on a patent system that can be used by all members; this enlarges the reach of protection for a European entrepreneur. In Japan, the assumption is made that government funds supporting industry are for the purpose of stimulating innovation, which also is aided by exclusive patent rights. The strongly held political view in the United States that invention supported by public funds should become general public property is not widely held in other industrialized western nations. In the United States, the belief persists that an invention has some intrinsic value. However, it is clear that an invention only becomes valuable after successful innovation. Public ownership and general licensing of a patent destroy the chances of successful innovation.

Another difference between the United States and many other countries is that seldom in other countries is the remedy sought in antitrust suits that requires the "guilty" firm to license any and all others to practice their invention (even when the licensed firm is partly or fully owned by foreign nationals and located outside the United States). Sometimes government actions initiated for security reasons prevent U.S. firms from filing abroad and may prevent exports from the United States.

Patents are more important to new venturesome firms during the introduction of radically new products than for firms producing mature products in large-scale integrated plants. Past studies of the effectiveness of the patent system have not distinguished between inventions that relate to different stages of the product life cycle. In a study of innovations in five countries, it was found that patents were much more important in the electronics industry than in the more mature automobile industry. [7] In the developing technologies like microprocessors, change is now so rapid that patents offer little protection. Rather, competitive position depends on the speed and effectiveness of the firms in developing and introducing new products, as Utterback's description of the innovation process in Chapter 2 would predict.

Basic Standards of Weights and Measures

Every industrial nation supports a system that ensures a consistent set of weights and measures. These systems are designed to ensure the uniformity of measures used throughout both the community of science and the economy, and to ensure their improvement with time. Uniform standards and measures simplify and reduce the costs of manufacturing by making it possible to use tools and parts having the same measures. Such systems make possible large potential markets and lower the costs of innovation. The research and development related to the standards system is aimed at the development of new standards since technology advances, and new products and processes, require more sophisticated measurements. In the United Kingdom, these functions are performed by the National Physical and Chemical Laboratories; in France, by several laboratories of the Centre National de la Recherche Scientifique; and in the United States, by one of its oldest scientific institutions, the National Bureau of Standards.

The basic physical standards of the world are based on the metric system. All basic standards bodies of the world exchange information and research results to ensure the reliability and uniformity of the measurements of the world. However, the fact that the United States does not use the metric system in most of its industrial and commercial activities requires a transfer of the basic metric standards to the "British" system. The continued use of the "British" system limits the market for U.S. products and increases the cost of product design and manufacture.

Industrial Standards

Within each country, industrial standards that relate to the products and processes of commerce are set by a cooperative arrangement among industry, government, and public representatives. Government representatives do not always participate, and the public is seldom involved. Setting standards too early in the life cycle of a product is likely to limit innovation and the dynamic development of adaptive and competitive products. Setting standards too late may restrict the market and deny the investor some of the fruits of his investment. The control of standards setting by a firm or group of firms may limit competition; and the control of standards setting by a nation may prevent other nations' products from entering its market.

Sometimes, industrial standards become nontariff barriers to trade and impede innovation that might otherwise occur. In West Germany, standards are set by the Deutsche Normennausschauft; in France, by the Association Franscais de

Normalization; in Britain, by the British Standards Institution; and in Japan, by the Japanese Industrial Standards Committee of the Ministry for International Trade and Industry. All of these are highly centralized bodies. The first three, though private, receive direct government subsidy, while Japan's standards setting committee is a quasi-government organization. In the United States, hundreds of technical organizations -- weakly coordinated by the American National Standards Institute -- are engaged in standards setting. The process in the United States is entirely voluntary and not supported by the government except for the participation of experts largely drawn from the Bureau of Standards. The U.S. Federal Trade Commission has proposed, under its antitrust authority, a set of "standards" for the standard setting process.

In recent years the European Common Market has issued about 100 standards and has about 100 more in the process, all of which will become mandatory for member states within the market. While the United States has been given the opportunity to comment on the proposed standards, there has been little U.S. participation in their development. There is no organized system in the United States for developing a strong industry-backed involvement in the European process.

International industrial standards are set through two organizations: the International Electrotechnical Commission for electrical and electronics standards and the International Standards Organization for all others. These standards are developed by representatives of the various national standards organizations and by special technical experts. Only the United States depends on the voluntary participation of industrial representatives. Other nations support their representatives' participation in whole or in part, and these representatives tend to reflect the viewpoint of their national governments and industries. The U.S. representatives tend to be from the large prosperous firms for which the particular standards may be of special commercial interest.

For ten years, the U.S. Congress has had before it a bill to legitimize and to support the international standards activities of the United States, but it has failed to act upon this bill. Even a cursory examination of the hearing record on this bill indicates there is little understanding of how modest difficulties with standards, or with the procedures for modifying products to meet the standards, will affect the innovation process and will enlarge or restrict the market for new or modified U.S. products. Restrictive standards may so raise the costs of developing or modifying a product that they prevent the use of innovative products or processes which could better or more cheaply serve the market.

In the current General Agreement on Tariffs and Trade (GATT) negotiations, agreement has apparently been reached for an international reciprocal arrangement that permits commentary on all proposed national or Common Market standards. If the United States vigorously pursues these new procedures, there may well be a significant positive benefit for U.S. trade. At present there are not even government funds to translate Japanese standards into English or to disseminate those standards to U.S. industry, whereas the reverse process is fully supported and highly developed in Japan. Standards setting, and the technical work involved, benefits all firms engaged in the related businesses, but these activities should be carried out with government participation and support since no single firm is likely to make the investment and still represent the public interest.

Support for Science

New scientific understanding may be needed for a specific applied goal; may be required to permit a broad range of technology to develop; or may be sought simply to satisfy the desire to understand. Science, while sometimes leading to new technological capability, does not itself have commercial value and will seldom be supported by a single firm or even a group of firms in a single industry. Science is supported by governments in the belief that it provides the knowledge and understanding that will permit the improvement of the defense, health, well-being, prestige, or the economy of the nation. [8] (See Chapter 7 for a description and comparison of government support for R&D.)

Some of the support for technical activity related to national purposes is in the form of support for basic science. It may be carried out in government laboratories, universities, or industry. Where such activities are performed, and how those involved in industrial innovation are connected with them, are perhaps more crucial questions than the amount of scientific research that takes place. In France, the emphasis is on work in government laboratories largely unconnected with industry. In the United States, scientific research takes place predominantly in academic institutions where connections with other than the most sophisticated large firms is mostly through students who later work in industry.

One reason for firms' and countries' support of basic research activities is that the results of basic scientific research are open to all those who participate in it. The results are immediately available to those working in a field and somewhat later to those who can use and interpret the results. World science is such a large endeavor that few firms can support enough effort to make significant continuing contributions to it. Only a few firms are large or diversified enough to support scientific research, and then it is largely in fields believed to be important for creating new technology of value to them. But even for these firms, the activity is primarily aimed at permitting entry into the world of science. Competent scientists working in a given field have access to the international scientific community and can bring knowledge of scientific advances that may contribute to innovation in the firm. Therefore, to establish mechanisms that connect the institutions performing basic science with industry has become a matter of policy concern for governments.

In the United States, before World War II, the support of science by both industry and government was insignificant compared to that of today. The large federal involvement grew out of the contribution of scientists to weapons development. While U.S. space and defense agencies still support a large portion of the basic science research, the Department of Health, Education, and Welfare (especially the National Institutes of Health), the National Science Foundation and the Department of Energy (and its predecessor, the Atomic Energy Commission) are also contributors.

The mode of government support for basic science in Europe and Japan differs from that of the United States. In West Germany, there is a long history of the support of science by the Lander (prefects) and the federal government (through the German Research Society) in universities and in the Technischer Hochschulen (technical institutes). A most important institution which led to Germany's preeminence in science in the early part of the century was the Max Planck Gesellschaft (formerly the Kaiser Wilhelm Gesellschaft). Funds are provided to the Max Planck Society by the central government, the Lander, and industry. The work is carried out in a large number of diverse institutes. In 1972, the Max Planck Society consisted of 49 research institutes and had a budget of about $200 million. The institutes have independent staff drawn in part from university faculty. The Max Planck Society operates quite independently and has shown great flexibility in entering new fields and in reinvigorating old ones.

Basic science in the United Kingdom is supported as a part of general allocations to the universities by the University Grants Commission. The National Research Councils also provide funds to universities and to their own research institutes. Basic scientific research is also performed in national laboratories such as the National Physical and Chemical Laboratories.

In France, while some scientific research is done in universities and in Les Grandes Ecoles, the national laboratories of the Centre National de la Recherche Scientifique (CNRS) have been a major vehicle for this activity. CNRS is responsible for several hundred laboratories whose staffs belong to the French Civil Service. Apparently, this system has difficulty in adapting to the changing emphasis in science and to the changing needs of the French people.

In Japan, the Ministry of Education fully supports the faculty of the state system of higher education and allocates research support in major fields of science. With this control by the minister of education, there is little possibility for the universities to carry on work for other government agencies or for industry. But special institutes near the universities are being organized to circumvent this difficulty. Unlike those in the United States, the universities in Japan are not a major element in the government's research activities directed toward major public needs (e.g., space, defense, environmental protection, and industrial development). In Japan, a large part of the scientific work is carried out by private firms. Less than 30 percent of the total R&D is supported by the state (compared to more than half in the United Kingdom, France, and the United States).

New science, and the awareness of the possibilities of new technology derived from science, may, like the patent system, be a necessary condition for radical new technological innovation. However, it is not sufficient.

Science is an "open" enterprise -- scientists know what other scientists do through informal and personal contacts. The literature of science is open, and the world of science is open to all. Therefore, the basic scientific activities of nations are not correlated with their economic and industrial conditions. The connections between the two are complex and indirect, and they occur over a very long period. Many important industrial innovations in a nation are based on science developed elsewhere, often on science developed years before. Therefore, it is important that an industrial nation have mechanisms, policies, institutions, and arrangements that permit the quick assessment of the potential of new science for innovation and that encourage the innovation process to proceed. The scope of the scientific activity in a country may give an indication of its potential for contributing to the advancement of science, but <u>not</u> of how its technology will advance or how its industrial enterprise will grow and adapt. In fact, a nation may provide so much support to science that few of its gifted scientific people are attracted to those institutions that actually develop industrial technology. Israel is currently reducing its support of academic science and increasing its support of industrial science for this very reason. [9]

The cases of the United Kingdom, Japan, and France illustrate the absence of a correlation between scientific research and industrial development. The United Kingdom has a long tradition of scientific achievement, nurtured by a distinguished and healthy academic environment. But its industry (with some spectacular exceptions) has a poor record of industrial innovation and a very low rate of productivity improvement. On the other hand, during their very rapid industrial growth in the 1950s and 1960s, the Japanese concentrated on technology transfer and adaptation and contributed relatively little to the world of science. French leaders, concerned with making the laboratories of the Centre National de la Recherche Scientifique more useful to the economy, created a special organization, Agence Nationale pour la Valorisation de la Recherche (ANVAR), to support the transfer of work in the laboratories to practical use.

Support for Industrial Technology

Technology is the body of art and science that permits, or is the basis for, the creation, design, and manufacture of useful products and processes. The development of new technology comes largely through advances in understanding and from expansion in the development, manufacture, and use of new products and processes. By basic industrial technology is meant the body of general knowledge of such subjects as catalysis, corrosion, and the strength of materials and the knowledge and understanding of manufacturing processes like casting, welding, and machining. Also included are new techniques of engineering design of products and processes.

The economic rationale for government support of basic industrial technology is similar to that for the support of basic science. Basic industrial technology does not usually apply only to single firms or to a class of products. The social benefits of the development of new technology exceed the private benefits to a single firm. Thus, most nations support the development of broad areas of technology to some degree. But, unlike the longstanding support of science, the institutions for assessing broad technological needs and opportunities for industrial development are only now being established in developed countries. Japan and West Germany have the most comprehensive and effective means for assessing and supporting broadly applicable industrial technology, as well as for supporting specific areas of technology believed to be important for future commercial innovations.

Until 1969, the West German Government had restricted most of its support of new technology to the very special areas of aeronautics, nuclear technology, computers, and electronics. [10] These have been the glamour fields in which the United States, through its defense and space programs, had apparently achieved technological leadership. In 1969, the German Government began a program of support of "new technologies." This program represents the political acceptance in Germany that the broad support of new technology leading to the development of new products for the ordinary commercial market was an important and appropriate function of a central government of a modern industrial state. A 1975 German Federal Government report justifies this program exclusively in terms of its potential for industrial innovation. [11] For example, the government now supports major activities in the use of computers in all phases of manufacturing. This program is growing rapidly, while the emphasis in most of the other European nations continues to be on aircraft, nuclear, and computer technology.

The industrial technology program of the Ministry of International Trade and Industry of Japan (MITI) is more clearly analogous to that of West Germany than to those of the United States, the United Kingdom, or France. Special areas of technology on which future innovation will be based are selected for national attention. Unlike most European countries, Japan has not concentrated its effort on aircraft, nuclear, and computer technology. [12] The Japanese government supports programs in such fields as: pattern recognition, combustion and jet engines, integrated circuits, and computer-aided manufacturing. Active support for this indigenous, industrially oriented technology began about a decade ago. Prior to that time, the Japanese largely obtained their technology by transfer and adaptation (which is still an important basis for Japanese industrial growth).

In France, the support for the development of industrial technology largely derives from the program called "Actions Concertees." This program is the responsibility of the Delegation Generale a la Recherche Scientifique et Technique (DGRST). It was originally designed to encourage coordinated action between government, academic, and industrial centers for research. The results of the work generally are available to the public.

There is no broad program in the United States like those in Europe and Japan for the development of industrial technology. The absence of such programs is somewhat surprising in light of the successful early support of mining technology by the Bureau of Mines and the successful history of the support for agricultural technology (both for development and extension) since the middle of the 19th century. Instead of using these successes as the basis for instituting broader programs, the United States has continued to respond to new opportunities with more specific programs such as the National Advisory Committee for Aeronautics (NACA), which aided the development of aeronautical technology and contributed to the continued leadership of the U.S. aircraft industry. The Atomic Energy Commission (AEC), created after World War II, and its successor (ERDA) are responsible for assisting the development of civilian nuclear technology. The huge technology support for weapons during World War II, and for space and defense during the 1950s and 1960s, contributed significantly to aeronautical, nuclear, and electronic technology. The apparent success of these activities, and the assertion that their by-products would substantially aid U.S. industry, diverted attention from broadly applicable industrial technology in the United States. Other countries, following the U.S. lead, invested heavily in these fields to close what some thought was a "technological gap" between Europe and the United States. There are indications that this stage is passing, and nations are now becoming more concerned with developing technology that will serve as a basis for broad industrial development.

DIRECT INTERVENTION IN THE INNOVATION PROCESS

Support for Specific Product and Process Development

Most European countries, and Japan, employ various schemes for reducing the cost to firms of the early stages of the innovation process. The purpose of all of these programs when they were begun was to encourage risky projects, particularly those that would lead to products for export. Since firms cannot appropriate all the benefits of innovation, the governments provide part of the funds for the early stages of innovation.

In West Germany, support for product and process development is provided by the Ministry for Science and Technology. In 1973, this ministry provided support of about $4 billion for industrial science and technology. [13] Rights to the results of this work are fully retained by industry if the government funding does not exceed 50 percent. If projects lead to commercially successful products, the government expects some repayment. Small- and medium-sized firms usually have the exclusive use of these results under all conditions. The economic ministry (through a complicated procedure involving the Lander) provides interest-free loans of up to 50 percent of the costs of technological innovation.

In Japan, the Japan Research and Development Corporation supports industrial development directly, at least in fields thought to be of importance to the future of the Japanese economy. Special low interest loans are also provided by the Japanese Development Bank.

The Aide de Development program of the DGRST in France receives proposals from industry for partial support of commercial developments based on new technology. In the beginning, this program was intended to aid small firms and to support projects too risky for industry to undertake. More recently, a few large, highly sophisticated firms have obtained a major share of the support. There is concern that a company with a large project portfolio may use government subsidies provided for sponsored projects to support other, marginal projects in the company's portfolios.

A similar program in the United Kingdom, called "Launching Aid," also tends to support large firms. As in the French program, there is a question as to whether government subsidies are being used by firms for the intended purposes.

A study by the Center for Policy Alternatives of four European countries and Japan indicates that direct support from programs of this type were involved with a larger number of commercial innovations than any other direct government intervention. [14]

Direct support of commercial projects in industry without some special public purpose is unusual in the United States. Obviously, when a space or defense product has civilian application, government funds have, in effect, subsidized a part of the cost of the commerical innovation. There is little doubt that the high expenditures for defense and space beginning during World War II created much of the new technology related to aeronautics, electronics, computers, and nuclear reactors. New firms and new ventures in older firms to bring new products to market were encouraged by the government R&D support and the procurements of the government agencies. However, these programs also diverted resources and attention from other requirements of industry and probably have obscured the need for government intervention generally now felt so strongly in other industrialized countries. More recently, as a result of the oil embargo and the high price of fuel, public subsidies for energy supply and for conservation products and processes have grown enormously. Most of these subsidies flow to industry, but there is apparently no consistent theory or procedure that governs the way they are provided. For energy the market is made up of private suppliers and users, unlike the situation for space and defense where the government is the buyer of the industrial products. The same schemes for government intervention cannot apply in the two cases, though the Department of Energy has tended to operate like a space agency. Presumably, the U.S. Government is providing these subsidies because the social benefit of new energy sources, or of energy conservation, is greater than the private benefits that would flow to private energy suppliers and users. Also, the subsidies are used to provide encouragement to firms to develop commercial products and services. The social benefit argument is used in Europe and Japan to justify government support of industrial technology and the subsidization of part of the costs of industrial innovation projects.

Support for Invention, Venture, and New Firms

Studies of the innovation process indicate that, at least in the United States, small, new, high-technology firms have spawned a disproportionate share of innovations. While there has been no explicit public policy to encourage individual inventors or the start-up of new firms in the United States, government contracting in the 1950s and 1960s stimulated new technological firms such as those along Route 128 (near Boston) and in Silicon Valley (near Palo Alto). The great growth of these companies occurred during the period of rapidly expanding government funding of advanced R&D and when venture capital was readily available. [15]

An important factor affecting the start-up of new firms is the tax treatment of capital gains, determined in part by the availability of capital for risky enterprises. A few years ago, in the United States, the taxes on capital gains were raised and then, following the tax reform act of 1978, were lowered slightly. In most of Europe and Japan, capital gains are either not taxed at all or at lower rates than in the United States.

Another factor influencing the ability of small firms is the tax treatment of the losses incurred in their early years. Limited partnerships permit these losses to be used to reduce ordinary income, and Chapter "S" corporations (limited to 12 incorporators) permit the losses as deductions to the shareholders. In the United States, early losses may also be carried forward for five years to offset future income. In France, the losses are not prorated; they may be carried forward indefinitely and used by the firm whenever desirable to offset income.

Japan has the most complete set of government programs to aid individual inventors, entrepreneurs, and new high technology firms (see Table 8.1). [16] Examples of all the Japanese programs are found in other nations, but in no other country is there such a complete array of programs to encourage the new. Perhaps the most important aspect of these programs is the variety of ways by which new small firms can obtain assistance for the start-up phases of new ventures.

Grants-in-aid to private inventors are provided in West Germany by the Fraunhofer Society. The Japanese Research and Development Corporation provides support by techniques much like the National Research and Development Corporation in the United Kingdom. The latter has a long tradition of supporting part of the costs of product innovation with arrangements for pay back (royalties) for successful projects. In France, a government funded organization called Agence Nationale pour la Valorisation de la Recherche (ANVAR) aids in early feasibility determination and exploitation of inventions (largely of inventions in government laboratories). In Sweden, the Ministry of Industry provides direct grants to inventors and new firms to support the critical early stages of the innovation process.

Spreading of Technical Information and New Techniques to Established Firms

In almost all large industrialized nations, except the United States, there is a system for cooperative industrial research associations that aid in the development and dissemination of technology to firms. [17] These institutions have a

TABLE 8.1

JAPANESE GOVERNMENT PROGRAMS TO AID INVENTORS, ENTREPRENEURS, AND NEW FIRMS

- Grants-in-aid for the implementation of inventions.
- Special development contracts and grants.
- Arrangements for commercialization of new developments.
- Support for capital equipment for new ventures.
- Rapid tax write-off for new developments.
- Low interest loans for high technology firms.
- Special analysis and advice for small firms.
- Special procurement policy for small business.
- "Open" research laboratories in all prefectures for the use of new and small firms.
- No interest loans for modernization of small firms.

Source: Nicholas Ashford, "National Support for Science and Technology: An Evaluation of Foreign Experiences -- Summary" (Report No. CPA 75-12, Center for Policy Alternatives, Massachusetts Institute of Technology, Cambridge, Mass., August 18, 1975).

long history in the United Kingdom, West Germany, and France, but are not as extensive in Japan.

There are now approximately fifty research associations in Great Britain serving almost all sectors of industry, ranging from pharmaceuticals to rubber and plastic products. Not only do they perform cooperative research, but they provide technical literature, answers to routine inquiries, and advice on specific problems. The government provides from 35 to 55 percent of the support for these industrial research associations, as well as giving grants for special equipment. [18]

A large number of similarly organized research associations operate in West Germany. In some industries such as mining, wood and wood products, textiles, and printing, a substantial fraction (20-50 percent) of the R&D of an industry is performed in this way.

The South Korean Government established a single industrial research laboratory, the South Korean Institute of Science and Technology, to serve all of industry. [19] This organization aided in technology transfer to South Korean industry. Now several such institutes are being formed, each dedicated to a specific industrial activity. These institutes are primarily aimed at adapting technology developed throughout the world. They are all partly supported by the Ministry of Science and Technology.

Cooperative research institutes, by their very nature, are seldom involved in innovation which would provide a firm with great competitive advantage. Instead, they develop techniques generally applicable to the industrial sector and transfer information and learning to the firms. Recently, some of these institutes have become "gatekeepers" for the firms, alerting them to new developments of possible use to them from around the world.

In France, industry is taxed a small percentage of its turnover and the funds distributed to the cooperative research institutes. In the United Kingdom and West Germany, the government provides a fraction of the support and industry provides the remainder. While Japan has a few such institutes organized by industrial sectors, each prefecture supports a research institute for industrial science and technology for the industry of the region. Conceptually, these latter institutes are not unlike the agricultural research and extension centers in each state of the United States. Separate from the cooperative institutes, the Japanese government supports "diagnosticians" (licensed by MITI) that analyze and teach modern techniques to small- and medium-sized firms. In West Germany, a system of industrial extension agents has now been authorized by the central government and will apparently be operated by local chambers of commerce. Many of the states of the United States operate modest but similar programs, usually through local universities. Most of these organizations were initiated as a result of the State Technical Services Act that provided matching funds for these services to industry. [20] Although federal funds for that program are no longer available, state governments and local industries have continued to support many of them.

There are only a few cooperative industrial research associations in the United States -- for example, in the textile, copper, lead, natural gas, and iron and steel industries. There has not been much enthusiasm for them and no federal support. The Electric Power Research Institute serves the broad technical needs of the electric utility industry. It came into being as a response to legislation proposed in the U.S. Senate to impose a tax for this purpose and to have the government allocate the funds. Fear of antitrust action seems to account for some of this lack of industrial cooperation in technical matters.

Reducing Market Risk

The introduction of new commercial products involves two types of risks -- one technical and the other having to do with uncertainties of the market. Programs that aid in the development of technology, the subsidy of R&D, and the dissemination of information and know-how tend to reduce technical risk. Programs of government that provide sure markets tend to reduce commercial risk. [21]

All governments procure products for their own use, particularly for defense. Some of these products are standard, off-the-shelf items, and others require R&D and may be highly innovative. Increasingly, government procurements have tended to use "performance" specifications which, in contrast to detailed specifications of product characteristics, may give some rein to the ingenuity and innovativeness of the supplier. The French central government has used such techniques for the development and construction of less expensive and more effective secondary schools. Recently, the U.S. General Services Administration has begun to use performance specifications to purchase improved commercial products. The specifications have been prepared for lawn mowers, air conditioners and even some government buildings. It is claimed by some critics that the government market for such goods in the United States is too small to affect industrial development generally. However, most product innovations at the beginning find a small, special use that eventually broadens to encompass a large market. For example, during the development of computer-controlled machine tools, the U.S. Air Force created a market by making a large purchase of tools to be used in the plants of its contractors. This government action, along with government support of the "software" for these machines, speeded the innovation process.

The United Kingdom has developed a unique program for preproduction order support. Here the government supports part of the purchase cost of new products to be used in industrial production. The government helps arrange for industrial use and guarantees a quantity of initial sales. It was through such arrangements that computer-controlled machine tools manufactured by British firms were first introduced into British industry.

Similar suggestions have been made for the development of new synthetic fuels in the United States. At present oil prices, these fuels are not economic. There is a need to determine the feasibility and the range of possible costs of new processes and to provide a means of reducing the demand for imported oil. As an example, the government, rather then supporting R&D and pilot demonstrations, could seek bids for a supply of synthetic fuels for a contract period. The supplier would then specify the technology, undertake the process development, construct the plant, and supply the fuel which the government would then sell at market prices. Procedures of this type create a market, remove the government from technical decision making, and spur the innovation process.

Restrictive national industrial standards and other nontariff barriers tend to assure markets for domestic suppliers. These arrangements, including restricted government procurement, are intended to provide stimulus to local innovation. However, protecting markets for domestic suppliers may restrict the entry of competitors from other nations and may counter the very stimulation that the government actions are aimed at producing. Recently, GATT negotiators have reached agreement that all countries will reduce the restrictions on government purchases from foreign firms. If the United States is technologically competitive, the implementation of such an agreement should be a spur to U.S. trade and innovation.

As part of their programs for stimulation of the computer industries, the governments of the United Kingdom, France, and West Germany gave preferences in their purchases to domestic suppliers, similar to the "buy America" conditions of U.S. procurements. Generally, national postal and telephone authorities use domestic suppliers almost exclusively. The Japanese government has for years restricted imports to its market, giving preference to domestic producers using technology purchased from other Western nations, notably the United States. In the recent GATT negotiations, the Japanese have apparently agreed to liberalize such arrangements, but complex import and distribution systems still effectively restrict entry of foreign products.

AMELIORATING THE ADVERSE EFFECTS OF INNOVATION

The discussion of the effects of government actions on the innovation process would be incomplete without consideration of the adverse effects of the introduction and spread of the use of new products and processes throughout the society. The introduction and diffusion of new products and processes can displace existing products, techniques, labor, management, firms, and sometimes whole industries. In the long history of industrialization, despite prediction to the contrary, new products and processes have not created vast unemployment. A rising standard of living, new purchasing power, reduced work, and new goods to purchase have been created along with the improvement in productivity. However, during the period of dynamic change, some are hurt and others benefit. The displacements can be so traumatic that governments must provide means of adjusting to them. Bourdon examines the relationship of productivity, unemployment, and innovation in Chapter 6.

As industrialization activities become widespread, and society lives with and uses its products, the conditions of the environment both within and without the workplace become of great concern. The costs of polluting air and water and of unsafe, unhealthy workplaces were not in the past borne by firms as a part of the costs of goods. Generally, these costs fall on the public and must, like worker displacement, be a major concern of public policy.

Displacement and Government Policy

Major displacements of firms and their workers occur as a result of the spread in the use of new products and processes. In rapidly growing firms, regions, and economies, shifts within a firm or an economy are relatively easy since new jobs are created and new investments made as the old becomes obsolete. However, even then adjustments may be traumatic; the locations of plants may change, and the new skills required may be very different from the old. These effects of innovation are harmful to workers and when little is done to ameliorate the hurt, resistance to innovation may increase. Steps may then be taken to protect affected workers and industries and to slow down the change. All of the European countries place major restrictions on the ability of firms to dismiss their workers. In Japan, the historic, paternalistic attitude of large firms, at least, is such that lifetime employment in a firm has been virtually guaranteed. When necessary, the firms provide retraining and relocation of their workers. For Japan, this employment policy, along with the paraphernalia of government protection of the domestic market and the stimulation of industry, has not retarded innovation or technological changes. In fact, this policy may have stimulated innovation because, to assure employment, the large firms must compete diligently in the world export market. With competition becoming stiffer from South Korea, Brazil, Taiwan, and Mexico (just as European and U.S. firms faced

Japanese competition in the past), the more mature Japanese industries must find means, besides expansion of their export market, to maintain productivity and employment, or there will be disruption in the traditional Japanese society.

Although in West Germany, as in all European countries, it is difficult to lay off workers, that country has a much more highly developed set of programs for assisting in the retraining and relocation of displaced workers than any other European nation with the possible exception of Sweden. Firms and individuals are partially reimbursed for the costs of retraining. Jobs are partly subsidized in geographical areas of high unemployment. In France, a few years ago, a special tax was levied on wages and salaries to be used for the sole purpose of the education and retraining of workers. The United Kingdom has often chosen to rescue major unprofitable aircraft, automobile, steel, and shipbuilding firms. These policies preserve jobs for a time, but they divert resources from industrial activities that are dynamic and growing.

Recently, European nations faced with high unemployment; competition with their traditional mature industries from progressive, less-developed countries; and dynamic competition in high technology products from other developed countries have become much more concerned with displacement. The tendency has been toward protective measures to prevent competitive imports, subsidize domestic competitors, and limit the scope of multinational firms.

Demands in the United States for protection of the textile, shoe, and steel industries are loud. Special quotas and "trigger" mechanisms have been established to slow the displacement resulting from the spread of technology and production to the lesser-developed countries. There is even talk of attempting to limit the flow of "U.S." high technology to the industrial competitors of the United States as the balance of trade becomes more unfavorable. The U.S. Economic Development Administration aids the economic development of distressed regions (after high unemployment is recognized). However, the modest assistance provided for job training in the United States does not compare with the programs of West Germany or Sweden.

Programs that encourage the location of industrial activity in depressed regions are supported in all European countries by the European Economic Community. Special capital grants and loans, as well as tax abatements, are used extensively. The agency responsible for regional development in France has initiated a new program to encourage the start-up of new firms in special regions outside of Paris. Ireland has perhaps the most successful programs for attracting manufacturing firms to locate in its country. The Irish Development Corporation and the Irish Government provide grants, low interest loans, tax abatements, and training subsidies for firms that locate there. Ireland views this activity as a healthy way to reduce unemployment and improve the Irish economy.

Ameliorating the Environmental and Health Consequences of Industrialization

In dealing with environmental problems, governments have given little consideration to the way regulation might impede or accelerate the development of new products and processes. However, government programs to improve the environment and working conditions do affect the innovation process.

While environmental regulations operate differently in the various countries belonging to the Organization for Economic Cooperation and Development (OECD) all are concerned with environmental degradation and the condition of the working environment. Improving worker satisfaction and health and safety conditions, have higher priority in Europe than in the United States, and industrially initiated programs to restructure work patterns, increase worker participation, and reduce worker alienation are much further advanced. In Japan, the long accepted relationship between the worker and the firm creates a very different work environment from anywhere else in the world.

The various countries have different histories of, and emphasis on, control of environmental pollution. Though the data are scanty and unreliable, it appears that most industrialized countries spend between .5 and 1.5 percent of their GNP's for pollution control -- West Germany having, perhaps, the highest rate of expenditure.

The relation between environmental regulations and the innovation process is extremely complex. The effects of these regulations tend to be concentrated in specific industries which, in turn, tend to be located in certain geographical regions. The basic materials, chemical, automobile, and energy-producing industries are strongly affected. In fact, it is often the older, more mature, industries with large production units that are the targets of environmental regulations and must make the abatement investments. And it is these industries that are facing competition from countries with lower labor costs, modern technology, and without extensive pollution-control programs.

Environmental regulations directly and indirectly affect the innovation process. New regulations generate markets for new products to perform the new functions that the regulations generate. These regulations also raise the costs of production, diverting capital from investments for other new products and businesses -- a serious complaint of some businessmen but not of those whose companies produce abatement equipment. Environmental regulations have an indirect effect on innovation. The regulations create new requirements for products or processes, often in an otherwise slowly changing industry. Firms producing mature products are likely to respond to regulations first by substantial resistance -- e.g., the U.S. automobile industry's posture toward environmental and safety regulations. Then, when that fails, they are forced to introduce radical innovation that meets the new requirements while, at the same time, providing an improved commercial product or process. The Center for Policy Alternatives analysis of commercial innovations in Europe and Japan showed that environmental regulations stimulated commercial innovation in the chemical and automobile industry. [22] These results are consistent with the current understanding of the way innovation proceeds in industry. New external forces on firms producing mature products stimulate and invigorate innovation.

SOCIOECONOMIC CLIMATE AND INNOVATION

The special government policies and programs that affect innovation directly all operate in a national social and economic climate. Even though there is little doubt that this climate is crucial to establishing the preconditions for industrial innovation, no studies exist that relate innovation activity to such matters as fiscal, monetary, and tax policies or to the cultural heritage of the particular nations. The general knowledge of economics and of the innovation process indicates that tax policies rewarding risk and investment, a high savings rate, low inflation, and active competition all contribute to a dynamic industrial economy. Klein discusses these matters more fully in Chapter 3.

CONCLUSIONS AND POLICY IMPLICATIONS

There is a growing acceptance throughout the industrialized world that technological innovation is crucial to social and economic development and to the amelioration of the adverse effects of the manufacture and use of industrial products.

The examination of the policies and programs of some European countries and Japan indicates that they recognize that the innovation process is complex -- requiring a number of conditions for it to occur readily. Reflected in these policies and programs is a growing recognition that the support of science and of R&D are alone insufficient to stimulate technological innovation. These policies and programs also reflect the growing appreciation that individual firms cannot and will not fully support all of the activities necessary to a healthy innovative environment. The most common activities in support of innovation in the market-oriented industrialized countries are:

- Support for the development of technology believed basic and important to future industrial progress and economic development through a widespread system of industrial research associations that develop technology commonly used in the industrial sectors and that provide training, advice, and information to their member firms.

- A variety of institutions concerned with broad technical fields that connect science to the needs of industry and perform some of the technical work to advance technology.

- Direct subsidies to firms to reduce the cost of innovation through grants or contracts and that favor projects that are technologially risky, lead to products for foreign markets, or ameliorate the adverse effects of industrial activity.

- Specific programs to aid the individual inventor or new firm through grants provided for the support of the early stages of the innovation process, the demonstration of feasibility, managerial and financial advice, and support for early manufacture.

- Activities that are aimed at easing labor displacement from innovation, new patterns of industrial activity, and of foreign trade through retraining and relocating workers.

- Programs and policies to ameliorate the adverse effects of industrial activities through regulation of toxic chemicals, worker health and safety, and air and water pollution. The form and severity of these regulations differ among the countries. In none of the countries have government regulatory policies been designed to consider their effects on innovation, although regulations have been found to stimulate commercially successful innovation.

The fact that one nation has developed a set of policies and programs that successfully influence innovation does not mean that they are necessarily applicable to another country. The size, the industrial system, and the political, social, and economic environments all may differ, and different actions may be appropriate to the differing circumstances. Granted these differences, the practices of other nations, coupled with a knowledge of economics and an understanding of the innovation process, do provide suggestions for new policies and programs that might be effective in improving the climate for innovation in the United States.

Other countries are beginning to recognize the central role of innovation and technology in meeting the needs of the public, improving productivity, competing in foreign markets, and adjusting to a changing world, and they are organizing to attend more effectively to these matters. It appears that the U.S. Government, in both its legislative and executive branches, should consider the organizational changes necessary to recognize and deal with these central issues. The United States should consider:

- Initiating a major program of support in both industry and the universities for the development of technology broadly applicable to industry (not for proprietary product or process development).
- Creating and partially supporting a broad system of industrial research associations to aid in the development of technology and to provide services, advice, and technical information to industrial firms.
- Providing partial support (perhaps with the states) for an extensive system of research institutes, possibly attached to universities, to connect science to broad industrial needs.
- Improving the climate for new ventures through changes in the tax laws, and providing subsidies to private inventors and small firms for the early risky stages of innovation.
- Developing a more complete system for anticipating industrial changes and displacements, and supporting more fully the retraining and relocation of workers.
- Improving regulatory techniques to encourage the innovation needed to ameliorate the harmful effects of industrial activity.

NOTES

1. This chapter is based primarily on analysis of some specific countries, on experiences and observations of the author, and on two broad studies: Nicholas Ashford, "National Support for Science and Technology: An Evaluation of Foreign Experiences -- Summary" (Report No. CPA 75-12, Center for Policy Alternatives, Massachusetts Institute of Technology, Cambridge, Mass., August 18, 1975), and Pierre L. Bourgault, "Innovation and the Structure of Canadian Industry," (Special Study No. 23, Science Council of Canada, October 1972).

2. Kenneth J. Arrow, "Economic Welfare and the Allocation of Resources for Invention," in _The Rate and Direction of Inventive Activity: Economic and Social Factors_, Report of the Conference of the Universities, National Bureau Committee for Economic Research and the Committee of Economic Growth of the Social Science Research Council, National Bureau of Economic Research (Princeton, N.J.: Princeton Universtiy Press, 1962), pp. 609-26.

3. J.H. Hollomon, et al., _Energy Research and Development_ (Cambridge, Mass.: Ballinger Publishing Co., 1975).

4. K. Pavitt and W. Walker, "Government Policies Toward Industrial Innovation: A Review," _Research Policy_ 5 (1976): 11-97.

5. P. Bourgault, "Innovation and the Structure of Canadian Industry."

6. Pavitt and Walker, "Government Policies Toward Industrial Innovation"; Ashford, "National Support for Science and Technology."

7. Ashford, "National Support for Science and Technology."

8. _Commission of the European Communities, Public Financing of Research and Development in the Community Countries, 1967-1971. Analysis by Objectives_, (Research and Development No. 3, May 1972); "Statistisches Amt der Europaischen Gemeinschaften," Die offentlichen Aufwendungen fur Forschung und Entwicklung in den Landern der Demeinschaft. Analyse nach Forschungszielen 1968-1972" (Statistiche Studien und Erhebungen No. 1, 1972).

9. Floyd R. Tuler (with the assistance of K. Nagaraja Rao), University and Industry, (Working Paper 78-30, Center for Policy Alternatives, Massachusetts Institute of Technology, Cambridge, Mass.).

10. O. Keck, "West German Science Policy Since the Early 1960s: Trends and Objectives," Research Policy 5 (1976): 116-57.

11. "Fifth Report of the Federal Government on Research," Federal Republic of Germany, Ministry for Research and Technology, 1976.

12. "The Role of Technology in the Change of Industrial Structure," Industrial Research Institute (Japan, April 1978).

13. "Fifth Report of the Federal Government on Research," (Germany).

14. Ashford, "National Support for Science and Technology."

15. Edward B. Roberts, "Entrepreneurship and Technology," Research Management, July 1978.

16. Ashford, "National Support for Science and Technology."

17. Industrial Research Associations in France, Belgium, and Italy, Organization for Economic Cooperation and Development (Paris, 1965), p. 117, and Research Associations: The Changing Pattern (London, 1972).

18. Ashford, "National Support for Science and Technology."

19. Policy and Strategy for Science and Technology, Ministry of Science and Technology, Republic of Korea, 1975, p. 8.

20. State Technical Services Act of 1965 (P.L. 89-182, U.S. 85th Congress, 1st Session).

21. Fritz Prakke, "Government Procurement Policies and Industrial Innovation" (Delft: TNO, 1978).

22. Ashford, "National Support for Science and Technology."

9 Summary and Policy Implications

Christopher T. Hill
James M. Utterback

THE CENTRAL ISSUE: A DYNAMIC ECONOMY

The authors of this book are concerned with a central issue of our times: how to ensure that the United States economy maintains its ability to respond to changing opportunities and to new challenges -- that is, that it remain a dynamic economy. The authors have shown that technological innovation is an important means of growth and change. While use of new technology undoubtedly creates new problems for the country, it can also help to control inflation, create jobs, enhance productivity, maintain growth, enhance environmental quality, and support a healthy balance of trade with other nations.

America currently faces numerous problems which a more rapid rate of technological innovation could help to resolve. The challenges of high inflation, declining productivity advance, slow growth, high trade deficits, high-priced energy, and a deteriorating environment can all be addressed through innovation. The authors of this book believe that the emergence of these problems can be traced, in part, to a declining rate of innovation in the United States over the last two decades. They believe that the United States could and should do more to ensure that a healthy rate and direction of technological innovation are maintained.

This chapter briefly summarizes the main ideas on the nature of the innovation process and what government can do to guide, direct, and stimulate it. It concludes with a summary of specific policies that should be considered in the ongoing discussions on industrial innovation in the United States.

THE NATURE OF THE INNOVATION PROCESS

In the market-oriented U.S. economy, technological innovation nearly always occurs in private firms, although government plays many roles in supplying the resources necessary for innovation and creating the environment within which it must happen. However, with rare exceptions, government itself does not innovate. Instead, it must find ways to influence the rate and direction of technological innovation in private firms. Similarly, only rarely do other institutions in our society, such as universities or special interest organizations, engage in technological innovation. When they do, it is nearly always by establishing or cooperating with a profit-making venture to commercialize an idea or an invention.

Technological innovation involves matching in a new way a social or economic need with capabilities drawn from science, technology, or craft. Innovative activity is risky business. It is not possible to know in advance whether any particular project will be successful. The greater the advance sought, the more uncertain the outcome.

According to the classic managerial theory of the firm, firms attempt to choose among the variety of alternative products, strategies, and technologies available to them in order to maximize some measure of profitability. The major concerns of such firms are to increase the probable rewards to success and to reduce the risk of failure in the marketplace. Therefore, firms must consider the profits likely to result from an innovation, and the technological risk involved, when deciding to attempt it.

However, there is another and more important risk -- the risk that in failing to innovate, an existing business may be taken over by a competitor who does innovate. Thus, an important incentive for firms to attempt risky technological innovation is their desire to survive in the face of effective competition, or rivalry, from other firms. Such rivalry can be especially effective if the competitor is a new entrant who has a new technological product or process that is superior to that of the existing firms in the industry.

Small new ventures and larger firms entering a business for the first time, introduce a disproportionate share of the innovations that create major competitive threats and rivalry. Established firms often respond to a new entrant's invasion of their product line with redoubled creative effort and investment in what they already know well. Even though it may be crude, the new technology may have great performance advantages in certain submarkets and gain ground by competing there first. Use of the new technology then expands as it captures a series of submarkets. The new technology often has a much greater potential for improvement and cost reduction than does the existing technology. Thus, price cutting by established units as a defense may be ineffective.

The continuing entry of new firms is an important determinant of the degree of rivalry in an industry, because new entrants, in hopes of becoming a major factor in an industry, have a greater incentive to take risks. In the absence of such new entrants, there is a tendency for the degree of risk taking in existing firms to decline. In existing firms, entrepreneurs' habits of thinking and their patterns of search are likely to be highly constrained by the forms of the technology and the organization they already have. Therefore, they are likely to search for new technologies more narrowly than do new entrants.

Firms must become more specialized and efficient to exploit innovations over an extended time. This drives them toward a more stable production process and a more structured organization. Most established large firms have evolved from small, disorganized, and highly innovative beginnings. Evolution starts through the origination of one or more major product innovations. These are usually stimulated by users' needs through frequent interaction between entrepreneurs and potential users of the innovation. Exploration of the product's potentials in different applications follows. Rising production volume may lead to the need for innovation in the production process. Demands for greater sophistication, uniformity, and lower cost in the product create an ongoing demand for development and improvement of both product and process. This means that product design and process design become more and more closely interdependent as a line of business develops. A shift from radical to evolutionary product innovation usually occurs as a result of this interdependence. This shift is accompanied by heightened price competition and by an increased emphasis on process innovation. Small-scale units that are flexible and highly reliant on manual labor and craft skills, and that use general-purpose equipment, develop into units that rely on automated, capital-intensive, high-volume processes. Thus, innovative patterns, production processes, and the level of production capacity all change together in a consistent, predictable way over time.

Technological innovations that lead to productivity advances depend not only on the degree of rivalry in particular industries, but also on the relative difficulty of bringing about productivity gains. Productivity advances typically come about as a result of broadening the definition of a particular technology. For example, diesel engines broadened the definition of railroad technology, jet engines broadened the definition of airplane technology, and semconductors broadened the definition of computer technology. How rapidly a particular technology will run into diminishing returns, as measured by the cost of bringing about advances, will depend on how easy or difficult it is to broaden the definition of the technology in question.

Highly innovative firms have much different characteristics from those that are long established. It is difficult for firms producing standardized products in high volume to respond to the challenges of major technological or market changes. Thus, they are vulnerable to invasion of their markets by other products or alternative solutions. For example, the major manufacturers of manual typewriters did not introduce the electric typewriter. Few major manufacturers of mechanical calculators now manufacture electronic calculators and few manufacturers of vacuum tubes were successful in making the shift to transistors. Unless firms that reach this mature stage are dealing with a long-lived product, they may eventually be forced out of business.

ENCOURAGING TECHNOLOGICAL INNOVATION

Government's most effective role in reversing the current decline in productivity gains is to create a more positive climate within which technological innovation can occur. This does not mean that the climate should be benign, especially toward entrenched firms which are slow to protect their positions by reducing costs or improving their products. In fact, creating a more dynamic economy depends on generating greater rivalry and competition between firms. This, in turn, depends on an increased rate of capital formation and, especially, on the formation and entry of new technology-based ventures by existing firms and by new firms aimed at invading established markets. In order to make the economy more dynamic, certain steps should also be taken to free some industries from the protective cocoon of economic regulation and to subject others to new and more vigorous antitrust standards, including limits on mergers and on vertical integration.

Protection of a domestic industry from foreign competition reduces the incentives to innovate in that industry. To get the country moving again, it is imperative to prod the less dynamic industries to move more rapidly than they have in many years, through exposure to foreign competition and through other means of creating rivalry. At the same time, more active government participation is needed in translating and informing U.S. firms of foreign business needs and opportunities.

Federal resources for industrial innovation should be directed toward creating new technological options through support of basic research, and toward broadening the scientific base of industrial technology, but not toward the commercial development of particular technologies, except to meet specific public needs. The private creation of new firms to participate in federally funded civilian R&D should be strongly encouraged. Public means for anticipating industrial changes and relocation of people should be greatly strengthened. Education of all citizens should be oriented toward developing a capability for problem solving and toward developing the skills that enable them to adapt to changing employment possibilities.

Regulations aimed at reducing hazards in the workplace, in the natural and human environment, and in product use are not major causes of the contemporary declines in innovation and in productivity improvements. Regulations can and should be designed to promote rivalry, the entry of new firms, and the development of new, more acceptable, technologies, rather than to protect vested positions in existing markets.

In general, a climate in which firms, their products, and their market shares are unchanging will lead to an increasingly unstable economy. Such an economy may feature rising levels of imports and rising inflation coupled with declining employment, reduced productivity improvements, and slow growth. On the other hand, an economy marked by sufficiently intense rivalry and frequent creation of new ventures, new firms, and innovative technologies will be characterized by rising productivity, by higher employment, by a favorable balance of trade, by stable prices, and by high rates of real economic growth.

Frequently, it is claimed that the risks of technological innovation are too great today, and that government should take actions to reduce that risk by, for example, subsidizing development of new technologies, weakening the antitrust laws, or rolling back environmental, health, and safety standards. The main message of this book is just the opposite -- government should take steps to encourage innovation by increasing the risks that firms face -- not by increasing the risk of failure of new technology, but by increasing the risk that a firm will fail if it does not innovate!

POLICY CONSIDERATIONS

Throughout this book, specific policies have been identified that the government might consider in encouraging technological innovation; they are summarized here. These policies are not all that government might do, but they are central to restoring and strengthening a dynamic economy. (Each of the preceding chapters contains a more detailed discussion of the policy issues and options in each area.)

Encouraging Domestic Rivalry

Rivalry among firms in an industry, and between those firms and new entrants, is essential to stimulating risk taking, technological innovation, and productivity advance. Specific actions to enhance such rivalry include encouraging new entrants, restructuring existing industries, and removing certain economic regulations.

- Encouraging new entrants by increasing the availability of venture capital.

The U.S. economy has been marked by low rates of new capital formation in recent years. Symptomatic of the problem are low rates of return on invested capital (according to U.S. Treasury estimates, the average rate has fallen from about 8 percent in the middle 1960s to about 4 percent at present), low rates of creation of new business ventures, and a decline in the public issuance of new equity securities. The problem is particularly acute for new business ventures in the high-technology industries. Small companies have contributed a disproportionate share of innovation in these industries, and the rivalry engendered by these companies serves to stimulate larger companies' innovative activities. The rate of formation of new, small, high-technology ventures has declined substantially in recent years, a condition that has been accompanied by the virtual "drying up" of the venture capital market.

In contrast, Japan has had high rates of capital formation. It has a broad complement of government programs to aid individual inventors, entrepreneurs, and new technology-based firms. In no other developed country is there such a complete array of programs to encourage innovation, including: grants-in-aid for the implementation of inventions, special development contracts and grants, arrangements for commercialization of new developments, support for capital equipment for new ventures, rapid tax write-off for new developments, low-interest loans for high-technology firms, special analysis and advice for small firms, special procurement policies for small businesses, "open" research laboratories in all prefectures for the use of new and small firms, and no-interest loans for modernization of small firms. The most important aspect of these programs is the diversity of ways in which new small firms can obtain assistance to enter a business and compete.

The United States should consider adopting new programs of the kind now found in Japan. In addition, venture capital might be made more available through such actions as eliminating the corporate income tax on dividends returned to shareholders and encouraging managers of employee retirement funds to invest in somewhat more risky ventures.

- Modify the structure of existing industries.

Both vertical integration and mergers among firms in dissimilar industries are designed to reduce business risks by controlling access to markets and suppliers, by limiting competition, or by diversifying risks in a portfolio of activities.

These business actions can distort static allocative efficiency, contribute to higher prices, and wield substantial political power. But, they have another, perhaps even more pernicious, effect on technological innovation and dynamic efficiency: the firm that minimizes its risks has less incentive to innovate.

Specifically, two kinds of actions should be considered. First, corporate mergers above a certain size should be prohibited. The exact nature of such a law -- with regard to the sizes of firms in various lines of business -- is not clear at this time, but it is clear that mergers that reduce risk are generally undesirable.

Second, the antitrust laws need to be changed to recognize the importance of stimulating rivalry and dynamic behavior, in addition to achieving improved static efficiency. Workable indicators of the failure of an industry to be dynamic might include unchanging market shares over a period of time, an inability to cope with foreign competition, or the lack of new entrants into the industry.

- Remove economic regulations where possible.

Economic regulation in such areas as transportation, communications, and electric power has removed much of the rivalry and the incentive to innovate from these industries. Therefore, the ongoing deregulation of passenger airlines and the proposals for deregulating the trucking industry should be supported as steps to encourage innovation and productivity improvement.

One proposal deserving consideration is requiring the separation of electricity generation firms from those in the electricity transmission and distribution businesses. There are no "natural monopolies" in generation, and generating plants are already operated by consortia of utilities in many cases. If new generating stations were individually owned, they would have to charge rates equivalent to the (rising) long-run marginal costs of power generation, but they would also face strong pressures to hold costs down. This scheme would also facilitate new entry by firms that develop such alternative sources of electric energy as solar and wind power.

Managing Rivalry by Foreign Firms

American competitiveness in international markets is to a large extent determined by the pace of technological innovation in the domestic economy. Any measure designed to bolster the nation's competitiveness in the world economy that does not work by improving the climate for domestic innovation is, at best, a measure of short-run or secondary importance.

- Avoid protecting U.S. firms from foreign competition.

Industrialization of developing nations presents the United States with a number of dilemmas and important choices. In countries where labor costs are low, industry can become highly competitive in trade in goods based on mature, standardized technologies. The more labor input required per unit of output of a good, the more competitive the developing nations will be as producers of that good. In the absence of governmentally imposed barriers to trade, some goods from developing nations will inevitably capture a growing share of world markets, including the U.S. domestic market.

Equally inevitable will be calls for import tariffs or import quotas to protect threatened U.S. industries. Such calls should not be heeded unless foreign industries are clearly subsidized by their governments, for example, by direct measures or by the absence of social controls over worker safety. To protect U.S. industries would be to force U.S. consumers to pay high prices for domestically produced goods that could be purchased at lower prices from abroad. This would be tantamount to taxing U.S. consumers in order to subsidize producers. Furthermore, protection of a domestic industry shields producers in that industry from the stimulus of foreign rivalry. The shield reduces the incentive to innovate within the industry, eventually allowing the industry to lag behind world standards in the introduction and use of new technology. Thus, protection of a domestic industry ultimately brings technological obsolescence in its wake.

Assisting Workers in Declining Industries

A vital component of the nation's economic policy is an effective adjustment assistance program. The need for adjustment assistance derives from the fact that in a dynamic economy the relative importance of some sectors of the economy inevitably declines. If the decline is rapid, there is loss of jobs in these sectors. Concurrently, other sectors gain in importance, and new job opportunities are created. If the rising sectors are marked by a high rate of technological innovation, as has generally been the case in the United States in the post-World War II era, the new jobs would typically be better in terms of compensation and working conditions than the jobs in the displaced sectors. It is, however, likely to be of little comfort to the worker who faces a job loss in a Massachusetts shoe factory to know that there are employment opportunities in a Texas semiconductor plant. The Massachusetts worker might not particularly care to move to Texas, and even if he did, he might not have the skills necessary to move into a desirable new position there.

- Avoid or limit temporary assistance efforts.

Whether some sort of adjustment assistance or relief should be given to firms or industries threatened by imports is a matter of controversy. Such assistance, as practiced by many nations, typically takes such forms as temporary tariffs or quotas on imports, or temporary government subsidies to the affected firms. In principle, the assistance should last only as long as is reasonably required for the affected firms to adjust to changing circumstances. In practice, however, firms more often than not are unwilling or unable to adjust. If this is the case, political pressure for the assistance to be extended will mount. Thus, a program of temporary assistance can evolve into a program of wholesale protection or subsidization of a noncompetitive industry. This will further weaken the incentives to innovate and lead to the ultimate demise of the industry.

- Develop an effective adjustment assistance program.

The U.S. labor movement rightfully scorns the present program in this area as constituting little more than a small "burial insurance." More imaginative and far-reaching programs are needed. In addition to enlargement of the existing program and liberalization of eligibility criteria, at least three additional steps should be taken by the government: (1) granting lifelong pensions to displaced workers who, because of age and skill, cannot reasonably expect to find new work; (2) creating incentives for business firms to locate plants in regions where competition has created pockets of severe unemployment; and (3) sponsoring extensive programs to retrain and relocate displaced workers who can reasonably expect to move into new careers. While it is doubtful that the government can do much to reduce the trauma of a job change, it can and should bear part of the cost. The alternative would be for the nation to subsidize and protect declining industries -- an alternative that would be far costlier and would ultimately stifle the vitality of the entire economy.

Creating New Scientific and Technological Options

- Support basic scientific and engineering research.

Government should support research and development in basic science and engineering and in technologies that are widely applicable in many areas of industry. Industry underinvests in these areas, which provide the basic knowledge upon which the next generation of technological innovation can be built.

By allocating its R&D money to a wide variety of firms and individuals, the federal government can enhance competition and rivalry in industry, particularly by smaller firms and by new entrants. If it concentrates its resources in a few established firms, or if it funds relatively few projects at levels beyond the reach of any but the largest companies, the government can actually reduce competition, raise barriers to entry, and have a negative effect on innovation.

- Avoid support for commercializing new technology.

Federally funded R&D should not be carried too far in the direction of commercialization. Resources may be diverted from other promising areas, and competition may be stifled by concentrating resources in a small number of large firms, or in government owned research facilities. Federal sponsorship of civilian industrial technology should place emphasis on creating an atmosphere in which such innovation can flourish, rather than on efforts towards bringing innovations to the marketplace.

Managing Noncommercial Risks of New Technology

This book speaks of the need to encourage greater risk taking in industry, as a means to increase the rate of technological innovation and, thereby, to enhance the strength of the economy. However, use of technological innovation can also create risks to the health and safety of workers, consumers, and the public, and to the quality and productivity of the natural environment.

Managing these noncommercial risks of new technology is quite a different matter from encouraging greater economic and financial risk taking on the part of managers and entrepreneurs. Yet, the two are connected in at least three ways. First, new technology is needed to identify and control many of the noncommercial risks, or to avoid them altogether. Second, programs to manage noncommercial risks can increase financial risks and inhibit new technology development. Third, programs to manage noncommercial risks can act to dislodge mature industries from established patterns of activity and can stimulate development of new technology in much the same way as the threat of rivalry from a new entrant or foreign competitor.

- Maintain strong programs to control noncommercial risks of new technology.

When innovations result in hazards in the workplace, in the general environment, or to the users of products, the resulting social costs are often of great public concern. Requirements that firms bear more of these costs have sometimes been decried as slowing productivity growth, sapping resources from innovation, and increasing industry vulnerability to foreign competition. But regulations are a forceful expression of a public demand for such public goods as clean air and water, regardless of the failure to account for these goods in the gross national product. Regulation may reduce innovation in some regulated industries, but encourage the development of less costly or more desirable substitute products in others. In the long run, failure to control the risks of new technology would threaten the very possibility of maintaining a viable economy.

- Design regulations to allow a variety of technological responses.

Regulatory constraints that tend to drive all products or processes in a market to the same design can severely constrain the arena within which innovative efforts can occur. Other regulatory thrusts that demand high performance and create markets for new products may spur innovation.

Regulations may tend to reinforce entrenched positions in an industry. Innovation may decrease as a result of entry barriers and decreased competition. The quality or novelty of innovations may also decline. To the extent that regulations contain provisions for variances and financial help, these effects will be reduced. On the other hand, regulation can effectively increase the degree of rivalry in an industry and provide a stimulus that encourages new entrants with new technology, especially if it is based on performance rather than design requirements. Regulation should be designed to encourage the best possible technological response from industry.

- Consider alternatives to direct regulation of noncommercial risks.

For a limited range of noncommercial risks of new technology, economic incentive approaches may be preferable to, or should be coupled with, regulation of the standards type in order to encourage technological innovation. There is little experience with such programs in the United States, and some pilot efforts would be needed to determine whether incentive approaches to regulation are workable and whether they, indeed, favor technological innovation. Wholesale change to the incentive approach is not now warranted.

CONCLUSION

Public and political interest in the state of U.S. technological innovation has been growing for the last two decades. Hardly a top priority concern for policy makers in the past, the priority assigned to technological innovation has risen in recent years as its connection to the vitality of the entire economy has come to be realized. Viewed properly, policy for technological innovation is not merely the province of pleaders for academic science, boosters of grandiose engineering feats, or advocates of "appropriate technology." Instead, policy towards technological innovation is coming to be viewed for what it actually is -- a key element in the long-term economic and social policies of the United States.

About the Authors

NICHOLAS A. ASHFORD is Assistant Director of the MIT Center for Policy Alternatives. He is the Chairman of the National Advisory Committee on Occupational Safety and Health and is a member of the Science Board of the Environmental Protection Agency.

CLINTON C. BOURDON is Assistant Professor in the Harvard Business School. He has been a consultant to the National Center for Productivty and Assistant Professor in the Department of Economics at Boston University.

EDWARD M. GRAHAM is currently on the Staff of the Office of the Secretary, U.S. Treasury Department. At the time he prepared his chapter, he was Assistant Professor in the Sloan School of Management at MIT.

GEORGE R. HEATON, JR. is a lawyer and Research Associate in the MIT Center for Policy Alternatives.

CHRISTOPHER T. HILL is a Senior Research Associate in the MIT Center for Policy Alternatives. He has worked for the Congressional Office of Technology Assessment, Washington University in St. Louis, and Uniroyal, Inc.

J. HERBERT HOLLOMON is Director of the MIT Center for Policy Alternatives. Previously, he was President of the University of Oklahoma, Acting Undersecretary of Commerce, and Assistant Secretary of Commerce for Science and Technology.

PAUL HORWITZ is a Principal Research Scientist and member of the Business Planning Group at Avco Everett Research Laboratory, Inc. He has been a Congressional Fellow in the office of Senator Edward M. Kennedy.

BURTON H. KLEIN is Professor of Economics in the California Institute of Technology. Previously, he was the head of the Economics Department at the Rand Corporation.

W. CURTISS PRIEST is a Research Associate in the MIT Center for Policy Alternatives. He has had extensive research experience in several firms in the areas of systems and economics.

JAMES M. UTTERBACK is a Research Associate in the MIT Center for Policy Alternatives. Previously, he was Assistant Professor in the Graduate School of Business at Indiana University.

Abstracts

TECHNOLOGICAL INNOVATION FOR A DYNAMIC ECONOMY

Chapter 1 TECHNOLOGICAL INNOVATION: AGENT OF GROWTH AND CHANGE by Christopher T. Hill

This chapter examines the importance of technological innovation for the economy and the society of the United States. Although it is not the only driving force, technological innnovation is a key agent of growth and of change in such important aspects of the society as economic growth, productivity, inflation, employment and the nature of work, international trade, and the quality of human life. A high rate of technological innovation can contribute to improvements in all these areas. Therefore, the rate and the direction of technological innovation should be of major concern to the United States. If the nation is to enjoy the benefits of a dynamic economy and rapid technological change, government, industry, and labor need to consider taking various actions outlined in this chapter and in the other chapters of this book.

Chapter 2 THE DYNAMICS OF PRODUCT AND PROCESS INNOVATION IN INDUSTRY by James M. Utterback

This chapter discusses the nature of the process of technological innovation within firms. Innovation is an intended act of a firm's management resulting from its decisions about the firm's competitive thrust and resource allocations. Both the type of innovation and the competitive circumstances condition the results of attempts to create and introduce change. The innovation process differs for new products, new production equipment, and incremental improvements.

The conditions necessary for rapid innovation are different from those necessary for high levels of output and production efficiency. The pattern of change in an organization often shifts from innovative and flexible to standardized and inflexible as output and productivity increase. Different creative responses to competitive and technological challenge are observed for different productive units, which provides a way of analyzing the effects of various policy options on the innovative process.

Chapter 3 THE SLOWDOWN IN PRODUCTIVITY ADVANCES: A DYNAMIC EXPLANATION by Burton H. Klein

Sustained productivity improvements can only be made if firms, industries, and the economy as a whole manifest a high degree of dynamic competition, or rivalry. The decline in the rate of advance of productivity in the United States can be traced to a decline in the willingness and ability of firms to take the risks associated with major technological advances. Such risk taking is encouraged not only by the opportunity to prosper and grow (Adam Smith's "invisible hand"), but also by the threat that a firm's rivals might make an advance if it doesn't (the "hidden foot").

The most effective way to ensure that the hidden foot functions to induce technological change is to ensure that firms in an industry face effective rivalry from other firms, both domestic and foreign, and especially from new entrants to the industry. However, the trend in the United States over the last two decades has been for firms' managements to become less responsive to feedback from rivals in the market. In addition, it has become much more difficult, in part for institutional reasons, for new firms to enter established markets. Pressures have also increased to isolate American firms from what are often their most effective rivals: foreign competitors. Each of these trends has contributed to chronic inflation, slowed productivity advance, and slowed economic growth.

Chapter 4 TECHNOLOGICAL INNOVATION AND THE DYNAMICS OF U.S. COMPARATIVE ADVANTAGE IN INTERNATIONAL TRADE by Edward M. Graham

This chapter reviews the major theories of the determinants of international trade, focusing on those theories and hypotheses that suggest that technological innovation is a source of U.S. comparative advantage in international trade. In light of these theories, the present position of the United States in the world economy is examined, and recommendations are presented for the international economic policies of the nation. It is argued that neither trade protectionism nor technological protectionism is likely to work to the nation's advantage at the present time, but rather that the nation must pursue policies that enable it to adjust to changing circumstances.

Chapter 5 ENVIRONMENTAL, HEALTH, AND SAFETY REGULATION AND TECHNOLOGICAL INNOVATION by Nicholas A. Ashford, George R. Heaton, Jr., and W. Curtiss Priest

This chapter reviews and synthesizes evidence on the effects of environmental, health, and safety regulation on technological innovation in industry. Care is taken to separate the assessment of regulation's effect on innovation from the valuation of regulation's costs and benefits. The variety of effects that occur in the regulation-innovation relationship are discussed within an analytic framework that distinguishes between short- and long-term effects and between adjustment effects and permanent changes. The framework also distinguishes the impact of regulation on innovation for ordinary business purposes from its impact on innovation for compliance purposes. Major policy issues are identified throughout the chapter and highlighted in the concluding section.

Chapter 6 LABOR, PRODUCTIVITY, AND TECHNOLOGICAL INNOVATION: FROM AUTOMATION SCARE TO PRODUCTIVITY DECLINE by Clinton C. Bourdon

Over the last century, workers have made gains in real wages and leisure time that are roughly commensurate with productivity increases. However, the distribution of the gains from productivity increases between capital and labor and between different skill groups is still not well understood. Case studies of technological change in particular firms and industries show that unions and workers have generally supported -- and contributed to -- technological change. Much technological change is incremental and the costs of adjusting to it can be absorbed by firms and workers. Contemporary management still encourages automation of production, even though this may not be competitive with less capital-intensive alternatives.

Social conflict will continue over the optimal levels of adjustment assistance, and over the degree of labor and management control over changes in production. Present public and vate policies provide quite disparate benefits to different groups of displaced workers. Continuing efforts are needed to articulate, fund, and manage a coherent adjustment strategy in the United States.

Chapter 7 DIRECT GOVERNMENT FUNDING OF RESEARCH AND DEVELOPMENT: INTENDED AND UNINTENDED EFFECTS ON INDUSTRIAL INNOVATION by Paul Horwitz

This chapter presents data on R&D support by the United States and foreign governments, describes the justifications in economic theory for such public support, and examines its intended and unintended effects on technological innovation. The findings indicate that governments have been most successful in promoting innovation when they have used a range of policies in addition to R&D support. To enhance the environment for innovation, the government should emphasize support for basic technology, maintain close contact with the potential users of research results, and promote competition between innovating firms. The government should avoid supporting the later stages of innovation and avoid premature concentration on single technical approaches.

Chapter 8 POLICIES AND PROGRAMS OF GOVERNMENTS DIRECTED TOWARD INDUSTRIAL INNOVATION by J. Herbert Hollomon

All industrialized nations carry out policies and programs that directly or indirectly are aimed at affecting the process of industrial innovation. This chapter examines selected government programs in several European countries and Japan.

In recent years, most of the nations studied have shifted their emphases from support of R&D, and from development of a few advanced technologies, to encouragement of the innovation process more generally. Neither West Germany nor Japan, the two countries most competitive with the United States, has followed the lead of the United States in emphasizing innovation for space and defense. Instead, each has actively promoted a wide variety of programs designed to stimulate both technological innovation and adaptation of new techniques in both new, small firms and mature, large ones.

While particular government programs are not directly transferable from one country to another, the widespread use of an instrument of intervention to support innovation in other countries raises questions about its absence in the United States. This examination of foreign experience concludes with suggestions for specific policies and programs that should be considered for the United States.

Chapter 9 SUMMARY AND POLICY IMPLICATIONS by Christopher T. Hill and James M. Utterback

This chapter summarizes the main themes, arguments, and policy inplications of the previous eight chapters.

Technological innovation, properly directed, can make major contributions to the vitality of the economy and the society of the United States. However, the United States is not now taking full advantage of the opportunities that a more rapid rate of innovation would offer in controlling inflation and the balance of trade; enhancing employment, economic growth, and productivity; resolving the energy dilemma; or improving the quality of life in general. Broad directions for U.S. policy toward technological innovation are suggested in light of these findings.

Index

Abernathy, William 40, 102, 165, 244
Action Concerties Program (France) 302
Advances in knowledge 7, 240, 265
Adjustment assistance 150, 250
Advent Corporation 49
Agence Nationale pour la Valorisation de la Recherche (ANVAR) (France) 305
Agriculture, Department of 279
Aide de Developement Program (France) 303
Air Mail Act of 1934 74
Air Force, U.S. 307
Allen, Thomas 181
Ambiguous feedback (see feedback)
American Airlines 75
American Association for the Advancement of Science 258
American Federation of Musicians 246
American National Standards Institute 297
American Telephone and Telegraph 91, 147
Ancillary innovation (see innovation)
Antitrust Division of the Justice Department 111
Antitrust 108, 110, 112, 187, 286
Association Francais de Normalisation (France) 297
Atomic Energy Commission 276, 283, 299, 302
Automation 222
Avery, Clarence W. 73

Balassa, Bela 125
Balance of payments 20
Bardeen, John 76
Barnett, Harold 31
Barriers to
    entry 181, 183, 206, 283
    innovation (see innovation)
    trade 141
Basic research (see research and development)
Basic R&D (see research and development)
Battelle Memorial Institute 103
Beckman Instruments 76
Belitsky, Harvey 248
Bell Laboratories 91, 186
Boeing 74, 75, 110

Brattain, Walter 76
Brazil 141, 310
Bretton Woods Agreement 136, 139
Bright, James 243
Britain (see United Kingdom)
British Standards Institution 246
Buick 72, 76

Canada 293
Cain, Bruce 114
California Institute of Technology 75
Center for Policy Alternatives, MIT 164, 185, 191, 193, 304, 313
Centre National de la Recherche Scientifique (France) 296, 299, 301
Chemical Institute of Toxicology 187
Chrysler 96
Clark, Kim 238
Claudon, M.P. 127
Clean Air Act 192
Cole, H.S. 31
Commission on Technology, Automation, and Economic Progress 222
Communications Workers of America 245
Comparative advantage 21, 118
Cost constraint 70, 73, 75, 77, 79
Council of Economic Advisors 227
Cummins Engine 247
Cyert R. 82

Davidson, William 128, 142
Defense, Department of 261, 275
Delay 94
Delegation Generale a la Recherche Scientifique et Technique (France) 302
Demand similarity model 132
Democracy 113
Denison, Edward 7
Denmark 115
Denver Research Institute 180, 187
Deutsche Normennausschauft (W. Germany) 296
Digital Equipment Corporation 49
Douglas Aircraft 52, 74, 75
Douglas, G.K. 124
DuPont 248
Dynamic
  behavior 70, 80, 87
  economy 32, 34, 68, 70, 97, 107
  efficiency 20, 81, 83
  equilibrium 95

Eads, George 164
Eastman Kodak 48, 49, 291
Economic Development Administration (U.S.) 311
Edison Thomas A. 49
Electric Power Research Institute 307
Employment 12, 17, 28, 40, 49, 57, 104, 150, 222
Energy 15
Energy, Department of 2, 261, 299, 302, 304
Entrepreneur 69
Environmental Protection Agency 189, 192
Equipment suppliers 51, 53
European Economic Community 133, 311
Experience curve 53

Factor proportions theory 120
Fairchild Camera and Instrument 49
Federal Aviation Administration 74
Federal Communications Commission 92
Federal Trade Commission 297
Federal Water Pollution Control Act 191, 192

Feedback
ambiguous 71, 85
unambiguous 71, 98
Fiorina, Morris 114
Food and Drug Administration 172
Ford, Henry 72
Ford Motor Company 50, 83, 96
Ford, Model T 52, 72, 77
Foreign competition (see Trade, international)
Forrester, Jay 31
France 25, 33, 135, 261, 265, 279, 293, 296, 298, 300, 305, 306, 309
Fraunhofer Society for the Advancement of Applied Research (Germany) 305
Freeman, Christopher 12, 123

General Agreement on Tariffs and Trade 148, 297
General Electric 50
General Motors 49, 96, 247
General Services Administration 307
Germany (see West Germany)
Gollop, F. 234
Government R&D (see research and development)
Great Britain (see United Kingdom)
Greenberg, Edward 171
Griliches, Zvi 7
Gross national product 2, 4, 8, 23, 108, 257
Growth 4, 81
Grubel, H.G. 133
Gruber, William 134

Hansen, Ronald 172
Health, Education and Welfare, Department of 261, 299
Hidden foot 69, 71, 85, 92, 107, 110
Hill, Christopher T. 171
Hirsch, Fred 31
Hirsch, Seev 124
Hirschmann, Albert 123
Hoffmeyer, Erik 123
Hollander, Stephen 244, 248
Honda Corporation 83, 108, 271
Hotelling, Harold 78, 96, 107
Hufbauer, Gary 123, 133

Inflation 14, 104
Innovation
ancillary 178
barriers to 46, 57, 103
compliance 163, 178, 183
main business 163, 171
negative effects of 29
Internal Revenue Code 197
International Business Machines 49, 147, 250
International trade (see Trade)
India 141
Industry structure 51, 206
Intergovernmental Regulatory Liaison Group 199
International Electrotechnical Commission 297
International Institute for Production Engineering Research (CIRP) 249
International Longshoremen's and Warehousemen's Union 245
International Standards Organization 297
Ireland 311
Irish Development Corporation 311
Islands of automation 43, 51
Israel 300
Italy 8, 25, 33
Iverstine, Joseph 180, 181

Japan 23, 25, 28, 33, 114, 135, 139, 142, 148, 150, 248, 265, 279, 295, 298, 300-303, 310, 313
Japan Industrial Standards Committee of the Ministry for International Trade and Industry 297

Japan Research and Development Corporation 303
Japanese Development Bank 303
Jefferson, Thomas 113
Jorgenson, Dale 7, 234

Katzell, Raymond 247
Kaysen, Carl 113
Keesing, Donald 124
Kendrick, John 237, 240
Kennedy, John F. 107
Kindleberger, Charles 123
Klein, Arthur L. 75
Klein, Burton 32, 127, 131, 138
Kneese, Allen 30
Kodak (see Eastman Kodak)
Korea (see South Korea)
Korean Institute of Science 306
Kravis, Irving 123, 127

Labor 49, 70, 122, 124, 222
  productivity 9, 12, 16, 222
  technological change 222
  unions 104, 238
Labor Statistics, Bureau of 105, 242
Lander (West Germany) 299
LaPierre, D. Bruce 192
Launching Aid Program (Britain) 303
Leibenstein, Harvey 81
Leisure time 230
Leone, Robert 172
Leontief, Wassily 21, 122, 124
Les Grandes Ecoles (France) 299
Lewis, John L. 245
Linder, Staffan 127, 132, 141
Linneman, H. 133
Littlewood, William 75
Lockheed 74
Luck 69, 70, 73, 74, 80, 74, 89, 101

Macrae, Norman 138
Macrostability 93
Majone, Giandomenico 202
Main business innovation (see innovation)
Malthus, Thomas 31
Manhattan Project 282
Mansfield, Edwin 98, 177, 269
March, J. 82
Marconi, G. 49
Market
  niche 47, 284
  pull 206
  rights 200
  shares 52, 56, 70, 78, 93, 98, 112, 167, 183
Marshall Plan 136
Marx, Karl 225, 223
Massachusetts Institute of Technology, (see Center for Policy Alternatives)
Max Planck Gesellschaft (West Germany) 299
Meadows, Donella 31
Mehta, Dileep 124
Mergers 101, 106
Mexico 141, 310
Microstability 93
Mines, Bureau of 302
Ministry for Science and Technology (West Germany) 303
Ministry of Education (Japan) 300
Ministry of Industry (Sweden) 305
Ministry of International Trade and Industry of Japan 302
Minnesota Mining and Manufacturing Company 80
Morse, Chandler 31
Multinational firms 27, 134, 140, 149, 182, 311
Multiplier effect 272
Musicians Guild 247

National Advisory Committee
on Aeronautics 275, 302
National Aeronautics and
Space Administration 258
National Bureau of Standards
270, 296
National Institutes of Health
299
National Physical and
Chemical Laboratories
(Great Britain) 296,
299
National Research and
Development Corporation
(Great Britain) 305
National Research Council
(Great Britain) 299
National Science Foundation
247, 258, 277, 281, 299
Negative effects of
innovation (see innovation)
Netherlands 125
Newburger, David J. 171
Nordhaus, William 31
Nuclear energy 109, 276

Office of Management and
Budget 258
Ohlin, Bertil 123
Occupational Safety and
Health Act 172, 180,
189, 191, 195
Organization for Economic
Cooperation and
Development 2, 33, 312

PCBs 164, 182
Patents 23, 25, 129, 266,
286, 292, 294
Pavitt, Keith 22
People's Republic of China 28
Pilkington 50, 55
Piore, Michael 20
Polaroid 49, 90
Polychlorinated biphenyls
(see PCBs)
Posner, Michael 123
Practolol 175
Process improvements 179
Product
cycle theory 21, 27
life cycle model 127, 131,
140, 143, 292
Productivity 9, 67, 77, 222
Public utility companies 109

Quality of life 29
Quinn, James 16

Rawls, John 202
Raymond, Arthur 74
Regulation 34, 41, 55, 271
and innovation 30, 43, 46,
50, 161
antitrust (see antitrust)
economic 83, 98, 109
environmental, health, and
safety 13, 108, 149, 161
fuel economy 184
tax incentives (see tax
incentives)
Research and Development 46,
255
basic 34, 47, 138, 265
government 255
Resource Conservation and
Recovery Act 199
Ricardo, David 119
Rivalry 57, 71, 78, 100,
103, 105, 106, 127,
142, 146
Rolls Royce 101
Rose-Ackerman, Sarah 200

Salter, W.E. 240
Sarett, L.H. 178
Scanlon Plan 247
Scherer, F.M. 83
Schmalensee, Richard 96
Schmookler, Jacob 144
Schultze, Charles 30
Schwartzman, David 172, 176
Schweitzer, Glenn 178
Scotland 103
Sears and Roebuck Co. 72
Securities and Exchange
Commission 256, 286
Sherman Antitrust Act
(see antitrust)

Shockley, William 76
Smith, Adam 81, 111
Social welfare 29, 233
Soete, Luc 22
Solow, Robert 4, 7
South Korea 306, 310
Soviet Union 146
State Technical Services Act 307
Static
  economy 68
  efficiency 20, 81, 113
  equilibrium 95
Stobaugh, Robert 127
Sweden 33, 279, 311

TRW 247
Taiwan 310
Tax incentives 199
Technological innovation
  definition of 2
Technology
  forcing 188, 191
  push 189
  transfer of 27, 142
Terleckj, Nestor E. 237, 269
Texas Instruments 49
3M Company (see Minnesota Mining and Manufacturing Co.)
Tilton, John E. 127
Timing of regulations 194
Toxic Substances Control Act 162, 196, 198
Trade Act of 1974 250
Trade secrets 25, 175, 199
Trade, international 20, 103, 112, 118
Trans-World Airlines 74
Treasury, Department of 144
Turner, Donald 113

Unambiguous feedback (see feedback)
Uncertainty,
  Type I 84, 89, 90, 101, 107, 131, 270
  Type II 87, 101, 131
Unemployment 20, 40, 106, 151, 223, 233, 240, 310, 311
Unions (see labor)
United Kingdom 25, 69, 100, 103, 151, 175, 265, 275, 279, 295, 297, 299, 300, 303, 306, 309
United Mine Workers 245
U.S. Government (see individual departments and agencies)
U.S. and foreign competition (see trade)
United Steelworkers 246
Utterback, James M. 244
University Grants Commission (Great Britain) 299

Vernon, Raymond 103, 124, 127, 130, 142

Wang 49
Weights and measures system 294, 296
Wells, Louis 130
West Germany 8, 25, 114, 135, 150, 265, 279, 295, 296, 299, 301, 303, 305, 306, 309, 311, 312
Western Electric Company 92
Work 16, 19

Xerox 250

Zenith 248

Pergamon Policy Studies

No. 1 Laszlo—*The Objectives of the New International Economic Order*
No. 2 Link/Feld—*The New Nationalism*
No. 3 Ways—*The Future of Business*
No. 4 Davis—*Managing and Organizing Multinational Corporations*
No. 5 Volgyes—*The Peasantry of Eastern Europe, Volume One*
No. 6 Volgyes—*The Peasantry of Eastern Europe, Volume Two*
No. 7 Hahn/Pfaltzgraff—*The Atlantic Community in Crisis*
No. 8 Renninger—*Multinational Cooperation for Development in West Africa*
No. 9 Stepanek—*Bangledesh—Equitable Growth?*
No. 10 Foreign Affairs—*America and the World 1978*
No. 11 Goodman/Love—*Management of Development Projects*
No. 12 Weinstein—*Bureacratic Opposition*
No. 13 De Volpi—*Proliferation, Plutonium, and Policy*
No. 14 Francisco/Laird/Laird—*The Political Economy of Collectivized Agriculture*
No. 15 Godet—*The Crisis in Forecasting and the Emergence of the "Prospective" Approach*
No. 16 Golany—*Arid Zone Settlement Planning*
No. 17 Perry/Kraemer—*Technological Innovation in American Local Governments*
No. 18 Carman—*Obstacles to Mineral Development*
No. 19 Demir—*Arab Development Funds in the Middle East*
No. 20 Kahan/Ruble—*Industrial Labor in the U.S.S.R.*
No. 21 Meagher—*An International Redistribution of Wealth and Power*
No. 22 Thomas/Wionczek—*Integration of Science and Technology With Development*
No. 23 Mushkin/Dunlop—*Health: What Is It Worth?*
No. 24 Abouchar—*Economic Evaluation of Soviet Socialism*
No. 25 Amos—*Arab-Israeli Military/Political Relations*
No. 26 Geismar/Geismar—*Families in an Urban Mold*
No. 27 Leitenberg/Sheffer—*Great Power Intervention in the Middle East*
No. 28 O'Brien/Marcus—*Crime and Justice in America*
No. 29 Gartner—*Consumer Education in the Human Services*
No. 30 Diwan/Livingston—*Alternative Development Strategies and Appropriate Technology*
No. 31 Freedman—*World Politics and the Arab-Israeli* Conflict
No. 32 Williams/Deese—*Nuclear Nonproliferatrion*
No. 33 Close—*Europe Without Defense?*
No. 34 Brown—*Disaster Preparedness*
No. 35 Grieves—*Transnationalism in Politics and Business*

No. 35 Grieves—*Transnationalism in World Politics and Business*
No. 36 Franko/Seiber—*Developing Country Debt*
No. 37 Dismukes—*Soviet Naval Diplomacy*
No. 38 Morgan—*Science and Technology for Development*
No. 39 Chou/Harmon—*Critical Food Issues of the Eighties*
No. 40 Hall—*Ethnic Autonomy—Comparative Dynamics*
No. 41 Savitch—*Urban Policy and the Exterior City*
No. 42 Morris—*Measuring the Condition of the World's Poor*
No. 43 Katsenelinboigen—*Soviet Economic Thought and Political Power in the U.S.S.R*
No. 44 McCagg/Silver—*Soviet Asian Ethnic Frontiers*
No. 45 Carter/Hill—*The Criminal's Image of the City*
No. 46 Fallenbuchl/McMillan—*Partners in East-West Economic Relations*
No. 47 Liebling—*U.S. Corporate Profitability*
No. 48 Volgyes/Lonsdale—*Process of Rural Transformation*
No. 49 Ra'anan—*Ethnic Resurgence in Modern Democratic States*
No. 50 Hill/Utterback—*Technological Innovation for a Dynamic Economy*
No. 51 Laszlo/Kurtzman—*The United States, Canada and the New International Economic Order*
No. 52 Blazynski—*Flashpoint Poland*
No. 53 Constans—*Marine Sources of Energy*
No. 54 Lozoya/Estevez/Green—*Alternative Views of the New International Economic Order*
No. 55 Taylor/Yokell—*Yellowcake*
No. 56 Feld—*Multinational Enterprises and U.N. Politics*
No. 57 Fritz—*Combatting Nutritional Blindness in Children*
No. 58 Starr/Ritterbush—*Science, Technology and the Human Prospect*
No. 59 Douglass—*Soviet Military Strategy in Europe*
No. 60 Graham/Jordon—*The International Civil Service*
No. 61 Menon—*Bridges Across the South*
No. 62 Avery/Lonsdale/Volgyes—*Rural Change and Public Policy*
No. 63 Foster—*Comparative Public Policy and Citizen Participation*
No. 64 Laszlo/Kurtzman—*Eastern Europe and the New International Economic Order*
No. 65 United Nations Centre for Natural Resources, Energy and Transport—*State Petroleum Enterprises in Developing Countries*